Lecture Notes in Computer Science 16521

Founding Editors

Gerhard Goos
Juris Hartmanis

Editorial Board Members

Elisa Bertino, *Purdue University, West Lafayette, IN, USA*
Wen Gao, *Peking University, Beijing, China*
Bernhard Steffen, *TU Dortmund University, Dortmund, Germany*
Moti Yung, *Columbia University, New York, NY, USA*

The series Lecture Notes in Computer Science (LNCS), including its subseries Lecture Notes in Artificial Intelligence (LNAI) and Lecture Notes in Bioinformatics (LNBI), has established itself as a medium for the publication of new developments in computer science and information technology research, teaching, and education.

LNCS enjoys close cooperation with the computer science R & D community, the series counts many renowned academics among its volume editors and paper authors, and collaborates with prestigious societies. Its mission is to serve this international community by providing an invaluable service, mainly focused on the publication of conference and workshop proceedings and postproceedings. LNCS commenced publication in 1973.

Luca Manzoni · Sylvain Cussat-Blanc · Qi Chen
Editors

Genetic Programming

29th European Conference, EuroGP 2026
Held as Part of EvoStar 2026
Toulouse, France, April 8–10, 2026
Proceedings

 Springer

Editors
Luca Manzoni
University of Trieste
Via A. Valerio, Italy

Sylvain Cussat-Blanc
Universite Toulouse
Toulouse, France

Qi Chen
Victoria University of Wellington
Wellington, New Zealand

ISSN 0302-9743   ISSN 1611-3349 (electronic)
Lecture Notes in Computer Science
ISBN 978-3-032-23004-1   ISBN 978-3-032-23005-8 (eBook)
https://doi.org/10.1007/978-3-032-23005-8

This Springer imprint is published by the registered company Springer Nature Switzerland AG
The registered company address is: Gewerbestrasse 11, 6330 Cham, Switzerland

If disposing of this product, please recycle the paper.

Preface

This volume contains the proceedings of EuroGP 2026, the 29th European Conference on Genetic Programming (GP). The conference was part of EvoStar 2026, the leading event on bio-inspired computation in Europe, and was held in Toulouse, France, between Wednesday, 8th April and Friday, 10 April, 2026.

EuroGP is the premier annual conference on GP, the oldest and the only meeting worldwide devoted specifically to this branch of Evolutionary Computation. At the same time, under the EvoStar umbrella, EvoAPPS focuses on the applications of Evolutionary Computation, EvoCOP targets Evolutionary Computation in combinatorial optimization, and EvoMUSART is dedicated to evolved and bio-inspired music, sound, art, and design. The proceedings for these co-located events are available in the LNCS series.

GP is a distinctive branch of Evolutionary Computation that aims to automatically solve design problems, in particular computer program design, without requiring the user to specify the form or structure of the solution in advance. It uses the principles of Darwinian evolution to approach problems in the synthesis, improvement, and repair of computer programs. The universality of computer programs, and their importance across a wide range of domains, make the automation of these tasks an exceptionally ambitious challenge with far-reaching implications. Over the years, GP has attracted a large and active research community, producing significant theoretical and practical advances.

Since the first EuroGP event in Paris in 1998, EuroGP has served as the primary conference dedicated exclusively to the evolutionary design of computer programs and symbolic models. EuroGP continues to play an important role in the development of the field by providing a forum for the exchange of ideas, fostering collaborations, and presenting new advances in theory, methodology, and applications.

EuroGP 2026 received 34 submissions from around the world. The papers underwent a rigorous double-blind peer-review process, each being reviewed by at least three members of the Program Committee. We selected 12 papers for full oral presentation, while 7 works were presented in short oral presentations and/or as posters. All accepted contributions, regardless of presentation format, appear as full papers in this volume.

An event of this kind would not be possible without the contributions of many individuals. We express our gratitude to the authors for submitting their work and to the members of the Program Committee and additional reviewers for their careful evaluations and constructive feedback. We also thank the EvoStar organizers and the local organizing team for their dedication in coordinating the conference.

Finally, we gratefully acknowledge Springer Nature for publishing these proceedings in the Lecture Notes in Computer Science series.

February 2026

Luca Manzoni
Sylvain Cussat-Blanc
Qi Chen

Organization

Program Chairs

Luca Manzoni   University of Trieste, Italy
Sylvain Cussat-Blanc   University of Toulouse, France

Publication Chair

Qi Chen   Victoria University of Wellington, New Zealand

Local Chairs

Dennis Wilson   University of Toulouse, France
Sylvain Cussat-Blanc   University of Toulouse, France

Publicity Chair

João Correia   University of Coimbra, Portugal

Conference Administration

Anna Esparcia-Alcázar   Universidad Politécnica de Valencia

Program Committee

Wolfgang Banzhaf   Michigan State University, USA
Ying Bi   Victoria University of Wellington, New Zealand
Stefano Cagnoni   University of Parma, Italy
Mauro Castelli   NOVA Information Management School (NOVA IMS), Portugal
Qi Chen   Victoria University of Wellington, New Zealand
Ernesto Costa   University of Coimbra, Portugal
Andrea De Lorenzo   University of Trieste, Italy

Mario Giacobini	University of Turin, Italy
Jin-Kao Hao	University of Angers, France
Erik Hemberg	MIT, USA
Malcolm Heywood	Dalhousie University, Canada
Ting Hu	Queen's University, Canada
Zhixing Huang	Victoria University of Wellington, New Zealand
Giovanni Iacca	University of Trento, Italy
Domagoj Jakobovic	University of Zagreb, Croatia
Krzysztof Krawiec	Poznań University of Technology, Poland
W. B. Langdon	University College London, UK
Yuri Lavinas	University of Toulouse, France
Nuno Lourenço	University of Coimbra, Portugal
Hervé Luga	IRIT, France
Evelyne Lutton	INRIA, France
James McDermott	University of Galway, Ireland
Eric Medvet	University of Trieste, Italy
Yi Mei	Victoria University of Wellington, New Zealand
Alberto Moraglio	University of Exeter, UK
Aidan Murphy	University College Dublin, Ireland
Fabrício Olivetti de França	Federal University of ABC, Brazil
Gisele Pappa	Federal University of Minas Gerais, Brazil
Stjepan Picek	Radboud University, The Netherlands
Gloria Pietropolli	University of Trieste, Italy
Peter Rockett	University of Sheffield, UK
Lukáš Sekanina	Brno University of Technology, Czech Republic
Sara Silva	University of Lisbon, Portugal
Moshe Sipper	Ben-Gurion University of the Negev, Israel
Ernesto Tarantino	ICAR-CNR, Italy
Alberto Tonda	INRIA, France
Leonardo Trujillo	Instituto Tecnológico de Tijuana, Mexico
Leonardo Vanneschi	NOVA University Lisbon, Portugal
Dennis Wilson	ISAE-SUPAERO, France
Man Leung Wong	Lingnan University, China
Mengjie Zhang	Victoria University of Wellington, New Zealand

Contents

Long Presentation

Short Presentation

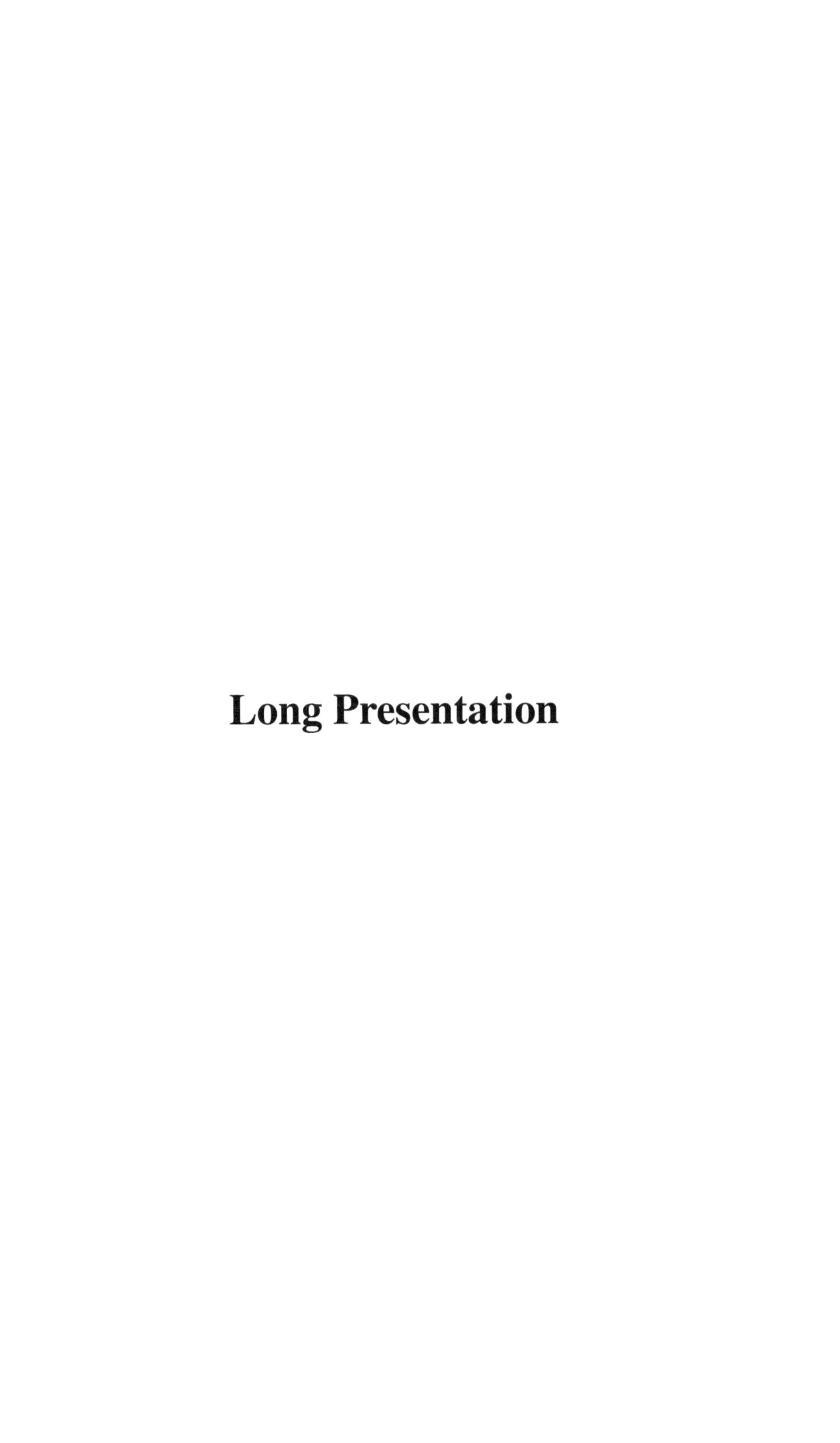

Long Presentation

On the Effects of Down-Sampling
for Tournament and Lexicase Selection
in Program Synthesis

Martin Briesch[(✉)][iD]

Johannes Gutenberg University, Mainz, Germany
`briesch@uni-mainz.de`

Abstract. In recent years random and informed down-sampling have
been used to increase the performance of genetic programming using both
tournament and lexicase selection. In the domain of symbolic regression
these down-sampling approaches have closed the gap between both selection methods. Building on this prior work we evaluate if these findings
transfer to the program synthesis domain, a problem domain mostly consisting of complex modal problems, a property that lexicase was specifically designed for. We conduct experiments for a diverse set of program
synthesis benchmark problems, using grammar guided genetic programming, and investigate the effects of different down-sampling schemes on
both tournament and lexicase selection. We find that while tournament
selection does indeed benefit more from down-sampling in terms of performance gains, generalization, code growth, and diversity, lexicase still
outperforms tournament selection due to its the superior promotion and
preservation of specialists in more complex problems.

Keywords: Selection · Program Synthesis · Down-Sampling · Genetic
Programming

1 Introduction

Selection is an integral part of genetic programming (GP) and over the years
many different selection methods have been proposed. One of these methods is
lexicase selection [35]. Instead of calculating an aggregated fitness value over
all training cases for selection (like in tournament selection), lexicase considers
individual training cases in random sequential order to filter the candidate pool.
This promotes diversity and specialists within a population and can improve GP
performance in both symbolic regression and program synthesis [26,29,30].

Lexicase can further be improved with random down-sampling (RDS), which
only uses a random subset of training cases at each generation for selection
[27]. This saves computational budget by requiring less evaluations per selection event which in turn can be used to run GP for more generations improving
the performance of lexicase even further [14,25,27]. Instead of naively down-
sampling the training cases at random, the down-sample can also be constructed
in a more sophisticated way: Informed down-sampling (IDS) leverages run-time

L. Manzoni et al. (Eds.): EuroGP 2026, LNCS 16521, pp. 3–18, 2026.
https://doi.org/10.1007/978-3-032-23005-8_1

population statistics to construct a more informative down-sample, further improving the performance in both program synthesis [3] as well as symbolic regression [16]. Both down-sampling methods do not only improve performance for lexicase selection, but also for other selection methods such as tournament selection [2,16]. However, down-sampling does not only influence the performance but also other factors like generalization, search convergence, code growth, diversity, and specialist preservation. These effects can either be negative or positive depending on the selection method [9,13,14,18,28,31,33].

With these effects in mind, a recent study by Geiger et al. [13] raised the question if down-sampling methods might bring the performance of tournament selection up to speed with lexicase selection [13]. They find that down-sampling reduces typical problems of tournament selection like generalization performance, code growth, and low diversity. This leads to comparable performance with lexicase at much lower run time complexity, making tournament selection the better choice. However, those results are limited to the domain of symbolic regression. In contrast, most program synthesis problems are modal in nature and might require specialists as stepping stones to be solved. Lexicase selection was specifically designed for such modal problems [35] and its performance gains are mostly attributed to higher diversity [21,22] and better preservation of specialists [23]. Therefore, it is unclear if the findings from Geiger et al. [13] generalize to the program synthesis domain.

In this work we investigate this question and analyze the effects of down-sampling for lexicase and tournament selection in the program synthesis domain. We perform experiments using grammar guided GP [10,11,36] on well known benchmark problems [20,24] for different down-sampling rates. We investigate both RDS and IDS and analyze the differences in performance as well as differences in search behavior in terms of generalization, convergence, code growth, diversity, and specialist preservation.

We find that both RDS and IDS improve the performance of tournament selection for program synthesis problems like in the domain of symbolic regression with similar effects like in Geiger et al. [13]. However, this only closes the gap between lexicase and tournament selection for some benchmark problems. Lexicase is still the better choice for the more complex modal problems in our experiments. Additionally, we find that this is mostly due to the superior promotion and preservation of specialists when using lexicase selection even when paired with down-sampling methods. Although down-sampling increases the ability to promote and preserve specialists for tournament selection while simultaneously impairing the ability of lexicase selection to do so, the difference between the two methods remains significant enough to give lexicase selection a clear advantage.

Overall, our contributions are a set of extensive experiments to extend the findings of Geiger et al. [13] to the program synthesis domain. Additionally, we analyze the effects of down-sampling in detail for different down-sampling rates, extending previous findings from RDS to IDS in a comprehensive way.

The paper is structured as follows: Sect. 2 presents the related work. Section 3 describes our methods followed by the results in Sect. 4. Section 5 discusses the results. Lastly, Sect. 6 concludes the paper.

2   Related Work

Down-sampling in GP is used to evaluate a population on fewer training cases to save computational resources or improve performance [12]. In recent years RDS has been used to improve GP performance in both program synthesis [9, 19,25,27] and symbolic regression [14,17,18]. RDS performs selection with a different random subset of the training data at each generation. This leads to fewer fitness evaluations required per generation and subsequently the possibility to increase the number of generations or population size within a GP run. This results in exploring more individuals during evolution, potentially improving the performance of GP [25].

The success of RDS has been improved with the introduction of IDS [3]. IDS constructs the down-sample in an informed way by leveraging run-time population statistics. First, the parent population is evaluated against the full training set to generate a 'solve vector' for each case, which characterizes how individuals perform on that specific case. These vectors are then used to compute a pairwise distance matrix between all cases. To construct the informed down-sample, the algorithm initializes with a random case and iteratively selects subsequent cases that maximize the minimum distance to those already in the down-sample. Finally, to maintain the computational efficiency of RDS, the distance matrix is updated only every few generations (distance update interval), and solve vectors are approximated using a sub-sample of the parent population (parent sampling rate). This informed down-sample results in less redundancy between cases and less test coverage loss, which in turn has improved the performance of lexicase selection on program synthesis benchmarks [3,4]. This performance increase has also been observed in the symbolic regression domain [16].

The performance increase of down-sampling has not only been observed for lexicase selection but also for tournament selection, both for symbolic regression [15,16] as well as for program synthesis [1,2].

In the symbolic regression domain, down-sampling can improve generalization for both lexicase and tournament selection [13,18]. Down-sampling methods also increase diversity for tournament selection [13] while decreasing diversity for lexicase selection [13,14]. Lastly, down-sampling can also prevent code growth [13,18].

Similarly, in the program synthesis domain, down-sampling can increase generalization and code growth for lexicase selection [33]. However, RDS in combination with lexicase selection has been found to reduce exploration capabilities, diversity, and specialist preservation [9,28,31].

The down-sampling rates can also have an effect. In prior work, lower down-sampling rates (meaning a smaller portion of cases is used for evaluation at each generation) result in better better performance. However, if the down-sample rate is too low, performance deteriorates again. A down-sampling rate between 10% and 25% achieves the best results in literature [14,25,27,31,33].

In addition to differences in performance, different down-sampling rates can have further effects. For lexicase paired with RDS lower down-sampling rates decrease specialist preservation and exploration capabilities even further [9,28].

Additionally, in the symbolic regression domain lower down-sampling rates lead to more hyper-selection events and consequently lower diversity when using lexicase selection [14].

With all these effects in mind a recent study raised the question if down-sampled tournament selection is the better choice considering better run-time complexity compared to lexicase selection. For the symbolic regression domain Geiger et al. [13] argue that down-sampling closes the gap between lexicase and tournament selection by increasing diversity, lowering the generalization gap, and decreasing code growth for tournament selection. This leads to equal performance between tournament and lexicase, making tournament selection the better choice. Additionally, the authors find that down-sampling helps performance even without making use of the saved evaluation budget.

However, it is unclear if those findings hold true for the program synthesis domain as lexicase was originally designed for modal problems [35], a property that program synthesis problems normally inhibit (in contrast to symbolic regression problems). Additionally, while the effects of down-sampling for RDS are well studied, for IDS these effects remain underexplored for both lexicase and tournament selection, especially in the program synthesis domain. Therefore, in this work we address these gaps.

3   Methods

In this section, we present our methods including benchmark problems, experimental setup, and investigated metrics.

3.1   Benchmark Problems

We focus on experiments in the program synthesis domain. More specifically, we perform experiments on six program synthesis problems from the established benchmark suites PSB1 [24] and PSB2 [20]: Count Odds, Fizz Buzz, Fuel Cost, Grade, Scrabble Score, and Small or Large. We select those problems in line with literature to cover a wide range of problem structure and difficulty [3,6,34].

Each problem is defined by a set of input/output pairs (cases) and consists of 200 training cases as well as $1,000$ test cases. We use binary error values in our experiments: Each unsolved case contributes an error of 1 to an individuals fitness and the total error is the number of all unsolved cases. Therefore, an individual that solves all training cases would have an aggregated fitness of 0.

To solve a problem a candidate solution must not only solve all training cases but is also evaluated on the test set. If it does generalize to those unseen test cases (meaning an aggregated error of 0 on the test set) it counts as a correct solution.

3.2   Experimental Setup

We use grammar guided GP [10,11,36] in this work, implemented in the PonyGE2 framework [8]. We use the same grammars as in Boldi et al. [3].

Table 1. Experimental settings

Parameter	Value
Initialization	Position-independent grow
Maximum initial tree depth	10
Population size	1000
Generations	2000
Selection	[Lexicase, Tournament $(n = 5)$]
Crossover probability	0.95
Crossover operator	Subtree crossover
Mutation probability	1
Mutation operator	Subtree mutation
Mutation events	1
Max. tree depth	17
Elite size	5
Number of training cases	200
Number of test cases	1000
Down-sampling methods	[RDS, IDS]
Down-sampling rates	[0.01, 0.05, 0.10, 0.25, 0.50, 1.00]
Parent sampling rate (IDS)	0.01
Distance update interval (IDS)	10

To better analyze the effect of down-sampling we study a wide range of down-sampling rates: 0.01, 0.05, 0.10, 0.25, 0.50, and 1.00. For each down-sampling rate we perform experiments with RDS and IDS, both for lexicase as well as tournament selection, to get a comprehensive overview of the effects.

All experiments employ a crossover rate of 0.95 and a mutation rate of 1.00 as suggested in [6] using subtree crossover and subtree mutation. We initialize our population using position-independent grow [7] with a maximum initial tree depth of 10. The maximum tree depth is set to 17 and we employ an elitism of 5. For informed down-sampling the parent sampling rate is set to 0.01 and the population dynamics are updated every 10 generations as suggested in [3]. For tournament selection the tournament size is set to $n = 5$. For both selection methods the population size is set to $1,000$ and the number of generations is set to $2,000$. Following the approach suggested in Geiger et al. [13] we keep the number of generations constant for all down-sampling rates to better isolate the effects of down-sampling. All settings are summarized in Table 1.

We perform 100 independent runs with different seeds for each configuration and benchmark problem.

3.3 Metrics

To analyze the effects of different down-sampling rates we record multiple metrics during our experiments.

First, we record the success rate of each approach as the number of successful runs, meaning those runs that produce a solution that solves both training and test set. We only consider the first candidate solution that solves the training set and evaluate this solution at the end of a run on the test set. Additionally, we calculate the generalization rate as the fraction of runs that produce a solution that solves both training and test sets divided by the number of runs that return an individual that only solves the training set but does not solve the test set.

Second, to get a better understanding of the effects of down-sampling we report the generation in which the first solution (solving both training and test set) was found for each run. Furthermore, to analyze code growth behavior we measure the size of individuals as the number of tree nodes and report the average number of tree nodes per individual in a population. Additionally, we measure the behavioral (or error) diversity defined as the ratio of unique error vectors within a population. An error vector describes the non-aggregated training error of an individual on a case-by-case basis, where each element corresponds to a specific fitness case and assumes a binary value indicating whether that case was solved correctly or incorrectly.

Lastly, we measure how different down-sampling schemes affect the promotion and preservation of specialists for both selection approaches. We define a specialists as an individual that performs poorly in terms of aggregated error across all training cases but solves cases that the rest of the population struggles with [9]. In this study, we consider an individual to be a specialists if it is in the bottom quartile of a population in terms of aggregated error, but can solve a case that only 10% or less of the population can solve. For each generation, we record the number of specialists within a population and the rate of unique specialists selected for the next parent population (specialist preservation rate).

4 Results

4.1 Performance Comparison

We present the performance of each selection method and down-sampling scheme. Table 2 displays the success rate using different down-sampling approaches and down-sampling rates for both lexicase and tournament selection respectively. We make multiple observations: First, we observe that lexicase selection without down-sampling outperforms tournament selection without down-sampling on all benchmark problems with a significant difference in 5 out of 6 problems according to a two-sided proportions z-test with a significance level of $\alpha = 0.05$.

Second, we find that both RDS and IDS can improve the performance of lexicase selection as well as tournament selection. For lexicase selection the best performance across all settings is achieved using down-sampling methods on 4

Table 2. Success rates for the different problems using lexicase selection and tournament selection with different down-sampling methods and down-sampling rates. The highest success rate per selection method for RDS and IDS respectively is **bold** and the highest success rate overall is **<u>underlined</u>**. A * marks a significant statistical difference between lexicase and tournament selection for the given down-sampling scheme and benchmark problem according to a two-sided proportions z-test with significance level $\alpha = 0.05$ and Bonferroni-Holm correction. A † indicates a significant difference between that down-sampling scheme and NDS using the same statistical testing.

		NDS	RDS					IDS				
	r	1.0	0.01	0.05	0.10	0.25	0.50	0.01	0.05	0.10	0.25	0.50
Lexicase	Count Odds	78*	80*	81*	88*	**<u>92</u>***†	83*	74	83*	74*	79*	87*
	Fizz Buzz	78*	51†	67*	72*	73*	**<u>83</u>***	50*†	76*	81*	74*	76*
	Fuel Cost	**<u>56</u>***	32†	45	46	43*	39*	33*†	24†	37*†	42*	53*
	Grade	41*	23†	35	**<u>51</u>**	**<u>51</u>**	36	27	44	41	47	41
	Scrabble Score	**<u>21</u>***	11	6†	6†	9	17	14	6†	4†	10	17
	Small or Large	55	51	56	59	57	50	60	62	**64**	51	51
Tournament	Count Odds	45	56	61	63	50	52	**69**†	56	54	42	52
	Fizz Buzz	30	38	47	36	36	42	32	43	**49**†	29	41
	Fuel Cost	20	23	**37**†	31	23	21	11	17	18	23	15
	Grade	26	38	37	39	38	**41**	36	31	32	39	38
	Scrabble Score	8	**11**	7	**11**	7	8	**11**	4	4	5	7
	Small or Large	51	57	57	47	46	50	49	**<u>65</u>**	56	56	51

out of 6 problems and for tournament selection the best performance is achieved using down-sampling methods on all benchmark problems. However, this performance increase is only significant in 1 setting for lexicase selection (using a two-sided proportions z-test and Bonferroni-Holm correction with a significance level of $\alpha = 0.05$) and only in 3 settings for tournament selection. Additionally, if the down-sampling rate is too low the performance of lexicase selection can decrease significantly. Meanwhile lower down-sample rates seem to be beneficial for tournament selection. We do not identify a clear winner between RDS and IDS in our experiments.

Third, for the Count Odds, Fizz Buzz, and Fuel Cost problems we observe that lexicase still outperforms tournament selection when using down-sampling and in most cases this performance difference is statistically significant (again using a two-sided proportions z-test and Bonferroni-Holm correction with a significance level of $\alpha = 0.05$). For the Grade, Scrabble Score, and Small or Large problems we find that down-sampling can close the gap between lexicase and tournament selection leading to no significant differences between the two selection methods and tournament selection even surpassing lexicase in solve rate in some settings and problems.

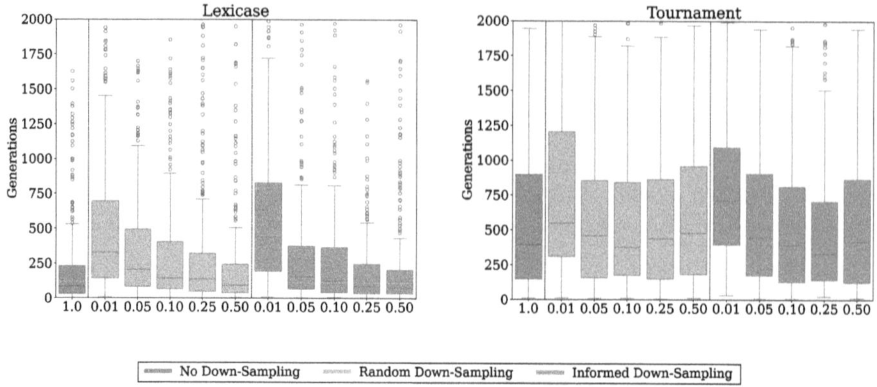

Fig. 1. Distribution of generations in which a solution was found that solves both training and test set using lexicase selection (left) and tournament selection (right) with different down-sampling methods and down-sampling rates aggregated over all 6 benchmark problems.

Overall, we find that down-sampling can be beneficial for lexicase selection depending on the benchmark problem and down-sampling rate. Additionally, we find that down-sampling is almost always beneficial for tournament selection in terms of success rate in our experiments. These findings are interesting considering that NDS runs for the same number of generation, therefore, down-sampling could improve the performance even further when using the additional program evaluations saved by down-sampling. Lastly, down-sampling can close the gap between lexicase and tournament selection, however, this seems to be problem dependent.

To better understand the difference in success rates we proceed to analyze generalization rates, convergence behavior, code growth, diversity, and specialist promotion.

4.2   Generalization Rates

Table 3 in Appendix A (found in our Zenodo repository [5]) displays the generalization rates across benchmark problems and down-sampling schemes. We find that down-sampling is clearly beneficial for tournament selection in terms of generalization and smaller down-samples lead to better generalization rates across all problems. However, this trend is less obvious for lexicase selection. While the best generalization rate is always observed with down-sampling, only for the Grade and Small or Large problem a clear benefit in terms of generalization rate can be observed. Overall we find that down-sampling does help with generalization, especially for tournament selection.

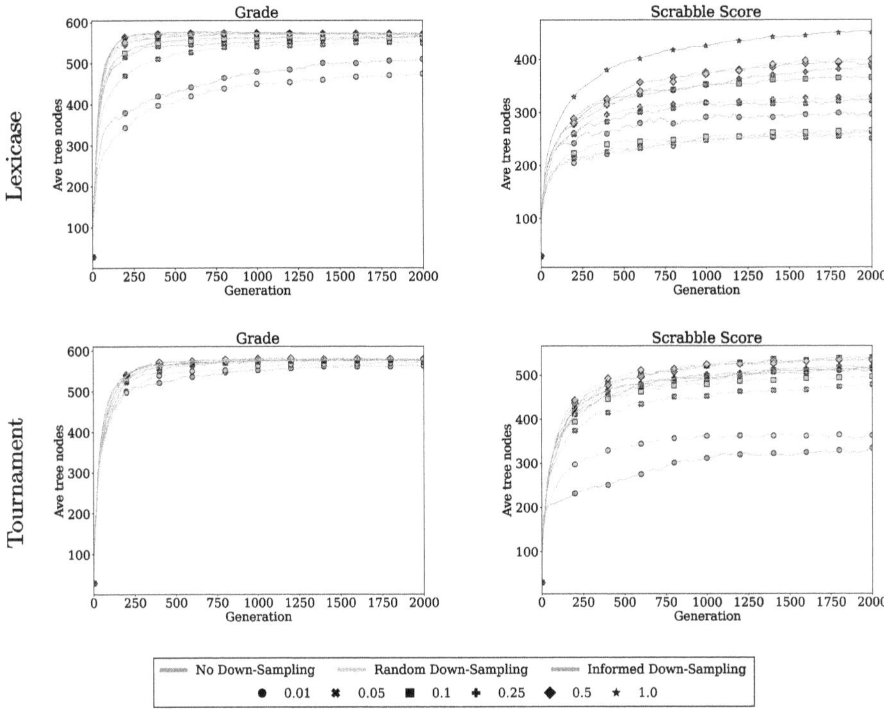

Fig. 2. Average number of tree nodes over generations using lexicase selection (top row) and tournament selection (bottom row) with different down-sampling methods and rates.

4.3   Convergence Behavior

Next, we investigate how the search converges depending on selection method and down-sampling scheme. Figure 1 displays the generations in which the first solution was found for each down-sampling scheme as box-plots for lexicase and tournament selection respectively. The results are aggregated over all 6 benchmark problems (see Appendix B for non-aggregated results). For lexicase we observe that with smaller down-sampling rates solutions are found at later generations for both RDS and IDS. We do not observe the same trend when using tournament selection, instead, there is no clear connection between down-sampling and convergence speed. When comparing lexicase and tournament selection, we find that tournament selection discovers solutions in later generations than lexicase selection, therefore, benefiting more from longer run times.

4.4   Code Growth Behavior

Furthermore, we investigate how down-sampling influences the code growth behavior. Figure 2 displays the average number of tree nodes of individuals in a population over generations using different down-sampling methods and down-sampling rates. For better readability all plots (and all following plots) are smoothed over 20 generations and markers are added every 200 generations. For reasons of space, we present only plots for the Grade and Scrabble Score problems in the main body of this paper and plots for the remaining problems in Appendix C. We chose Grade and Scrabble Score as those problems best reflect the effects for all 6 benchmark problems.

We find that both selection approaches produce smaller trees when paired with down-sampling methods and lower down-sampling rates result in smaller trees. RDS tends to produce slightly smaller trees than IDS except for a down-sampling rate of 0.01. Additionally, we observe that trees evolved using lexicase selection are either of the same size or smaller than using tournament selection. Overall, this shows that both RDS and IDS can slow down code growths for both lexicase and tournament selection with smaller down-sample rates having a higher effect.

4.5   Influence on Diversity

Next, we investigate the influence of down-sampling on the diversity of individuals in a population. Figure 3 displays the behavioral diversity over generations for different down-sampling methods and down-sampling rates. We initially observe the most diversity for NDS when using lexicase selection across all benchmark problems and small down-sample rates lead to lower diversity. However, over generations the diversity decreases for 4 out of 6 problems and smaller down-sample rates start to overtake larger ones in diversity by generation 500. This might be an effect of the early convergence of lexicase without down-sampling as observed before. Only for the Fuel Cost and Scrabble Score problems the initial difference in diversity is present over all 2,000 generations. Interestingly, these two problems are also those problems that achieve the highest success rate with NDS. Additionally, we find that that IDS promotes more diversity in these two problems as well, while RDS promotes more diversity in the other problems. For tournament selection we observe that down-sampling generally increases diversity, however, a down-sample rate of 0.01 produces a less diverse population for the Fuel Cost and Scrabble Score problems. In general RDS seems to produce slightly more diverse populations than IDS. Only for the Count Odds problem we find different results. Here the diversity decreases with down-sampling and RDS leads to slightly less diverse populations. When comparing lexicase and tournament selection directly, we find that lexicase selection initially has more diverse populations, however, over generations this evens out with tournament selection. Only for the Fuel Cost and Scrabble Score problems lexicase selection promotes a higher diversity when compared to tournament selection for larger down-sample rates.

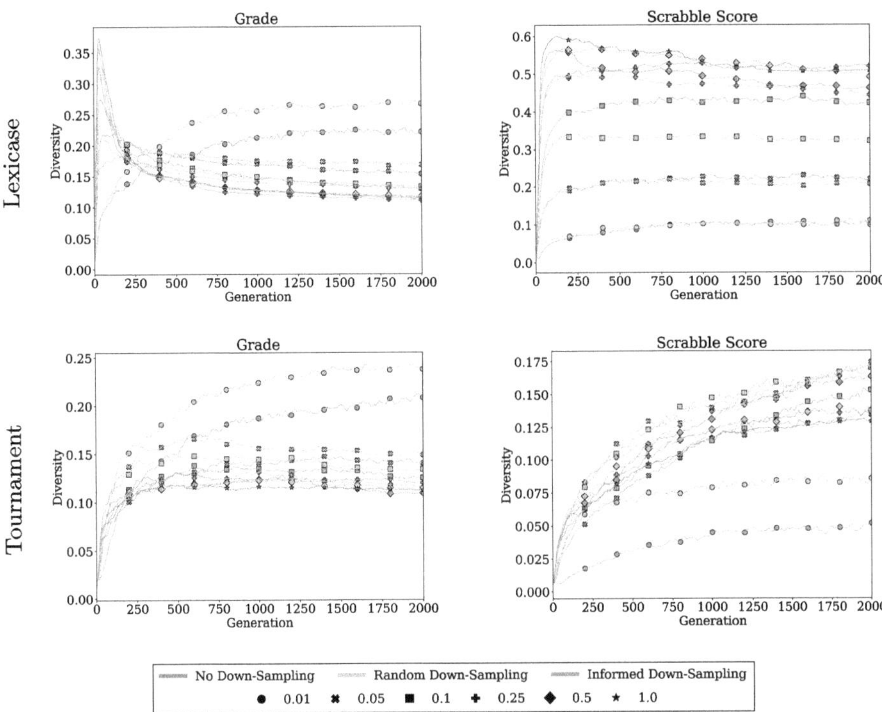

Fig. 3. Population diversity over generations using lexicase selection (top row) and tournament selection (bottom row) with different down-sampling methods and rates.

4.6   Influence on Specialist Promotion

Lastly, as the performance of lexicase selection is often attributed to better promotion and preservation of specialists [23] we analyze the influence of down-sampling on those factors. Figure 4 displays the number of specialists in a population over generations with different down-sampling methods and down-sampling rates. First, we observe that for lexicase selection down-sampling tends to reduce the number of specialists in a population and smaller down-samples result in less specialists. We do not find a clear trend if RDS or IDS promote more specialists for lexicase selection. Second, we find that for tournament selection down-sampling can increase the number of specialists per generation, however, this depends on the problem and down-sampling rate. For example, for the Small or Large problem we observe a decrease in specialists with down-sampling, while we observe an increase for Grade and Count Odds if the down-sampling rate is not too small. For the other problems we observe mixed results. RDS seems to produce slightly more specialists than IDS for tournament selection. When comparing lexicase and tournament selection directly, we find that lexicase has a high number of specialists in the population from the start while the number of specialists grows over generations when using tournament selection (except

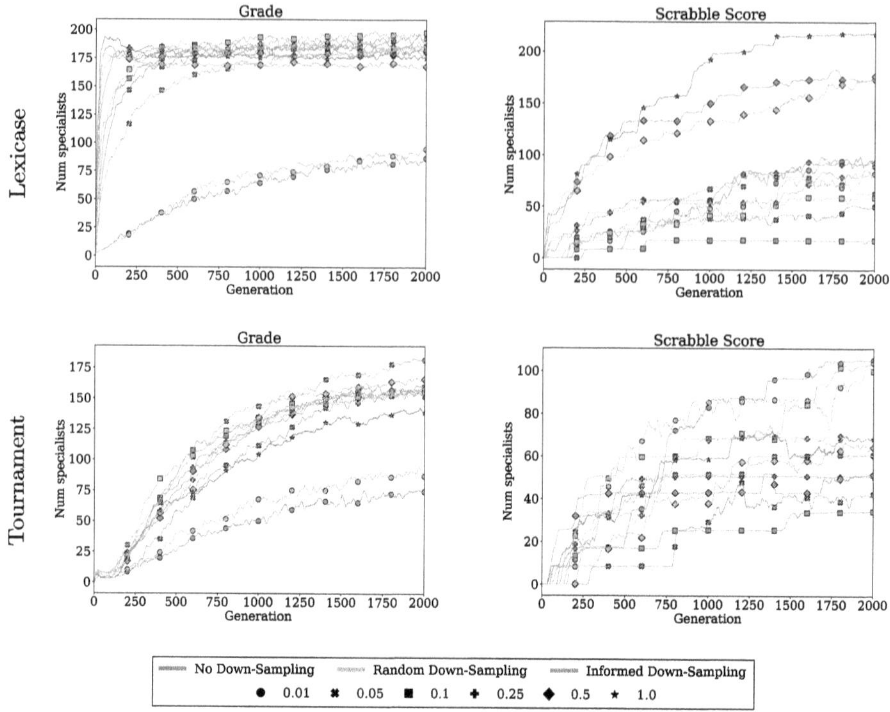

Fig. 4. Number of specialists over generations using lexicase selection (top row) and tournament selection (bottom row) with different down-sampling methods and rates.

for Small or Large). This results in an equal amount of specialists over long generations for the Fizz Buzz, Fuel Cost, and Grade problems. However, lexicase produces substantially more specialists initially and for the Count Odds and Scrabble Score problems the number of specialists remains higher.

We also investigate if those specialists are preserved at a different rate depending on selection method and down-sampling scheme. Figure 5 displays the preservation rate of unique specialists over generations. For lexicase selection we observe that down-sampling leads to lower specialist preservation rates and smaller down-sampling rates result in less specialist preservation. For tournament selection we observe the opposite: Down-sampling increases the specialist preservation rate when compared to NDS, however, we do not observe a clear correlation with down-sampling rates. For both selection methods we do not notice a clear difference between RDS and IDS. Overall, the specialist preservation rate is higher for 4 out of 6 problems (Count Odds, Fizz Buzz, Fuel Cost, Scrabble Score) when using lexicase selection and about even for the other 2 problems (Grade, Small or Large). Interestingly, those problems where the specialist preservation rate is higher are also those problems where lexicase still performs better even with down-sampling (apart from Scrabble Score where both selection methods find very little solutions). Similarly, as smaller down-

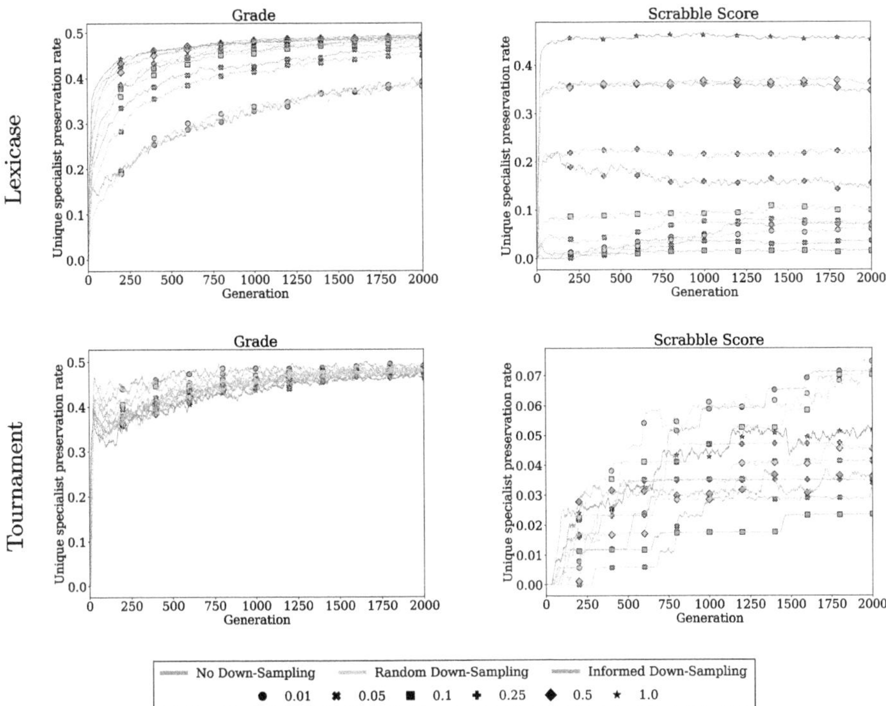

Fig. 5. Unique specialist preservation rate over generations using lexicase selection (top row) and tournament selection (bottom row) with different down-sampling methods and rates.

sampling rates decrease the specialist preservation rate for lexicase selection we also observed a decrease in performance for these settings (see Table 2). This indicates that the promotion and preservation of specialists is the main driver behind the performance differences between lexicase and tournament selection.

5 Discussion and Implications

In our experiments we found similar results as in Geiger et al. [13] for the program synthesis domain in terms of generalization, code growth, and diversity when pairing down-sampling methods with lexicase and tournament selection. This partially closes the gap between lexicase and tournament selection for some benchmark problems, however, in other benchmark problems lexicase is still significantly better than tournament even with down-sampling. Our results indicate that this is caused by the superior promotion and preservation of specialists when using lexicase selection. These specialists serve an important role in solving complex program synthesis problems. Therefore, while down-sampling closes the gap between both selection approaches for the easier benchmark problems, it fails to do so for the more difficult problems that require loops or complex control structures. In the domain of symbolic regression those modal problems rarely occur. However, we suspect that we would observe similar findings when analyzing piecewise functions for symbolic regression.

Future work should focus on what problem structures benefit the most from these specialist promotion effects to make informed decisions about the selection scheme. Additionally, future work should investigate how the effects of down-sampling observed in our study can be used to further improve down-sampling and selection techniques, e.g., by adapting the down-sampling rate during an evolutionary run.

Finally, our results are limited to a single GP system and two selection approaches. However, we know from literature that representations have significant effects on GP systems [32]. This has also been observed for down-sampling methods like IDS [3]. Therefore, expanding these results to other GP systems and selection approaches could further strengthen the results.

6   Conclusion

In this work we investigated the effects of down-sampling for tournament and lexicase selection for the program synthesis domain in terms of performance gains, generalization, convergence, code growth, diversity, and specialist preservation, to answer the question if down-sampling can close the gap between the two selection methods in the program synthesis domain similar to the symbolic regression domain. We performed extensive experiments for a grammar guided GP system using well known benchmark problems.

We found that down-sampling indeed improves the performance of tournament selection with similar effects as in prior work in the symbolic regression domain. However, for more complex program synthesis problems lexicase still significantly outperforms tournament selection, mostly due to its superior promotion and preservation of specialists.

Acknowledgments. The author would like to thank Dominik Sobania, Alina Geiger, and Franz Rothlauf, as well as the anonymous reviewers, for their discussions and feedback, which helped shape this work.

References

1. Boldi, R., et al.: The problem solving benefits of down-sampling vary by selection scheme. In: Proceedings of the Companion Conference on Genetic and Evolutionary Computation, pp. 527–530 (2023)
2. Boldi, R., et al.: Untangling the effects of down-sampling and selection in genetic programming. In: Artificial Life Conference Proceedings 36, vol. 2024, p. 88. MIT Press, Cambridge (2024)
3. Boldi, R., et al.: Informed down-sampled lexicase selection: identifying productive training cases for efficient problem solving. Evol. Comput. **32**(4), 307–337 (2024)

4. Boldi, R., Lalejini, A., Helmuth, T., Spector, L.: A static analysis of informed down-samples. In: Proceedings of the Companion Conference on Genetic and Evolutionary Computation, pp. 531–534 (2023)
5. Briesch, M.: Supplementary material of on the effects of down- sampling for tournament and lexicase selection in program synthesis (2026). https://doi.org/10.5281/zenodo.18360803
6. Briesch, M., Sobania, D., Rothlauf, F.: On the trade-off between population size and number of generations in GP for program synthesis. In: Proceedings of the Companion Conference on Genetic and Evolutionary Computation, pp. 535–538 (2023)
7. Fagan, D., Fenton, M., O'Neill, M.: Exploring position independent initialisation in grammatical evolution. In: 2016 IEEE Congress on Evolutionary Computation (CEC), pp. 5060–5067. IEEE (2016)
8. Fenton, M., McDermott, J., Fagan, D., Forstenlechner, S., Hemberg, E., O'Neill, M.: Ponyge2: grammatical evolution in python. In: Proceedings of the Genetic and Evolutionary Computation Conference Companion, pp. 1194–1201 (2017)
9. Ferguson, A.J., Hernandez, J.G., Junghans, D., Lalejini, A., Dolson, E., Ofria, C.: Characterizing the effects of random subsampling on lexicase selection. In: Genetic Programming Theory and Practice XVII, pp. 1–23. Springer, Heidelberg (2020)
10. Forstenlechner, S., Fagan, D., Nicolau, M., O'Neill, M.: A grammar design pattern for arbitrary program synthesis problems in genetic programming. In: European Conference on Genetic Programming, pp. 262–277. Springer, Heidelberg (2017)
11. Forstenlechner, S., Nicolau, M., Fagan, D., O'Neill, M.: Grammar design for derivation tree based genetic programming systems. In: European Conference on Genetic Programming, pp. 199–214. Springer, Heidelberg (2016)
12. Gathercole, C., Ross, P.: Dynamic training subset selection for supervised learning in genetic programming. In: International Conference on Parallel Problem Solving from Nature, pp. 312–321. Springer, Heidelberg (1994)
13. Geiger, A., Briesch, M., Sobania, D., Rothlauf, F.: Was tournament selection all we ever needed? A critical reflection on lexicase selection. In: European Conference on Genetic Programming (Part of EvoStar), pp. 207–223. Springer, Heidelberg (2025)
14. Geiger, A., Sobania, D., Rothlauf, F.: Down-sampled epsilon-lexicase selection for real-world symbolic regression problems. In: Proceedings of the Genetic and Evolutionary Computation Conference, pp. 1109–1117 (2023)
15. Geiger, A., Sobania, D., Rothlauf, F.: A comprehensive comparison of lexicase-based selection methods for symbolic regression problems. In: European Conference on Genetic Programming (Part of EvoStar), pp. 192–208. Springer, Heidelberg (2024)
16. Geiger, A., Sobania, D., Rothlauf, F.: A performance analysis of lexicase-based and traditional selection methods in gp for symbolic regression. ACM Trans. Evol. Learn. (2025)
17. Gonçalves, I., Silva, S.: Experiments on controlling overfitting in genetic programming. In: 15th Portuguese Conference on Artificial Intelligence (EPIA 2011), pp. 10–13 (2011)
18. Gonçalves, I., Silva, S., Melo, J.B., Carreiras, J.M.: Random sampling technique for overfitting control in genetic programming. In: European Conference on Genetic Programming, pp. 218–229. Springer, Heidelberg (2012)
19. Helmuth, T., Abdelhady, A.: Benchmarking parent selection for program synthesis by genetic programming. In: Proceedings of the 2020 Genetic and Evolutionary Computation Conference Companion, pp. 237–238 (2020)

20. Helmuth, T., Kelly, P.: Psb2: the second program synthesis benchmark suite. In: Proceedings of the Genetic and Evolutionary Computation Conference, pp. 785–794 (2021)
21. Helmuth, T., McPhee, N.F., Spector, L.: Effects of lexicase and tournament selection on diversity recovery and maintenance. In: Proceedings of the 2016 on Genetic and Evolutionary Computation Conference Companion, pp. 983–990 (2016)
22. Helmuth, T., McPhee, N.F., Spector, L.: Lexicase selection for program synthesis: a diversity analysis. In: Genetic Programming Theory and Practice XIII, pp. 151–167. Springer, Heidelberg (2016)
23. Helmuth, T., Pantridge, E., Spector, L.: Lexicase selection of specialists. In: Proceedings of the Genetic and Evolutionary Computation Conference, pp. 1030–1038 (2019)
24. Helmuth, T., Spector, L.: General program synthesis benchmark suite. In: Proceedings of the 2015 Annual Conference on Genetic and Evolutionary Computation, pp. 1039–1046 (2015)
25. Helmuth, T., Spector, L.: Problem-solving benefits of down-sampled lexicase selection. Artif. Life **27**(3–4), 183–203 (2022)
26. Helmuth, T., Spector, L., Matheson, J.: Solving uncompromising problems with lexicase selection. IEEE Trans. Evol. Comput. **19**(5), 630–643 (2014)
27. Hernandez, J.G., Lalejini, A., Dolson, E., Ofria, C.: Random subsampling improves performance in lexicase selection. In: Proceedings of the Genetic and Evolutionary Computation Conference Companion, pp. 2028–2031 (2019)
28. Hernandez, J.G., Lalejini, A., Ofria, C.: An exploration of exploration: measuring the ability of lexicase selection to find obscure pathways to optimality. In: Genetic Programming Theory and Practice xviii, pp. 83–107. Springer, Heidelberg (2022)
29. La Cava, W., Helmuth, T., Spector, L., Moore, J.H.: A probabilistic and multi-objective analysis of lexicase selection and ε-lexicase selection. Evol. Comput. **27**(3), 377–402 (2019)
30. La Cava, W., Spector, L., Danai, K.: Epsilon-lexicase selection for regression. In: Proceedings of the Genetic and Evolutionary Computation Conference 2016, pp. 741–748 (2016)
31. Lalejini, A., Moreno, M.A., Hernandez, J.G., Dolson, E.: Phylogeny-informed fitness estimation for test-based parent selection. In: Genetic Programming Theory and Practice XX, pp. 241–261. Springer, Heidelberg (2024)
32. Rothlauf, F.: Representations for genetic and evolutionary algorithms. In: Representations for Genetic and Evolutionary Algorithms, pp. 9–32. Springer, Heidelberg (2006)
33. Schweim, D., Sobania, D., Rothlauf, F.: Effects of the training set size: a comparison of standard and down-sampled lexicase selection in program synthesis. In: 2022 IEEE Congress on Evolutionary Computation (CEC), pp. 1–8. IEEE (2022)
34. Sobania, D., Schweim, D., Rothlauf, F.: A comprehensive survey on program synthesis with evolutionary algorithms. IEEE Trans. Evol. Comput. **27**(1), 82–97 (2022)
35. Spector, L.: Assessment of problem modality by differential performance of lexicase selection in genetic programming: a preliminary report. In: Proceedings of the 14th Annual Conference Companion on Genetic and Evolutionary Computation, pp. 401–408 (2012)
36. Whigham, P.A., et al.: Grammatically-based genetic programming. In: Proceedings of the Workshop on Genetic Programming: From Theory to Real-World Applications, Tahoe City, California, USA, vol. 16, pp. 33–41 (1995)

Comparison of Parent and Environmental Selection Schemes in Genetic Programming

Vladimir Stanovov[(✉)]

Reshetnev Siberian State University of Science and Technology, Institute of Informatics and Telecommunications, Krasnoyarsk 660037, Russia
vladimirstanovov@yandex.ru

Abstract. The selection mechanism is one of the crucial parts of any genetic programming algorithm, which determines its efficiency. In this paper 23 different variants of selection schemes for genetic programming are compared, including tournament, lexicase, and Friedman ranking-based, where first and second parent can be chosen differently. Four environmental selection, also known as replacement schemes are compared, including elitism, pairwise comparison, fitness sorting-based and Friedman ranking-based. The genetic programming has two populations with different replacement schemes. The comparisons are made for standard and down-sampled lexicase selection on 96 regression datasets. The analysis of the results shows that some of the typically applied population management schemes are suboptimal and can be easily changed to more efficient ones. The combination of current individual for first parent and lexicase selection for second, as well as soring-based replacement scheme demonstrates the highest performance among tested configurations.

Keywords: Genetic programming · Symbolic regression · Parent selection · Lexicase selection · Differential Evolution · L-NTADE

1 Introduction

The area of evolutionary computation (EC) has several directions, which focus on different research areas. These include optimization (pseudo-boolean, numerical, combinatorial), machine learning (regression, classification, clustering) [29] and their variants. The genetic programming (GP) [12] algorithms, which include tree-based GP (TGP) [1], linear GP (LGP) [3,6], Cartesian GP (CGP) [13], grammatical evolution (GE) [17,18] and Push GP [21,22] have found various applications in both optimization and learning. One of the most important parts of any GP is the parent selection operation for further crossover, which is independent of the solution representation.

Since the early stages of EC development as a field, several classical selection schemes were proposed, which are still in use today. These include fitness proportionate selection, also known as roulette wheel, tournament selection and rank-based selection [30,31]. Most of the other schemes are in fact variations of one of these, which change their parameters, but not the mechanism behind. However, the latter proposed lexicase selection [20] is significantly different from them, as it considers every training case separately, without aggregation of performance values. Its limitation, however, is that it is only applicable to the problems, where the fitness values are composed of several performance measures across different cases [10].

Other variants of selection mechanisms, specific for genetic programming, were also proposed, for example batch-tournament selection, where the training cases are combined into batches, and the fitness values during tournament are compared according to current batch. The ε-lexicase selection [4,5] is an extended variant, which is used for regression datasets, or cases where the performance values are numeric. The ε-plexicase selection is another modification, where the individuals are chosen based on approximated probabilities of being selected by ε-lexicase. Other modifications include down-sampling [8,9], which evaluates individuals by a fraction of the training set, resulting in a significant performance improvement, while decreasing compute time. Informed down-sampling [2] further improves this modification by creating a subset, composed of diverse cases.

Although there are many new approaches to select parents in genetic programming, in most studies the environmental selection stage, also known as replacement of parents with offspring, is often not even mentioned in the papers. The typically used replacement scheme is the elitism, where all individuals in the current population are replaced by offspring, with the exception of one or several best individuals, which are always preserved. Other evolutionary algorithms sometimes use different strategies, for example in differential evolution (DE) [15,26] the offspring is directly compared to its parent, and replaces it only if there is an improvement. Another known replacement scheme can be called "best of parents and offspring" - in this case they for a joined population, which is then sorted by fitness and the best individuals are to end up in the next generation. Similar schemes are sometimes used in multi-objective optimization, for example in a classical NSGA-II [7] algorithm.

Another important part of parents selection is the difference between the mechanisms for choosing first and second parent. In most genetic programming variants they are chosen using the same scheme, just applied twice. But, for example, in DE the basic vector to be changed is always the i-th individual in the population, $i = 1, 2, ...N$. That is, it is the current individual, which is combined with others using specialized mutation mechanisms. Such mechanism can be described as "give everyone a chance", as every individual participates in creating an offspring, and it has been proven to be an efficient strategy for DE. Another modification, recently proposed for DE is the usage of two populations, one with best solutions, and another with the newest ones. All of these mechanisms can be quite easily applied to tree-based and other variants of GP, and significantly improve the performance.

Considering all of the above, this study focuses on testing how selecting first and second parent with different mechanisms affects the performance of the tree-based GP. In particular, 23 variants of selection are considered, including selection from one of the two populations of best and newest solutions. Four replacement schemes are also compared, and the experiments are repeated for the standard and down-sampled ε-lexicase selection.

The rest of the paper is organized as follows. The next section describes the related work and used selection mechanisms, third section presents proposed selection and replacement mechanisms, the fourth section contains experiments and results, and the last section concludes the paper.

2   Related Work

The general workflow of most genetic programming algorithms includes the following steps: initialization, parents selection, crossover, mutation and replacement, the last four steps are repeated until a stopping condition is met. This means that there is a current population with solutions p_i, $i = 1, 2, ..., N$, where N is the population size, and the offspring population p_i of the same size. As we consider only the tree-based GP in this study, each individual is a tree, consisting of functional and terminal nodes. When evaluating a tree on the set of training cases t_j, $j = 1, 2, ..., |T|$, the error values are e_t are recorded for every individual of the population. The fitness f_i values are usually calculated as mean squared error (MSE) or coefficient of determination R^2, which has the best possible value of 1, making it easier to compare results on different datasets.

Some of the changes to the structure of selection in GP are inspired by another evolutionary algorithm, namely differential evolution, and its many variants. The reason for this is that DE has a flipped order of operations: the main loop starts with mutation, followed by crossover and then selection. Although this is mainly a terminology issue, as the selection in DE is in fact replacement, and for mutation several vectors should be selected, but this is not stated explicitly in most papers. The success of DE algorithms for numerical optimization, where they won most competitions leaves one wondering if some part of its structure can be efficiently applied in GP. In particular, in DE mutations usually the current vector is used, i.e. the i-th solution from the population is the basic vector to generate a new one. The newly generated trial vector is then evaluated and compared to the basic vector by their fitness values. Following this scheme, a variant of GP can be derived, where to produce an offspring two parents are chosen, one of which is the i-th individual p_i, and another is selected with any of the used selection mechanisms. If the fitness of the newly generated offspring o_i is better, then it replaces its parent. Attempts to change the workflow of genetic algorithm and genetic programming has already been made in [23] and [19].

Another feature of differential evolution is the application of more than one population. For example, in a popular SHADE algorithm [27] and its successors like L-SHADE [28] the archive set is introduced, which consists of the individuals, which were removed from the population during replacement. These vectors are then used in the mutation, increasing the diversity of generated trial solutions. An even more general approach was considered in [11], where all generated individuals were saved in the population with no size limitations. From this unbounded population vectors were chosen for mutation using different selection mechanisms, based on their fitness and novelty. Based on this study in [24] the dual-population framework was proposed, where one population contained the best vectors from the whole search process, and the other kept the latest individuals. The resulting L-NTADE algorithm was shown to be competitive with the best DE variants.

Regarding the selection mechanisms, in addition to the classical selections like tournament and lexicase selection variants, there some other approaches, which were applied to genetic programming. For example, in [25], GP was used to design parameter adaptation for DE, and the solutions were compared by their results on several functions and runs. For every run and function the whole population was ranked, and then the ranks were summed - this procedure is taken from the Friedman statistical test, which compares the differences across several tests. Although Friedman ranking aggregates performance into a single value, it is less sensitive to the distribution of values, unlike, for example, MSE, which can be skewed by an outlier. In contrast to lexicase selection, the Friedman ranking-based selection (different from rank-based selection) prefers individuals which are generalists, not specialists.

In the next section the details of implementing the mentioned selection and ranking mechanisms is described in application to genetic programming.

3   Proposed Approach

In order to describe all variants of selection and replacement used in this study, first the structure of the GP should be presented. Following the ideas of L-NTADE, the GP in this paper has two populations, the population of best (top) solutions p_i^t, and the population of latest (newest) solutions p_i^n, $i = 1, 2, ..., N$. These populations have the same size, and the offspring population o_i is also of size N. The second population p_i^n will be updated in the same way as in L-NTADE, but the update method for p_i^t will be changed. The main steps of GP are shown in Algorithm 1. The parameters such as population size, tournament size, maximum tree depth are set to typical values, which are used in other studies.

In line 6 of algorithm 1 the notation $p_i^{n/t}$ means either individual from the top population or the newest population. As at initialization these population are same, it does not matter which individual to evaluate. As for lines 9 and 10, the selected individual can be from either from the population of top or newest individuals, and in this case there is a difference. Note that in line 14 the update strategy for the population of newest individuals is fixed, and it is the same as in L-NTADE. The newly generated offspring replaces one of the individuals in the newest population if it is better than the $i - th$ individual from the population of top individuals.

The naming of the first population can be a bit misleading, as it does not always contain the best individuals in terms of R^2 metric - this will be the case only if the replacement sorting of parents and offspring is applied in line 19. The four different strategies used in this study for replacement are the following:

Algorithm 1. GP with two populations

1: Input: training set T
2: Output: best individual p^t_{best}
3: Set population size $N = 100$, max depth $D_{max} = 7$, max length $L_{max} = 75$
4: Set resource $NFE_{max} = 100000$, tournament size $ts = 3$, newest index $ni = 1$
5: Initialize population $\{p^n_i, ..., p^n_N\}$ and copy to $\{p^t_i, ..., p^t_N\}$, $NFE = 1$
6: Evaluate individuals, calculate error values $e_{t_j}(p^{n/t}_i)$, $j = 1, 2, ..., |T|$ and $R^2(p^{n/t}_i)$
7: **while** $NFE \leq NFE_{max}$ **do**
8:     **for** $i = 1$ to N **do**
9:         Select first parent $p^{n/t}_{s1}$ with first selection method
10:         Select first parent $p^{n/t}_{s2}$ with second selection method
11:         Apply crossover to produce o_i
12:         Mutate o_i
13:         Calculate error values $e_{t_j}(o_i)$ and $R^2(o_i)$, $NFE = NFE + 1$
14:         **if** $R^2(o_i) > R^2(p^t)$ **then**
15:             Copy o_i to p^n_{ni}
16:             $ni = mod(ni + 1, N)$
17:         **end if**
18:     **end for**
19:     Create new population p^t from o and p^t
20: **end while**
21: Return best solution p^t_{best}

---

1. Elitism - the offspring population o replaces all individuals in p^t, except for the best one, if it is in p^t_{best};
2. Pairwise comparison - p^t_i is replaced by o_i if $R^2(o_i) > R^2(p^t)$;
3. Sorting - best N individuals from p^t and o are chosen after ordering them by R^2 values;
4. Friedman ranking - same as sorting, but based on ranks r_i, calculated on joined population of p^t and o;

As for the selection mechanisms, there were four main selection methods, and all except one were be applied to both populations:

1. Current - the i-th individual from population p^t is chosen;
2. Tournament - three individuals are chosen and compared by R^2 metric, the best is returned. The choice can be made from either p^t or p^n, but always from the same population;
3. Lexicase - the ε-lexicase selection is used, with relaxation threshold chosen based on median deviation from the median, calculated for p^t and p^n separately;
4. Friedman - all population is ranked for every case separately, and ranks are summed for each individual. These ranks are used in tournament selection with $ts = 3$ instead of R^2 metric. Ranks are calculated separately for p^t and p^n.

The tournament, lexicase and Friedman selections can be applied to either p^t or p^n. In case if the down-sampling is used, the Friedman ranks are calculated based only on the cases present in the sample. Note that Friedman ranking based selection is different from typically used rank-based selection, as the latter uses aggregated fitness values, whereas Friedman ranking utilized information about performance on every training case separately.

As mentioned before, there were 23 different selection combinations tested, which differ in the way the first and second parent were chosen. In Table 1 these combinations are presented, together with their shorter notation used later.

Table 1. Selection combinations

#	First parent	Second parent	Notation
1	Current	Tournament(p^t)	C,T(B)
2	Current	Tournament(p^n)	C,T(N)
3	Current	Lexicase(p^t)	C,L(B)
4	Current	Lexicase(p^n)	C,L(N)
5	Current	Friedman(p^t)	C,F(B)
6	Current	Friedman(p^n)	C,F(N)
7	Tournament(p^t)	Tournament(p^t)	T(B),T(B)
8	Tournament(p^t)	Tournament(p^n)	T(B),T(N)
9	Tournament(p^t)	Lexicase(p^t)	T(B),L(N)
10	Tournament(p^t)	Lexicase(p^n)	T(B),L(B)
11	Tournament(p^t)	Friedman(p^t)	T(B),F(N)
12	Tournament(p^t)	Friedman(p^n)	T(B),F(B)
13	Tournament(p^n)	Lexicase(p^t)	T(N),L(N)
14	Tournament(p^n)	Lexicase(p^n)	T(N),L(B)
15	Tournament(p^n)	Friedman(p^t)	T(N),F(N)
16	Tournament(p^n)	Friedman(p^n)	T(N),F(B)
17	Lexicase(p^t)	Lexicase(p^t)	L(B),L(B)
18	Lexicase(p^t)	Lexicase(p^n)	L(B),L(N)
19	Lexicase(p^t)	Friedman(p^t)	L(B),F(B)
20	Lexicase(p^t)	Friedman(p^n)	L(B),F(N)
21	Friedman(p^t)	Friedman(p^t)	F(B),F(B)
22	Friedman(p^t)	Friedman(p^n)	F(B),F(N)
23	Friedman(p^n)	Friedman(p^n)	F(N),F(N)

The functional set of the GP included addition ($x + y$), subtraction ($x - y$), multiplication ($x * y$), analytic quotient (AQ [14], $\frac{x}{\sqrt{y^2}}$) instead of division, logarithm ($ln(1 + x)$), square root ($sqrt(x)$), sine ($sin(x)$), cosine ($cos(x)$), absolute value ($|x|$), negative value ($-x$), square (x^2), minimum ($min(x,y)$) and maximum ($max(x,y)$). All operations are protected, i.e. they return either minimum or maximum floating point value on overflow, or zero in case of NaN. The constants are generated with Cauchy distribution with a scale parameter of 5. During mutation, previous constant values are used as location parameter to generate new values.

The algorithm parameters and details about the experimental setup are provided in the next section.

4 Experiments and Results

The approach described here was implemented in C++ using only standard libraries, compiled with GCC 13.3.0 and ran on a cluster of 16 machines with AMD Ryzen 5700X CPU each working under Ubuntu Linux 20.04. The parallel processing was implemented using OpenMPI 4.0.3, and the network file system was used to store the datasets and results. The post-processing of results, statistical tests and visualizations were implemented in Python.

The regression datasets used in to evaluate the performance of the proposed approach were taken from the Penn Machine Learning Benchmark dataset collection [16]. In order to speed up the calculations, the datasets with more than 2000 instances were filtered out, giving a total of 96 datasets. The GP ran 25 times for each dataset and each parameter combination, 90% went to the training set, and the rest - to the test set. The down-sampling factor was set to 5, i.e. the training sample consisted of 20% of the cases, chosen randomly.

As there were 23 different selection combinations, 4 replacement combinations and two sampling strategies (standard and down-sampled), this resulted in a total of 184 different experiments for each of the 96 datasets.

Figure 1 shows the results of the Friedman ranking of the best R^2 on the test set. The results on each dataset were ranked separately, and then summed together. If several algorithms had same performance, the fractional ranks were assigned to them. On the horizontal axis the types of sampling (standard or down-sampled) are shown, as well as replacement scheme, while the vertical axis shows different selection strategies. Smaller ranks are better.

As can be seen from Fig. 1, the elitism replacement strategy performs worse than others, while the sorting is the best, both with and without down-sampling. The down-sampling performs better then standard method, as expected. As for the selection methods, the first six combinations where the first parent is simply the current individual, performed better than other strategies. However, the lexicase selection also performed well, while tournament selection and Friedman ranking had similar, but worse results.

In order to compare each couple of strategies, the Mann-Whitney statistical test was applied, with normal approximation of the U statistics, tie-breaking and significance level set to $p = 0.01$. Figure 2 shows the results of pairwise comparison of all selection schemes, with each cell containing the number datasets where the scheme on the vertical axis was significantly better then the one on the horizonal axis, minus the number of times when it was significantly worse. In other words, Fig. 2 shows the number of wins minus the number of losses, out of 96 datasets.

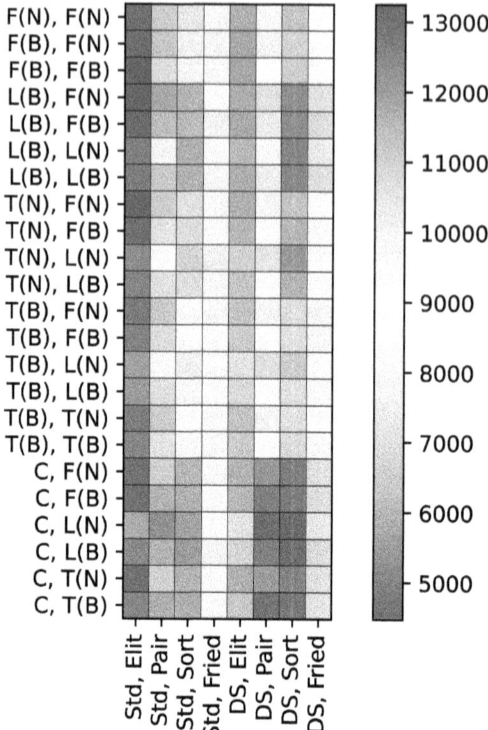

Fig. 1. Friedman ranking of the results of all experiments. The ranks are summed across all datasets and runs for every combination of settings, and smaller ranks are assigned to better R^2 values on the test set. The large number of experiments leads to high total ranks, i.e. from around 5000 to 13000. The labels on the X axis encode the sampling method, standard (Std) or down-sampled (DS), as well as replacement scheme: elitism (Elit), pairwise comparison (Pair), sorting (Sort) and Friedman ranking (Fried).

According to Fig. 2, the worst combinations are tournament selection from the top population and either Friedman ranking-based selection from top population, or lexicase selection from top population. If, however, the lexicase selection is applied to both individuals, no matter from which population they are chosen, the performance becomes comparable to the configurations, where the current individual is used as a parent. If, however, the tournament selection is applied to the population of newest individuals, its performance is significantly improved, especially if the second individual is chosen with lexicase selection from the population of newest. The best results are demonstrated by the combination of current and lexicase election from either top or newest population.

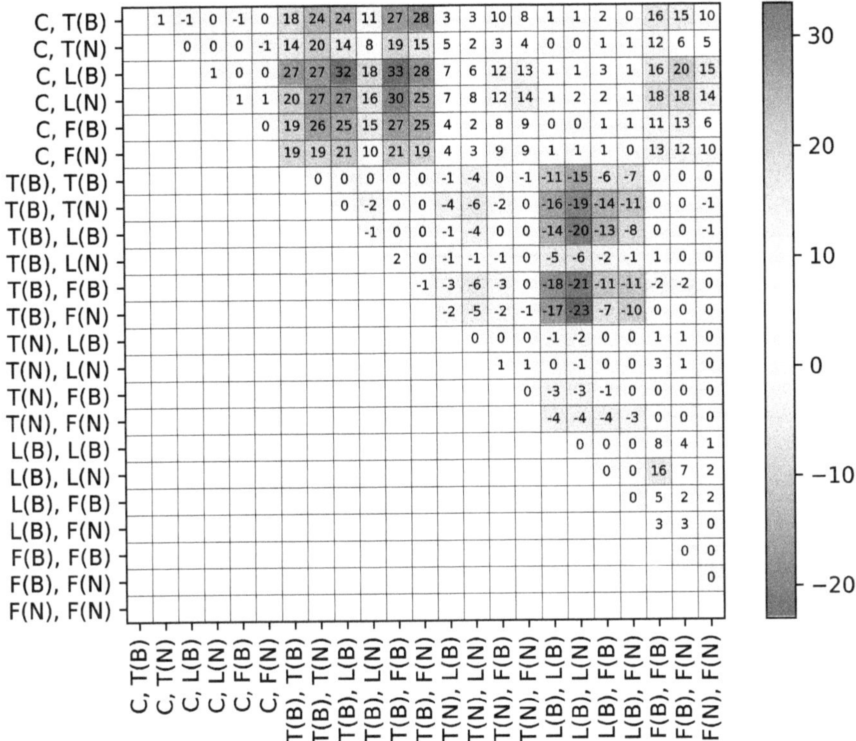

Fig. 2. Comparison with Mann-Whiney statistical tests. The values in the cells show the total score, calculated as the number of datasets, on which the selection method on the Y axis outperformed the selection method on the X axis. This number is calculated as number of wins minus number of losses, larger values mean that the method on the vertical axis is better.

To compare the performance of replacement and sampling strategies Fig. 3 demonstrates box plots of R^2 values on both training and test sets based on the results of the third selection configuration (C, L(B)) on all datasets.

As Fig. 3 shows, the best two replacement strategies are the sorting and pairwise comparison - they have the best values on the training set. The Friedman ranking here is slightly better than the elitism strategy. Similar trends are observed on the test set, with the only exception that the standard strategy and down-sampling have very similar performance.

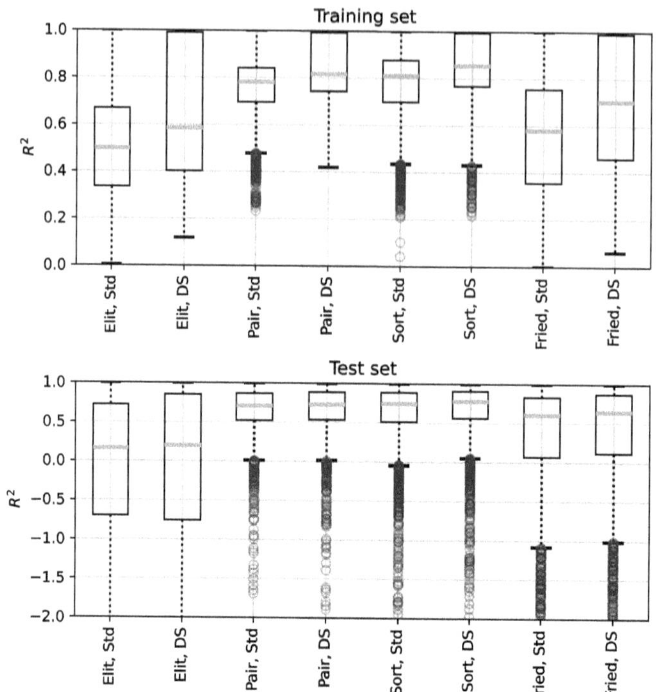

Fig. 3. Box plots of R^2 values with different replacement and sampling strategies, selection configuration C, L(B), higher is better.

Figure 4 compares the performance of selection configurations for the case of sorting-based replacement and down-sampling.

The first six selection schemes, which use the current individual as one of the parents, demonstrate much better results on both training and test samples, in terms of median and standard deviation. The only two schemes which have similar performance are the ones which use lexicase selection for both parents. As for the differences between choosing an individual from population of newest or top solutions, they seem negligible in most cases.

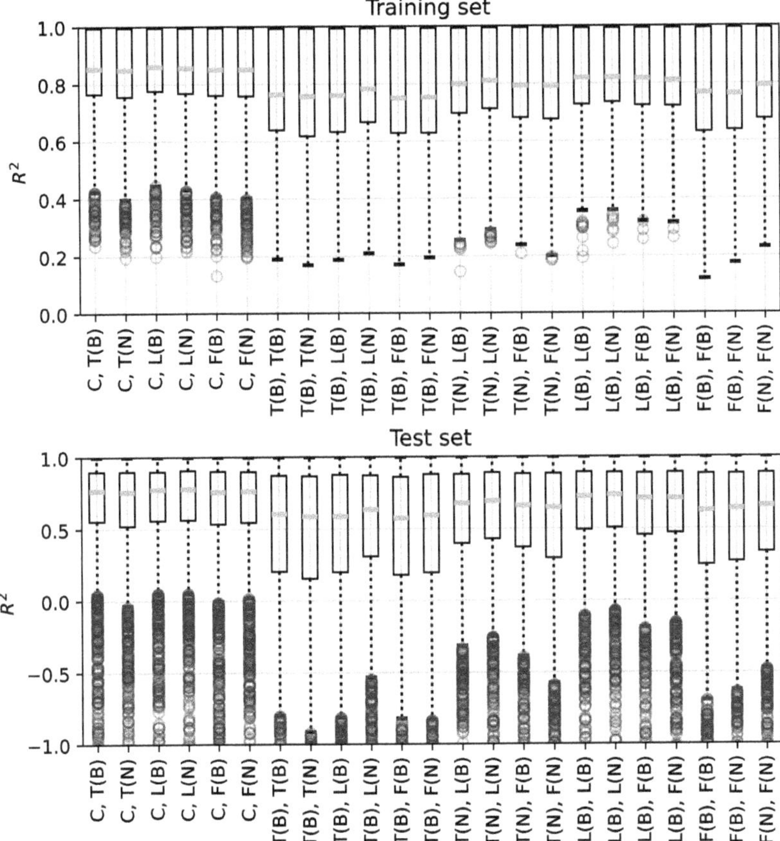

Fig. 4. Box plots of R^2 values with different selection configurations, sorting-based replacement and down-sampling, higher is better.

Figure 5 shows similar comparison, but based on the ranking instead of the raw R^2 values. In particular, all selection, replacement and sampling strategies are compared on the test set.

The advantage of ranking is that it is less sensitive to outliers in the data, and allows to see small differences more precisely. Again, it can be seen that the first six selection schemes have the best performance, but schemes with lexicase selection have worse results on average. This means that selecting one of the parents as current i-th individual is a more stable strategy, as it has smaller dependence on the replacement and sample handling techniques. As for the lower part of Fig. 5, the sorting of individuals by their fitness combined with down-sampling is the best strategy, followed by pairwise comparison. The Friedman ranking demonstrates relatively stable results compared to other strategies, but still worse.

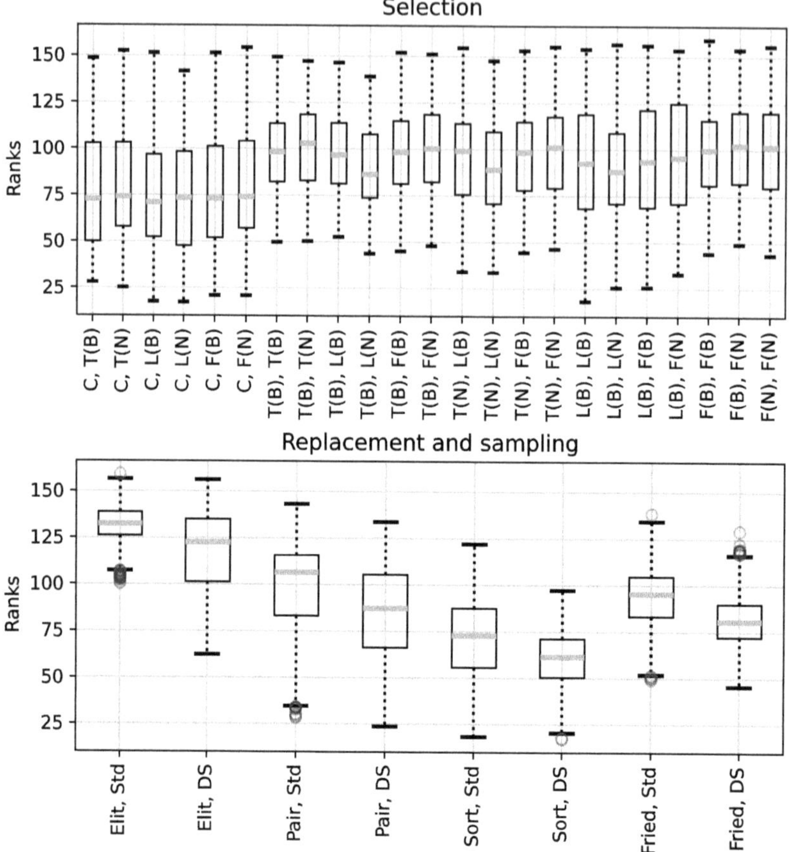

Fig. 5. Box plots of Friedman ranking with different selection, replacement and down-sampling, lower is better.

5   Conclusion

The analysis of the results of the experiments performed in this study shows that the typically used selection and replacement strategies are suboptimal - the best results were achieved by the cases where one of the parents was simply the current individual, and the other is chosen by lexicase selection. This means that many implementations of genetic programming can be significantly improved by using this simple strategy. Another advantage is that there is no need to perform lexicase selection twice, so the computational resources will be saved as well. As for the replacement strategy, the pairwise comparison, widely used in the differential evolution has proved itself not the best, but the fitness sorting-based replacement outperformed all other methods significantly. This method is a bit more complicated than elitism, but given relatively small (hundreds or thousands) population sizes in most GP implementations, should not add a significant additional computational burden.

The reasons for the higher efficiency of the tested schemes can be the following. The choice of current individual as the first parent makes sure that each individual participates in the crossover, no matter how good it is, whereas choosing both parents with selection makes some of the individuals in the population unused. Hence, this must increase the diversity of the generated solutions. As for the replacement scheme, sorting makes sure that no valuable individuals are lost in the process, so that they can participate in the crossover in the next generation.

The population management schemes considered here are not the only ones possible, and many other selection and replacement strategies can be proposed. For example, future works may focus on using multi-objective selection and replacement schemes, which use several criteria; another opportunity is the selection of complementary pairs of parents, i.e. parents who make errors on different cases - their offspring should have higher chance of inheriting the strength of both parents.

Disclosure of Interests. The authors have no competing interests to declare that are relevant to the content of this article

References

1. Banzhaf, W., Francone, F.D., Keller, R.E., Nordin, P.: Genetic programming - an introduction: on the automatic evolution of computer programs and its applications (1998)
2. Boldi, R., et al.: Informed down-sampled lexicase selection: identifying productive training cases for efficient problem solving. Evol. Comput. **32**, 307–337 (2023). https://doi.org/10.1162/evco_a_00346
3. Brameier, M., Banzhaf, W.: Evolving teams of predictors with linear genetic programming. Genet. Program Evolvable Mach. **2**, 381–407 (2001). https://doi.org/10.1023/A:1012978805372
4. Cava, W.L., Helmuth, T., Spector, L., Moore, J.H.: A probabilistic and multi-objective analysis of lexicase selection and ϵ-lexicase selection. Evol. Comput. **27**, 377–402 (2019). https://doi.org/10.1162/evco_a_00224
5. Cava, W.L., Spector, L., Danai, K.: Epsilon-lexicase selection for regression. In: Proceedings of the Genetic and Evolutionary Computation Conference 2016 (2016). https://doi.org/10.1145/2908812.2908898
6. Ciesielski, V.: Linear genetic programming. Genet. Program Evolvable Mach. **9**, 105–106 (2007)
7. Deb, K., Agrawal, S., Pratap, A., Meyarivan, T.: A fast and elitist multiobjective genetic algorithm: NSGA-II. IEEE Trans. Evol. Comput. **6**, 182–197 (2002). https://api.semanticscholar.org/CorpusID:9914171
8. Ferguson, A.J., Hernandez, J.G., Junghans, D., Lalejini, A., Dolson, E.L., Ofria, C.: Characterizing the effects of random subsampling on lexicase selection. In: Genetic Programming Theory and Practice (2019). https://doi.org/10.1007/978-3-030-39958-0_1

9. Geiger, A., Sobania, D., Rothlauf, F.: Down-sampled epsilon-lexicase selection for real-world symbolic regression problems. In: Proceedings of the Genetic and Evolutionary Computation Conference (2023). https://doi.org/10.1145/3583131.3590400

10. Helmuth, T., Spector, L., Matheson, J.E.: Solving uncompromising problems with lexicase selection. IEEE Trans. Evol. Comput. **19**, 630–643 (2015). https://doi.org/10.1109/TEVC.2014.2362729

11. Kitamura, T., Fukunaga, A.S.: Differential evolution with an unbounded population. In: 2022 IEEE Congress on Evolutionary Computation (CEC) (2022). https://doi.org/10.1109/CEC55065.2022.9870363

12. Koza, J.R.: Genetic programming: on the programming of computers by means of natural selection (1992)

13. Miller, J.F., Thomson, P.: Cartesian genetic programming. In: European Conference on Genetic Programming (2000). https://doi.org/10.1007/978-3-540-46239-2_9

14. Ni, J., Drieberg, R.H., Rockett, P.I.: The use of an analytic quotient operator in genetic programming. IEEE Trans. Evol. Comput. **17**, 146–152 (2013). https://api.semanticscholar.org/CorpusID:30086468

15. Reyes-Dávila, E., Haro, E.H., Casas-Ordaz, Á., Oliva, D., Ávalos, O.: Differential evolution: a survey on their operators and variants. Arch. Comput. Methods Eng. **32**, 83–112 (2024). https://doi.org/10.1007/s11831-024-10136-0

16. Romano, J.D., et al.: Pmlb v1.0: an open source dataset collection for benchmarking machine learning methods. arXiv preprint arXiv:2012.00058v2 (2021)

17. Ryan, C., O'Neill, M., Collins, J.J.: Introduction to 20 years of grammatical evolution. In: Handbook of Grammatical Evolution (2018). https://doi.org/10.1007/978-3-319-78717-6_1

18. Ryan, L., Collins, J., NeillDept, M.O.: Grammatical evolution: evolving programs for an arbitrary language. In: European Conference on Genetic Programming (1998). https://doi.org/10.1007/BFb0055930

19. Sherstnev, P.A., Semenkin, E.S.: Self-configuring genetic programming algorithms with success history-based adaptation. Siberian Aeros. J. (2025), https://api.semanticscholar.org/CorpusID:277880572

20. Spector, L.: Assessment of problem modality by differential performance of lexicase selection in genetic programming: a preliminary report. In: Proceedings of the 14th Annual Conference Companion on Genetic and Evolutionary Computation (2012). https://doi.org/10.1145/2330784.2330846

21. Spector, L., Klein, J., Keijzer, M.: The push3 execution stack and the evolution of control. In: Annual Conference on Genetic and Evolutionary Computation (2005). https://doi.org/10.1145/1068009.1068292

22. Spector, L., Robinson, A.J.: Genetic programming and autoconstructive evolution with the push programming language. Genet. Program Evolvable Mach. **3**, 7–40 (2002). https://doi.org/10.1023/A:1014538503543

23. Stanovov, V., Akhmedova, S., Semenkin, E.: Genetic algorithm with success history based parameter adaptation. In: International Joint Conference on Computational Intelligence (2019). https://doi.org/10.5220/0008071201800187

24. Stanovov, V., Akhmedova, S., Semenkin, E.: Dual-population adaptive differential evolution algorithm L-NTADE. Mathematics (2022). https://doi.org/10.3390/math10244666

25. Stanovov, V., Semenkin, E.: Genetic programming for automatic design of parameter adaptation in dual-population differential evolution. In: Proceedings of the

Companion Conference on Genetic and Evolutionary Computation (2023). https://doi.org/10.1145/3583133.3596310

26. Storn, R., Price, K.V.: Differential evolution - a simple and efficient heuristic for global optimization over continuous spaces. J. Global Optim. **11**, 341–359 (1997). https://doi.org/10.1023/A:1008202821328
27. Tanabe, R., Fukunaga, A.S.: Success-history based parameter adaptation for differential evolution. In: 2013 IEEE Congress on Evolutionary Computation, pp. 71–78 (2013). https://doi.org/10.1109/CEC.2013.6557555
28. Tanabe, R., Fukunaga, A.S.: Improving the search performance of SHADE using linear population size reduction. In: 2014 IEEE Congress on Evolutionary Computation (CEC), pp. 1658–1665 (2014). https://doi.org/10.1109/CEC.2014.6900380
29. Telikani, A., Tahmassebi, A., Banzhaf, W., Gandomi, A.H.: Evolutionary machine learning: a survey. ACM Comput. Surv. (CSUR) **54**, 1–35 (2021). https://doi.org/10.1145/3467477
30. Whitley, L.D.: The GENITOR algorithm and selection pressure: why rank-based allocation of reproductive trials is best. In: International Conference on Genetic Algorithms (1989)
31. Whitley, L.D.: A genetic algorithm tutorial. Stat. Comput. **4**, 65–85 (1994). https://doi.org/10.1007/BF00175354

A Comparative Study on Robustness in Evolved Image Classifiers

Camilo De La Torre[1,2], Stéphane Treillard[1,2,3], Camille Franchet[4], Hervé Luga[2,5], Dennis Wilson[6], and Sylvain Cussat-Blanc[1,2,7(✉)]

[1] University Toulouse Capitole, Toulouse, France
sylvain.cussat-blanc@ut-capitole.fr
[2] IRIT - CNRS UMR5505, Toulouse, France
[3] Centre Hospitalier Universitaire de Toulouse, Toulouse, France
[4] Oncopole Claudius Regaud, Toulouse, France
[5] Université Toulouse Jean Jaures, Toulouse, France
[6] ISAE-SUPAERO, University of Toulouse, Toulouse, France
[7] Institut Universitaire de France, Paris, France

Abstract. Deep Learning (DL) models, despite high accuracy, often degrade under domain shifts and image perturbations, which are common in medical imaging. This brittleness hinders reliable deployment in real-world applications. Some solutions, such as Lipschitz-constrained networks, offer formal robustness guarantees at the cost of added complexity. We explore program synthesis as an alternative, examining the inherent robustness of image classification programs evolved via Multi-Modal Adaptive Graph Evolution (MAGE). Using the PatchCamelyon histopathology dataset, we compare evolved programs against a standard ResNet-18 and a 1-Lipschitz ResNet-18 under Gaussian noise, Poisson noise, and brightness perturbations. On unperturbed test data, all models achieve competitive balanced accuracy. Remarkably, MAGE programs, evolved without data augmentation or explicit robustness objectives, exhibit inherent resilience comparable to, and often surpassing, the 1-Lipschitz ResNet-18 model. Moreover, as the evolutionary search yields a diverse population, it enables the selection of highly robust programs suited to specific perturbations.

Keywords: Machine Learning · Genetic Programming · Robustness

1 Introduction

Deep Learning (DL) has driven transformative advances in image analysis [20], enabling many applications in computer vision [22,30] and frequently matching or surpassing human performance [9,15]. Despite these achievements, a critical barrier to real-world deployment is their well-documented brittleness under common image perturbations or distribution shifts [11,17,33]. Such shifts occur when models encounter data drawn from a distribution differing from their training set. This vulnerability is especially pronounced in high-stakes domains like medical imaging, where acquisition devices, protocols, and disease prevalence vary widely across institutions [52]. In histopathology, for example, differences in tissue preparation and staining between laboratories significantly impair model generalization [37, 42]. These distributional shifts can lead to severe performance degradation, since deep learning models may rely on spurious correlations such as data provenance [52], patient's race [12] and even random labels [53], undermining their reliability in practice.

To address this, the research community has developed techniques to improve model robustness. When the amount of data is fixed, a common approach is data augmentation, where models are exposed to simulated input variations during training to improve generalization [20, 35]. A parallel approach is to build robustness into the model's architecture itself. A prominent example is the use of Lipschitz-constrained neural networks, which bound the distance between outputs in terms of the distance between inputs [51]. A 1-Lipschitz network ensures that the output distance between any pair of inputs is at most the distance between them, providing a certifiable robustness guarantee [34].

In this paper, we employ Program Synthesis (PS) via Genetic Programming (GP) as an alternative approach to image classification. Instead of optimizing a black-box Neural Network (NN), our method evolves populations of computer programs composed of well-established computer vision and mathematical functions, such as filters and feature extractors. We conduct the search for these multimodal programs using Multi-Modal Adaptive Graph Evolution (MAGE) [44], a typed extension of Cartesian Genetic Programming (CGP) [25, 26]. The resulting classifiers are white-box models, meaning they are explainable [8] and interpretable [45], properties of paramount importance in critical domains like medical imaging.

This paper examines the emergent robustness of programs evolved through GP to common image perturbations. Our empirical evaluation on the Patch-Camelyon (PCAM) histopathology dataset compares MAGE programs against standard ResNet-18 and a 1-Lipschitz ResNet-18 models. Importantly, both the MAGE programs and the Lipschitz network attain comparable Balanced Accuracy (BA) on the unperturbed test set. In robustness assessments, we demonstrate that MAGE programs can rival the resiliency of the Lipschitz network against Gaussian and Poisson noise. Notably, MAGE exhibits superior resistance to brightness variations, with performance degrading more symmetrically than the Lipschitz network, even when the latter incorporates additional brightness data augmentation. By contrast, the standard ResNet-18 baseline suffers substantial performance declines across all perturbations.

A key outcome of the evolutionary search is the generation of a distribution of high-performing yet diverse solutions. This diversity facilitates post-hoc selection of a program exhibiting exceptional resilience to a particular anticipated or known perturbation, providing a practical strategy for deploying robust models. These results illustrate that PS offers a viable means of generating performant and robust classifiers.

2   Related Work

2.1   Genetic Programming for Image Classification

GP has been applied to various problems in image analysis [19]. Early research often employed GP to evolve classifiers that used manually extracted features [29] or to design specialized image filters [32]. A prominent research direction has been to use GP for automatic feature engineering [6,10]. In this paradigm, GP evolves programs that transform images into feature vectors for a final separate and more conventional classifier algorithm. Moreover, CGP [25,26] have also been successfully applied to complex vision tasks, including the segmentation of Martian geological terrain [21], filter optimization [14], as well as, state of the art cell instance segmentation [7].

Additionally, GP has garnered increasing attention in two emerging areas: multimodal learning and interpretability. First, GP and CGP have been applied to evolve multimodal programs for image classification [8,49], which integrate or transform data across modalities, for example, converting raw image pixels into scalar features used for decision logic. Second, there is growing interest in the inherent explainability and interpretability obtained with GP based approaches [24,28,45].

Our work is situated within the more recent paradigm of end-to-end program synthesis, where a single evolved program handles the entire pipeline from raw pixels to classification. MAGE [44] has been used to synthesize multimodal, interpretable and high-performing policies for Atari games [45] and for biomedical image classification [8]. While these studies demonstrated the ability to synthesize programs that are both effective and interpretable for vision tasks, in this paper we explore an additional algorithmic trait. We present an empirical analysis of the resilience exhibited by MAGE programs, a critical attribute for trustworthy Artificial Intelligence (AI) deployed in real-world settings [23].

2.2   Robustness in Deep Neural Networks

The primary goals in developing Machine Learning (ML) models are high predictive performance and generalization to unseen data. However, the high capacity of Deep Neural Networks (DNNs) makes them prone to overfitting the training distribution [36] and learn spurious correlations [52,53]. Such models can also learn unexpected regularities that are hard to understand, even by experts [12]. A related challenge is their sensitivity to minor input perturbations, whether adversarial or stochastic, that can alter their predictions [13,39].

A common and practical approach to mitigate this sensitivity is data augmentation. This involves creating synthetic training examples by applying semantically-invariant transformations to the original inputs, such as small rotations, crops, or noise injections. This process enhances the model's ability to handle variations it is likely to encounter at test time [35]. Data augmentation is often used for DL applied to medical image data [18,42,50]. In histopathology, for example, using noise as an augmentation strategy has also been proven to improve the models' performance [18]. Nevertheless, augmentation may not cover the full spectrum of possible real-world perturbations and can be computationally demanding. Furthermore, training on specific perturbations may fail to generalize, and even be detrimental, to other type of perturbations [11].

For stronger assurances, more formal methods seek to provide mathematical guarantees of robustness. Lipschitz continuity provides a rigorous framework for this purpose. A function f is L-Lipschitz if the distance between its outputs is bounded by at most L times the distance between its inputs, formally expressed as:

$$\|f(x_1) - f(x_2)\| \leq L \cdot \|x_1 - x_2\|$$

for any inputs x_1 and x_2. By constraining a network to have a small Lipschitz constant (e.g., $L = 1$), we can certify its robustness against any perturbation within a certain radius ϵ [38]. Early techniques encouraged this through regularization, either by penalizing an upper bound on L [51] or by adding a margin term to the logits based on a Lipschitz estimate [38]. More recently, the dominant approach enforces the constraint architecturally using spectral normalization [27], which normalizes weights by an estimate of their spectral norm at each access. Many standard components, such as ReLU, Sigmoid, and Max-Pooling, are inherently 1-Lipschitz [38]; for convolutional layers, tighter bounds are achievable through architecture modification [34]. Libraries like deel-torchlip implement 1-Lipschitz versions of common layers (linear, convolutional, normalization, etc.) [34].

While effective, these methods impose strong architectural constraints. Adversarial training, for example, might lead to a robustness-accuracy trade-off, where improving robustness against perturbations may lead to a decrease in performance on non adversarial examples [46]. Although 1-Lipschitz networks can match the accuracy of unconstrained ones, there is an inherent trade-off between accuracy and robustness which must be controlled via loss function hyperparameters [5].

2.3   The PCAM Dataset

We evaluate our approach on the PCAM dataset [47], a balanced collection of 96×96-pixel patches extracted from lymph node Whole-slide imagess (WSIs) in the Camelyon16 (CAM16) dataset [4]. Each patch is labeled as malignant if its central 32×32 region contains at least one metastatic pixel according to PCAM's documentation[1]. The dataset comprises 327,680 images with no scan overlap across splits: 262,144 for training, 32,768 for validation, and 32,768 for testing. Prior work has achieved strong performance on PCAM. A P4M-DenseNet exploiting rotational symmetry reached 89.8% accuracy [47]. Distance Metric Learning approaches using ResNet-34 backbones reported 90.47% [41] and 90.36% [40] test accuracy.

---

[1] https://github.com/basveeling/pcam.

3   Methods

In this section, we detail our methods for training MAGE programs and DL models on the PCAM dataset. First we will introduce the approach that we took for synthesising programs for binary image classification with MAGE. Then we will expand into the DL (unconstrained) baseline as well as the 1-Lipschitz DL model.

3.1   Program Synthesis via MAGE

In this paper, we are interested in end-to-end binary image classification. As such, the evolved programs should take raw images in the RGB color space and predict labels. To infer the label of an instance, evolved programs must create relevant image features and a decision logic based on these features. We use MAGE [44], an extension of CGP [25,26], which is particularly suited for this task because of its multimodal capacities. Multimodal programs use and produce data of different types, such as images and scalar values. MAGE ensures that during the evolutionary process, only *type-safe* mutations are done (i.e., mutations that produce inputs of the types needed by the functions used in the program) [43,44].

For binary image classification, a MAGE genome is structured into four distinct chromosomes, each dedicated to producing outputs of a specific type: 2D intensity images, 2D binary images, 2D segmented images, and scalar values, respectively. Accordingly, each chromosome has access to a corresponding function library. These libraries include 2D image-to-2D image operations (e.g., Sobel filters), 2D image-to-scalar operations (e.g., mean of an image), and scalar-to-scalar operations. We would like to emphasize that a function within a given library produces an output matching the chromosome's type but may accept inputs from other chromosomes. Each evolved program processes an input image $x \in \mathcal{X}$ and generates two scalar outputs, interpreted as support scores for the two target classes. The predicted class is determined by the argument of the maximum score: $\hat{y} = \arg\max_{i \in \{0,1\}} f_i(x)$, where $f_0(x)$ and $f_1(x)$ denote the program's scalar outputs.

Similar to [8], we employ a Genetic Algorithm (GA) to optimize MAGE programs. The process involves evolving a population of program candidates. In each iteration, the fittest individuals are designated as elite and carried to the next generation. New candidate programs are generated by mutating the winners of a tournament selection, which is based on fitness. Due to the large dataset size, fitness for all candidates (elite and newly generated) is calculated using a non-fixed sampled subset of the training data at each iteration. At the end of every iteration, the best solution identified among the elite is evaluated on the full validation set, which for the PCAM dataset consists of 32,768 images. Parameters governing this evolutionary search are given in Sect. 4.

In this work, our primary metric of interest is the final BA of the classification model. As BA can be a coarse optimization signal, we also incorporate Negative Log-Likelihood (NLL) as a complementary metric. For the binary case, BA is defined as the mean of sensitivity (True Positive Rate) and specificity (True

Negative Rate): $BA = \frac{1}{2}\left(\frac{TP}{TP+FN} + \frac{TN}{TN+FP}\right)$, where TP, FN, TN and FP denote the number of true positives, false negatives, true negatives and false positives, respectively. To align with our minimization objective, we optimize its complement, the Balanced Error Rate (BER) which is given by $BER = 1 - BA$. The NLL is a standard loss function for binary classification, expressed as: $NLL = -\sum_{i=1}^{N}[y_i \log(\hat{y}_i) + (1 - y_i)\log(1 - \hat{y}_i)]$. Where, for each instance i, y_i represents the ground truth class $y \in \{0, 1\}$ and $\hat{y}_i$ is the predicted probability (e.g., softmax score) for the positive class. The fitness function ($\mathcal{F}$) that we minimize for each individual is an affine combination between both metrics $\mathcal{F} = \alpha \cdot BER + (1 - \alpha) \cdot NLL$. Additionally, individuals are also given a small time cost penalty p to discourage very slow programs.

3.2   Deep Learning Baselines

To compare the quality of our evolutionary approach, we establish two distinct DL baselines, both based on the ResNet-18 architecture [16], a widely adopted model for image classification. Our first baseline is a standard, non pre-trained, ResNet-18 model [16] from PyTorch [3,31]. This network comprises 18 layers and repeatedly uses residual connections to facilitate gradient flow and mitigate the vanishing gradient. Each layer reduces the field of view while increasing the number of filters. Our second baseline is a 1-Lipschitz ResNet-18 variant. We replaced every layer in the standard ResNet-18 architecture by their homologous 1-Lipschitz constrained layers available in the Python library *deel-torchlip*[2] [34].

4   Experimental Setup

This section details the training hyperparameters employed for both MAGE and the DL baselines. Experiments' code and examples of the noise regimes are available online[3].

4.1   Model Training

For MAGE programs, we used the GA described in Sect. 3.1. Parameters were chosen manually and also based on [8]; we did not perform additional hyperparameter tuning. An elite population, with a size denoted as E, of $E = 20$ candidate solutions used. At each iteration, $O = 124$ new candidate solutions were generated by mutation. Parent selection was performed using tournament selection, with a tournament size $T = 7$. For evaluating the fitness of each solution, a subset of $S = 300$ images was randomly sampled from the training dataset at each iteration. The evaluation time penalty parameter was set to $p = 100$. The evolutionary search was constrained by a time budget, with runs executed

---

[2] https://github.com/deel-ai/deel-torchlip.
[3] https://github.com/camilodlt/evo_star_robustness_image_classification_submission.

for as many generations as possible within a limit of 48 h. For each genome, the number of nodes was set to $N = 100$. Note that since MAGE genomes have multiple chromosomes, four in this case, then a full genome has a grid of $4 \times N$ nodes. Finally, the fitness hyperparameter α was set to 0.8. We performed 16 independent runs for MAGE and in the following section we will focus on the top 5 programs out of all the runs.

The DL baselines partially shared a set of common parameters. Both models were trained for 100 epochs using the Adam optimizer [1] with a learning rate of $1.0e - 5$. A batch size of 64 was used for ResNet-18 model and 1024 for the 1-Lipschitz ResNet-18 models mainly for faster training. Model selection was based on the epoch achieving the highest BA on the validation set. No early stopping was applied. For data augmentation, we adopted two distinct strategies to assess different profiles of robustness. The standard ResNet-18 and one instance of the 1-Lipschitz ResNet-18 were trained using random horizontal, vertical flips and normalization. Indeed, in principle, model's prediction on histopathological images should be rotation invariant [47]. To specifically evaluate the impact of training with brightness augmentation, a second instance of the 1-Lipschitz ResNet-18 was trained. This particular model used the same augmentations as before but additionally incorporated a random brightness adjustment, where the brightness factor varied uniformly from -0.3 to 0.3.

4.2 Robustness Evaluation

After training, all models were initially evaluated on the original, unmodified PCAM test set to establish a baseline performance. Subsequently, we thoroughly evaluated their robustness by applying three distinct types of image perturbations to the entire validation and test set at varying intensities. For all experiments, performance was measured using BA. The noise application procedures were as follows:

- **Gaussian Noise:** Additive Gaussian noise was applied to each pixel of each color channel in the images. The noise was drawn from a Normal distribution with a mean of 0.0 and a standard deviation (σ) ranging from 0.01 to 0.2 by steps of 0.01. The σ values express percentages relative to the maximum pixel intensity (255), meaning that the actual σ range was 0.01×255 to 0.2×255. After adding noise, each pixel was rounded to the nearest $Uint8$ in $[0, 255]$.
- **Poisson Noise:** Simulates photon shot noise, which is common in low-light conditions and arises from the inherent fluctuations in photon count in light [2]. This type of noise is common in histopathology [18] and in general in microscopy [48]. For each pixel, its intensity (denoted as x) was first normalized by the maximum intensity of its respective image (n). This normalized intensity was then multiplied by a specified *photons* level to determine the rate parameter (λ) for a Poisson distribution, i.e., $\lambda = (x/n) \times photons$. A random sample ($K$) was subsequently drawn from Poisson(λ) for each pixel. This sampled count K was then re-scaled back to an intensity value: $x_{sampled} = (K \times n)/photons$. We evaluated robustness across seven distinct

photons counts: $\{2000, 1000, 500, 255, 100, 20, 10\}$. A higher *photons* count leads to a larger λ. The variance of the Poisson noise is λ itself, so the amount of noise increases with the *photons* count [2]. However, the signal-to-noise ratio (SNR) $(photons/\sqrt{\lambda})$ improves with higher photon count [48]. Consequently, higher *photons* counts result in an improved SNR and perceptually less noise. After noise application, pixel values were clamped and rounded as with gaussian noise.

- **Brightness Changes:** Brightness perturbations were implemented by applying a fixed additive adjustment to all pixel values of each image. This additive amount ranged from -0.3 (to make images darker) to 0.3 (to make images brighter). The range used was: $\pm\{0.3, 0.2, 0.15, 0.1, 0.05, 0.01\}$. After the adjustment, the pixel values were clamped, scaled and rounded as with the other noises regimes.

5   Results

We analyze models, both deep learning and MAGE, for their robustness to different image perturbations. For the DL baselines (standard ResNet-18 and 1-Lipschitz ResNet-18), we retained the model checkpoint with the highest validation BA during training. For MAGE, we conducted multiple independent evolutionary runs and selected the top five programs based solely on their validation BA.

In this section, we report MAGE performance in three following ways:

- Table 1 presents the validation and test BA of the single best program (by validation BA) and the average across all five (unperturbed data).
- Figs. 1, 2 and 3 display individual curves for each of the five programs, along with their average.
- Within each figure, we highlight the program that maximizes the area under the validation BA curve across the noise axis. This serves to identify a potentially robust program for that perturbation that might generalize to unseen noisy test data.

5.1   Performance on Unperturbed Data

First, we establish the baseline performance of all models on the original, unperturbed PCAM validation and test set. Table 1 shows that MAGE and DL baselines are competitive and perform similarly. We can also see that the best MAGE model, selected based on validation performance, achieves a competitive test performance compared to DL models.

5.2   Robustness to Gaussian Noise

Figure 1 shows model performance across increasing intensities of Gaussian noise applied to validation and test images. On validation data, ResNet-18 achieves

Table 1. Balanced Accuracy (BA) on unperturbed data for each model.

Model	Validation BA (%)	Test BA (%)
ResNet-18	81.72	77.26
ResNet-18 1-Lipschitz	79.08	79.97
Best MAGE (Val.)	80.25	79.67
Avg MAGE (top 5 Val.)	79.1	78.22

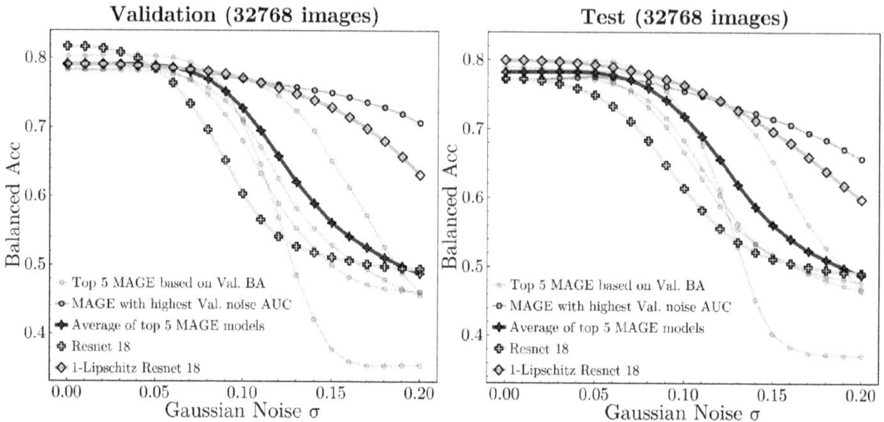

Fig. 1. Model's performance across different levels of Gaussian noise on validation and test data. Higher σ results in noisier images.

the highest BA. However, even without noise, it generalizes worst to test data. Its performance degrades sharply on both sets, revealing high sensitivity to Gaussian noise.

Although some individual MAGE models lacked robustness to high levels Gaussian noise, the average of the top five outperformed ResNet-18 on both validation and test data. All MAGE models proved resilient to low and moderate Gaussian noise. Their performance remained stable up to $\sigma = 0.08$. At low and moderate noise levels, MAGE and 1-Lipschitz ResNet-18 are competitive. However, at higher intensities, four of the top five MAGE models prove less robust than 1-Lipschitz ResNet-18, which retains discriminative ability even at $\sigma = 0.2$.

To identify the most robust MAGE model, we computed the AUC on unperturbed and noisy validation data. This yielded a MAGE model more resilient to high Gaussian noise than the 1-Lipschitz ResNet-18. For $\sigma > 0.14$, this MAGE model exceeded the 1-Lipschitz counterpart in discriminative ability. At the highest Gaussian noise level, it achieved 65.73% test BA compared to 59.78% for the 1-Lipschitz Resnet-18.

In summary, MAGE models demonstrate strong resilience to low and moderate Gaussian noise, matching the 1-Lipschitz ResNet-18 in these regimes.

On average, they outperformed ResNet-18 across all noise levels, except at $\sigma = 0.2$, where ResNet-18 held a marginal edge. Note that, on this balanced dataset, a BA of 50% can be the result of always predicting one class. While 1-Lipschitz ResNet-18 surpasses four of the top five MAGE models at high noise, the MAGE model maximizing validation AUC under noise generalizes best to noisy test data, exceeding 1-Lipschitz ResNet-18 by almost 6% BA at $\sigma = 0.2$.

5.3   Robustness to Poisson Noise

Figure 2 presents the models' performance under Poisson-distributed noise. The ResNet-18 achieved the highest BA on validation data. However, it generalized poorly to test data, even without noise. Among the models, ResNet-18 degraded faster and to a greater extent than MAGE and its 1-Lipschitz counterpart. Thus, ResNet-18 proved to be the least robust to Poisson noise.

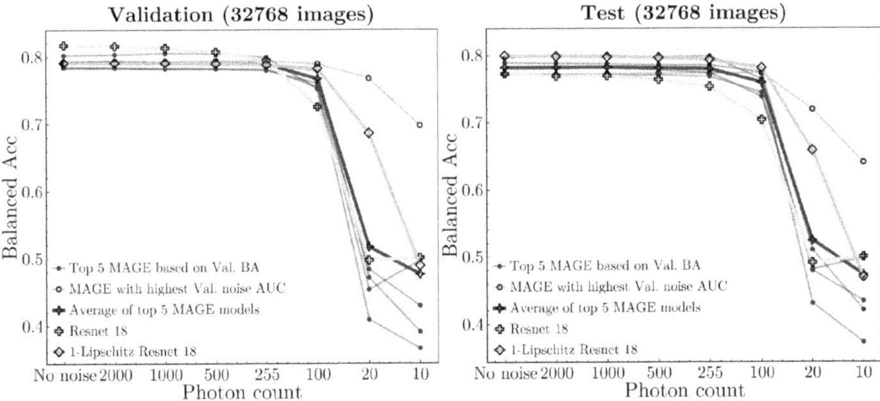

Fig. 2. Model's performance across Poisson-distributed noise levels on validation and test data. Lower photon count results in noisier images.

Figure 2 also shows the performance of the top five MAGE models, their average, and the 1-Lipschitz ResNet-18. To select the most robust MAGE model, we computed the area under the curve (AUC) on validation data using constant step sizes along the noise axis (no noise, 2000, 1000, 500, 255, 100, 20, and 10 photons). We treated these levels as equally spaced positions (1 through 8) for the AUC calculation. This approach accounts for the lack of a true photon count in the unperturbed condition. It also prevents disproportionately underweighting intervals at low photon counts (e.g., 10–20 photons) relative to higher ranges (e.g., 1000–2000 photons). The MAGE model with the highest AUC is highlighted in blue.

1-Lipschitz ResNet-18 and average MAGE models competed closely in low Poisson noise regimes, both on validation and test data. In noisier regimes (fewer than 100 photons), 1-Lipschitz ResNet-18 showed greater robustness than the average of the top five MAGE models or four of them individually. Yet, the MAGE model with the highest AUC on noisy validation data outperformed the 1-Lipschitz ResNet-18 by a wide margin in very noisy regimes. At the lowest photon count, this MAGE model retained discriminative power with a BA of 64% while all other models reached a BA of 50% or less at that level. We would like to emphasize that since the dataset is balanced, in all the splits, a constant prediction of a single class guarantees 50% BA.

In summary, on the PCAM dataset under Poisson noise, averaged MAGE models proved more robust than the standard ResNet-18. They matched ResNet-18 1-Lipschitz in moderate noise but fell short in high noise. Still, evaluating robustness of MAGE models on noisy validation data revealed one that generalized well to unseen data and surpassed ResNet-18 1-Lipschitz in very noisy Poisson regimes.

5.4   Robustness to Brightness Changes

Figure 3 shows the models' performance across varying perturbations to image brightness for validation and test data. Although this type of perturbations does not constitute noise, it is common and highlights robustness to uniform changes in lighting [17].

Brightness adjustments during training represent a common data augmentation strategy in DL. Accordingly, we included a third DL model: a 1-Lipschitz ResNet-18 trained with random brightness changes in the same range as the levels in Fig. 3 (i.e., $[-0.3 : 0.3]$), alongside the horizontal and vertical flips already applied to the other DL models.

A surprising observation from Fig. 3 is that the 1-Lipschitz ResNet-18 remains robust to minor brightness changes (i.e., less than 5% for brighter images and less than 1% for darker ones) but emerges as the least resilient model under more substantial alterations, whether darkening or lightening the images. This outcome is unexpected, given the model's resilience to Gaussian and Poisson noise. Under pronounced brightness shifts, the standard ResNet-18 preserves greater discriminative ability than its 1-Lipschitz counterpart. Furthermore, the 1-Lipschitz network exhibits inconsistent robustness between validation and test data: it demonstrates resilience to brighter images on validation data but fails to generalize this property to test data.

Incorporating brightness augmentation during training clearly benefited the 1-Lipschitz model. However, it uncovered an asymmetric robustness profile. Specifically, the 1-Lipschitz ResNet-18 with brightness augmentation proved highly resilient to brighter images, even at extreme levels, yet failed to sustain this resilience for darker images.

At moderate and high brightness alteration levels, the average performance of MAGE models substantially exceeds that of both the standard ResNet-18 and the 1-Lipschitz ResNet-18 on test data. Notably, MAGE models exhibit symmetric resilience on average, degrading homogeneously across darker and brighter images. The 1-Lipschitz ResNet-18 trained with brightness changes

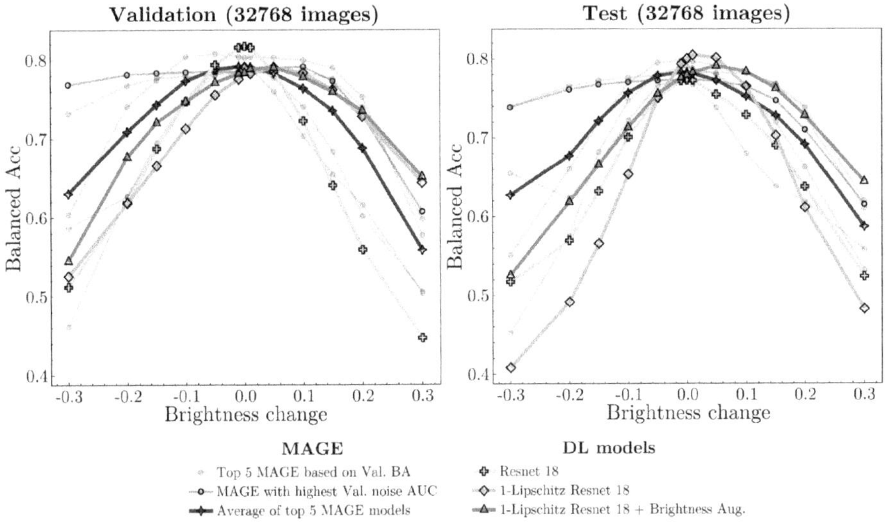

Fig. 3. Model's performance across different levels of uniform brightness change. Low values result in darker images, high values in lighter ones.

demonstrated greater resilience than the average of MAGE models on brighter images, but this trend reversed for darker images.

Finally, focusing on the MAGE model with the highest AUC along the brightness axis on validation data yields a model that competes effectively with the 1-Lipschitz ResNet-18 trained with brightness changes on very bright data ($\Delta = -2.8\%$) and substantially outperforms it on very dark data ($\Delta = 21.2\%$).

5.5   The Value of Model Diversity in Program Synthesis

The pipeline of model training in DL typically yields a single model. Such a model may lack resilience to noise on both validation and test data, as observed with the ResNet-18 in Fig. 1 and Fig. 2. Alternatively, a singular model may display signs of robustness when evaluated on noisy validation data but fail to generalize to noisy test data, as seen with the 1-Lipschitz ResNet-18 in Fig. 3.

In contrast, MAGE programs arise from a stochastic GA, resulting in a population of models with varying performances on unperturbed validation data. From this distribution, we select the top fraction, which provides multiple model options. This availability of choices allows us to identify one model or an ensemble that maximizes a desired property. In this study, we selected the top MAGE model that maximizes the AUC over unperturbed and noisy validation data. As evident from Figs. 1, 2 and 3, properties observed in MAGE models on validation data transfer effectively to unseen data.

In Figs. 1 and 2, the MAGE program that maximizes the AUC along the noise axis on validation data is the same one, indicating that a single MAGE model can exhibit robustness across multiple noise types. In Fig. 3, however, prior trends reversed for all models, and a different MAGE program demonstrated superior resilience.

In summary, having a distribution of high-performing models empowers the modeler to enhance the likelihood that a single MAGE model will perform as desired on unseen data, particularly when a specific perturbation is known. When the nature of real-world perturbations remains unknown, MAGE models, on average, exhibited clear robustness against all perturbations examined in this study. This finding highlights the potential of MAGE model ensembles for real-world applications, where data distributions may shift.

6   Limitations and Future Work

In this study, we compared MAGE with several DL models on binary image classification using a histopathology dataset of 327,680 images. We evaluated individual and average MAGE performance against a standard ResNet-18 and a certifiably robust (1-Lipschitz) ResNet-18 across varying levels and types of image noise. While our analysis was comprehensive and thorough, it is limited to a single dataset. Comparisons on additional datasets are left for future work. The primary goal of this paper was to highlight an emergent robustness property in MAGE models, critical for developing trustworthy AI systems in real-world settings.

A further limitation is that, while we demonstrated this robustness, we did not investigate its underlying causes. We defer this analysis to future work but propose, for the time being, two hypotheses. First, the evolutionary search imposes strong generalization pressure. As detailed in Sect. 3, at each iteration, the top program on a small training sample ($S = 300$) is evaluated on the full validation set (32,768 images). The final model is selected based on maximum validation BA. This process discourages overfitting: poor generalizers rely on chance alignment with S, while strong generalizers extract truly discriminative features transferable to the full validation set. Second, the function set in MAGE, which comprising operations like Sobel filters and morphological opening/closing, may inherently be less sensitive to image perturbations than learned convolutional filters, which can be highly specialized due to the number of parameters they contain.

7   Conclusion

In this paper, we compared MAGE-evolved programs for image classification against several DL models. The focus was to characterize, for the first time to our knowledge, an emergent robustness property of evolved image classification programs. Notably, MAGE training incorporated no data augmentation or fitness function tuning for robustness. We used the PCAM dataset, a histopathological binary classification task comprising 262,144 training images, 32,768 validation images, and 32,768 test images. On unperturbed data, MAGE and DL models proved competitive: the top performer was a 1-Lipschitz ResNet-18 at 79.97% BA, followed by the best MAGE model (selected on validation data) at 79.67%.

We subsequently evaluated the models under varying levels of perturbations: Gaussian noise, Poisson noise, and brightness changes. The standard ResNet-18 exhibited poor robustness on test data across all perturbation types. A 1-Lipschitz ResNet-18 demonstrated strong resilience to Gaussian and Poisson noise at test time but failed surprisingly under brightness changes, despite promising validation robustness signs. Training an additional 1-Lipschitz ResNet-18 with brightness augmentation enhanced its robustness, though asymmetrically across the perturbation range.

We trained 16 MAGE models and selected the top five based on validation metrics. On average, these MAGE models proved robust against all perturbations, particularly in low and moderate regimes. Across all noises, average MAGE performance surpassed the standard ResNet-18 on test data. Leveraging the distribution of high-performing MAGE models, we identified, for each perturbation, the program maximizing AUC along the noise axis on validation data. For Gaussian and Poisson noise, the same MAGE program outperformed the 1-Lipschitz ResNet-18 in high-noise regimes. For brightness perturbations, the MAGE model with the highest noise AUC competed effectively with the brightness-augmented 1-Lipschitz ResNet-18 under very bright conditions and substantially outperformed it under dark conditions. We demonstrate that MAGE models exhibit an emergent robustness property against multiple image perturbations, surpassing DL counterparts in numerous scenarios.

Acknowledgments. This project used compute resources from the CALMIP projects P21049 and P24042. This work has been supported by funds from Institut Carnot CALYM (DIALYM project) and IA pour la santé (Région Occitanie).

Disclosure of Interests. The authors have no competing interests to declare that are relevant to the content of this article.

References

1. Adam, K.D.B.J., et al.: A method for stochastic optimization. arXiv preprint arXiv:1412.6980 **1412**(6) (2014)
2. Alter, F., Matsushita, Y., Tang, X.: An intensity similarity measure in low-light conditions. In: Leonardis, A., Bischof, H., Pinz, A. (eds) ECCV 2006. LNCS, vol. 3954, pp. 267–280. Springer, Heidelberg (2006). https://doi.org/10.1007/11744085_21
3. Ansel, J., et al.: PyTorch 2: faster machine learning through dynamic python bytecode transformation and graph compilation. In: 29th ACM International Conference on Architectural Support for Programming Languages and Operating Systems, vol. 2 (ASPLOS 2024). ACM (2024). https://doi.org/10.1145/3620665.3640366. https://docs.pytorch.org/assets/pytorch2-2.pdf
4. Bejnordi, B.E., et al.: Diagnostic assessment of deep learning algorithms for detection of lymph node metastases in women with breast cancer. J. Am. Med. Assoc. (JAMA) (2017). https://doi.org/10.1001/JAMA.2017.14585

5. B'ethune, L., Boissin, T., Serrurier, M., Mamalet, F., Friedrich, C., González-Sanz, A.: Pay attention to your loss: understanding misconceptions about lipschitz neural networks. Neural Inf. Process. Syst. (2021)
6. Bi, Y., Xue, B., Zhang, M.: Genetic programming with image-related operators and a flexible program structure for feature learning in image classification. IEEE Trans. Evol. Comput. (2021). https://doi.org/10.1109/TEVC.2020.3002229
7. Cortacero, K., et al.: Evolutionary design of explainable algorithms for biomedical image segmentation. Nat. Commun. (2023). https://doi.org/10.1038/S41467-023-42664-X
8. De La Torre, C., et al.: Evolved and transparent pipelines for biomedical image classification. In: Xue, B., Manzoni, L., Bakurov, I. (eds.) Genetic Programming, pp. 173–189. Springer, Cham (2025)
9. Esteva, A., et al.: Dermatologist-level classification of skin cancer with deep neural networks. Nature (2017). https://doi.org/10.1038/NATURE21056
10. Fan, Q., Bi, Y., Xue, B., Zhang, M.: A multi-tree genetic programming-based ensemble approach to image classification with limited training data [research frontier]. IEEE Comput. Intell. Mag. (2024). https://doi.org/10.1109/MCI.2024.3446148
11. Geirhos, R., Temme, C.R.M., Rauber, J., Schütt, H.H., Bethge, M., Wichmann, F.A.: Generalisation in humans and deep neural networks. Neural Inf. Process. Syst. (2018). https://doi.org/10.15496/PUBLIKATION-30814
12. Gichoya, J.W., et al.: AI recognition of patient race in medical imaging: a modelling study. Lancet Digit. Health $4(6)$, e406–e414 (2022)
13. Goodfellow, I., Shlens, J., Szegedy, C.: Explaining and harnessing adversarial examples. In: International Conference on Learning Representations (2015)
14. Harding, S., Leitner, J., Schmidhuber, J.: Cartesian genetic programming for image processing. Sci. Eng. Faculty (2013). https://doi.org/10.1007/978-1-4614-6846-2_3
15. He, K., Zhang, X., Ren, S., Sun, J.: Delving deep into rectifiers: surpassing human-level performance on imagenet classification. In: International Conference on Computer Vision (2015). https://doi.org/10.1109/ICCV.2015.123
16. He, K., Zhang, X., Ren, S., Sun, J.: Deep residual learning for image recognition. In: Proceedings of the IEEE Conference on Computer Vision and Pattern Recognition, pp. 770–778 (2016). https://doi.org/10.1109/CVPR.2016.90
17. Hendrycks, D., Dietterich, T.G.: Benchmarking neural network robustness to common corruptions and surface variations. arXiv: Learning (2018)
18. Jena, P., Mishra, D., Das, K., Mishra, S.: NoiseAugmentNet-HHO: enhancing histopathological image classification through noise augmentation. IEEE Access (2024). https://doi.org/10.1109/ACCESS.2024.3518575
19. Khan, A., Qureshi, A.S., Wahab, N., Hussain, M., Hamza, M.Y.: A recent survey on the applications of genetic programming in image processing. In: International Conference on Climate Informatics (2021). https://doi.org/10.1111/COIN.12459
20. Krizhevsky, A., Sutskever, I., Hinton, G.E.: ImageNet classification with deep convolutional neural networks. Commun. ACM (2012). https://doi.org/10.1145/3065386
21. Leitner, J., Harding, S., Förster, A., Schmidhuber, J.: Mars terrain image classification using cartesian genetic programming. In: International Symposium on Artificial Intelligence (2012)
22. Litjens, G., et al.: A survey on deep learning in medical image analysis. Med. Image Anal. (2017). https://doi.org/10.1016/J.MEDIA.2017.07.005

23. Liu, H., et al.: Trustworthy AI: a computational perspective. ACM Trans. Intell. Syst. Technol. (2021). https://doi.org/10.1145/3546872
24. Mei, Y., Chen, Q., Lensen, A., Xue, B., Zhang, M.: Explainable artificial intelligence by genetic programming: a survey. IEEE Trans. Evol. Comput. (2023). https://doi.org/10.1109/TEVC.2022.3225509
25. Miller, J.F.: An empirical study of the efficiency of learning Boolean functions using a cartesian genetic programming approach. In: Proceedings of the Genetic and Evolutionary Computation Conference, vol. 2, pp. 1135–1142 (1999)
26. Miller, J.F.: Cartesian genetic programming: its status and future. Genet. Program Evolvable Mach. (2020). https://doi.org/10.1007/S10710-019-09360-6
27. Miyato, T., Kataoka, T., Koyama, M., Yoshida, Y.: Spectral normalization for generative adversarial networks. In: International Conference on Learning Representations (2018)
28. Nadizar, G., Medvet, E., Wilson, D.G.: Naturally interpretable control policies via graph-based genetic programming. In: European Conference on Genetic Programming (2024). https://doi.org/10.1007/978-3-031-56957-9_5
29. Nandi, R.J., Nandi, A.K., Rangayyan, R.M., Scutt, D.: Classification of breast masses in mammograms using genetic programming and feature selection. Med. Biol. Eng. Comput. (2006). https://doi.org/10.1007/S11517-006-0077-6
30. Noor, M.H.M., Ige, A.O.: A survey on state-of-the-art deep learning applications and challenges. Eng. Appl. Artif. Intell. **159**, 111225 (2025)
31. Paszke, A., et al.: PyTorch: an imperative style, high-performance deep learning library. arXiv: Learning (2019)
32. Poli, R.: Genetic programming for image analysis. Cognitive Science Research Papers-University of Birmingham CSRP (1996). https://doi.org/10.7551/MITPRESS/3242.003.0053
33. Recht, B.: Do imagenet classifiers generalize to imagenet? arXiv: Computer Vision and Pattern Recognition (2019)
34. Serrurier, M., Mamalet, F., González-Sanz, A., Boissin, T., Loubes, J., Barrio, E.d.: Achieving robustness in classification using optimal transport with hinge regularization. In: Computer Vision and Pattern Recognition (2021). https://doi.org/10.1109/CVPR46437.2021.00057
35. Shorten, C., Khoshgoftaar, T.M.: A survey on image data augmentation for deep learning. J. Big Data (2019). https://doi.org/10.1186/S40537-019-0197-0
36. Srivastava, N., Hinton, G.E., Krizhevsky, A., Sutskever, I., Salakhutdinov, R.: Dropout: a simple way to prevent neural networks from overfitting. J. Mach. Learn. Res. (2014). https://doi.org/10.5555/2627435.2670313
37. Stacke, K., Eilertsen, G., Unger, J., Lundström, C.: A closer look at domain shift for deep learning in histopathology. arXiv: Computer Vision and Pattern Recognition (2019)
38. Sugiyama, M., Sato, I., Tsuzuku, Y.: Lipschitz-margin training: scalable certification of perturbation invariance for deep neural networks. In: NeurIPS (2018)
39. Szegedy, C., et al.: Intriguing properties of neural networks. arXiv preprint arXiv:1312.6199 (2014)
40. Teh, E.W., Taylor, G.W.: Metric learning for patch classification in digital pathology (2019)
41. Teh, E.W., Taylor, G.W.: Learning with less data via weakly labeled patch classification in digital pathology. In: IEEE International Symposium on Biomedical Imaging (2020). https://doi.org/10.1109/ISBI45749.2020.9098533

42. Tellez, D., et al.: Quantifying the effects of data augmentation and stain color normalization in convolutional neural networks for computational pathology. Med. Image Anal. (2019). https://doi.org/10.1016/J.MEDIA.2019.101544
43. Torre, C.D.L., Cortacero, K., Cussat-Blanc, S., Wilson, D.G.: Multimodal adaptive graph evolution. In: Proceedings of the Genetic and Evolutionary Computation Conference Companion (2024). https://doi.org/10.1145/3638530.3654347
44. Torre, C.D.L., Lavinas, Y., Cortacero, K., Luga, H., Wilson, D.G., Cussat-Blanc, S.: Multimodal adaptive graph evolution for program synthesis. In: Affenzeller, M., et al. (eds.) PPSN 2024. LNCS, vol. 15148, pp. 306–321. Springer, Cham (2024). https://doi.org/10.1007/978-3-031-70055-2_19
45. Torre, C.D.L., Nadizar, G., Lavinas, Y., Luga, H., Wilson, D.G., Cussat-Blanc, S.: Evolution of inherently interpretable visual control policies. In: Proceedings of the Genetic and Evolutionary Computation Conference (2025). https://doi.org/10.1145/3712256.3726332
46. Tsipras, D., Santurkar, S., Engstrom, L., Turner, A., Mądry, A.: Robustness may be at odds with accuracy. In: International Conference on Learning Representations (2018)
47. Veeling, B.S., Linmans, J., Winkens, J., Cohen, T., Welling, M.: Rotation equivariant CNNs for digital pathology. In: Frangi, A., Schnabel, J., Davatzikos, C., Alberola-López, C., Fichtinger, G. (eds) MICCAI 2018. LNCS, vol. 11071, pp. 210–218. Springer, Cham (2018). https://doi.org/10.1007/978-3-030-00934-2_24
48. Waters, J.: Accuracy and precision in quantitative fluorescence microscopy. J. Cell Biol. (2009). https://doi.org/10.1083/JCB.200903097
49. Wu, Z., Xue, B., Zhang, M.: Multitree genetic programming for multimodal learning in multimodal medical image classification. IEEE Trans. Evol. Comput. **PP**, 1 (2025). https://doi.org/10.1109/TEVC.2025.3619532
50. Xue, W.: Deep learning framework for enhanced MRI analysis in healthcare diagnosis. Expert Syst. Appl. (2025). https://doi.org/10.1016/J.ESWA.2025.128487
51. Yoshida, Y., Miyato, T.: Spectral norm regularization for improving the generalizability of deep learning. arXiv: Machine Learning (2017)
52. Zech, J.: Variable generalization performance of a deep learning model to detect pneumonia in chest radiographs: a cross-sectional study. PLoS Med. (2018). https://doi.org/10.1371/JOURNAL.PMED.1002683
53. Zhang, C., Bengio, S., Hardt, M., Recht, B., Vinyals, O.: Understanding deep learning requires rethinking generalization. In: International Conference on Learning Representations (2016)

Syntactic Flexibility Enables Compact Solutions in Transformer Semantic GP

Philipp Anthes[(✉)] [iD]

Johannes Gutenberg University, Mainz, Germany
anthes@uni-mainz.de

Abstract. In genetic programming, semantic variation operators aim to produce new solutions that are semantically similar to their parents. Geometric semantic operators (GSOs) use linear combinations to generate semantically similar offspring; however, this easily leads to bloated solutions. In contrast, Transformer Semantic Genetic Programming (TSGP) uses a pre-trained transformer model as a semantic mutation operator to generate offspring with similar semantics. Experiments on four symbolic regression problems reveal that TSGP's variation operator exhibits significantly greater syntactic flexibility: while still producing semantically similar solutions, it performs a greater number and more diverse edit operations, frequently replacing nodes and modifying multiple parts of the program tree simultaneously. Unlike GSO-based methods, which rely on local additive or reductive changes, TSGP is able to perform more holistic changes of a program while preserving semantics, which explains its ability to evolve effective and compact solutions.

1 Introduction

Semantic variation operators in Genetic Programming (GP) aim to incorporate program behavior (semantics) into the variation process, enabling controlled and weighted behavioral changes of the parent program. In contrast to standard GP (stdGP), which relies on purely syntactic modifications that often lead to unpredictable semantic effects [14], semantic-aware GP guides the search directly in the semantic space. This is advantageous as it induces a smoother, often unimodal error landscape [21], eliminating local optima that can trap traditional syntactic based search approaches. As a result, semantic-based methods have proven to be effective search strategies [2,5,15,22,23].

One of the most popular approaches is geometric semantic genetic programming (GSGP) [15], which was recently extended by SLIM_GSGP (SLIM) [22]. Both methods are based on Geometric Semantic Operators (GSOs), which generate offspring through linear combinations of program structures, ensuring that the semantics of the offspring lie within a defined region of the parent's semantic neighborhood. In GSGP, each variation step adds new syntactic material to the parent program, leading to rapid and uncontrolled program growth (bloat). SLIM addresses this issue by additionally employing a deflation operator, which

L. Manzoni et al. (Eds.): EuroGP 2026, LNCS 16521, pp. 51–66, 2026.
https://doi.org/10.1007/978-3-032-23005-8_4

removes previously added program components that were added in earlier generations and thus generates offspring with similar semantics but reduced size. Experiments show that this deflation operator significantly mitigates the excessive growth of GSGP [22].

However, despite improved size control, the core limitation of GSO-based methods remains: they rely on fixed, pre-defined syntactic transformation rules. By only performing local additive (in GSGP) or also reductive (in SLIM) variations, they are constrained in their syntactic expressiveness. Additionally, they do not take the syntactic form of the parent into account, which makes adaptive syntactic modifications that preserve semantics impossible. This limits GSO-based methods from evolving truly compact programs.

The recently proposed approach Transformer Semantic GP (TSGP) [2] may have the potential to overcome this limitation. Instead of relying on fixed transformation rules, TSGP uses a pre-trained transformer model as a semantic mutation operator. The model is once trained in a single, offline pre-training step using millions of syntactically diverse program pairs of high semantic similarity. During evolutionary search, the transformer model proposes a semantically similar offspring to a given parent. Studies show that the transformer model of TSGP successfully learns to create offspring with similar semantics in a controlled way and can transfer this ability to diverse test problems. That leads to high performance, but also to compact solutions compared to GSO-based methods [1,2]. However, the underlying mechanism responsible for this improved compactness remains unclear.

Therefore, this study examines how the semantic variation operators of GSGP, SLIM and TSGP express semantic similarity syntactically and how this impacts program size. We include stdGP as a syntactic baseline. In extensive experiments on four symbolic regression (SR) problems, we focus on the syntactic differences between parent and offspring solutions, particularly using the Levenshtein distance, a measure of the minimum number of edit operations (insertion, deletion, substitution) required to transform one program sequence into another [13].

Our results reveal that TSGP exhibits significantly greater syntactic flexibility than GSO-based methods: it performs a higher number and more diverse edit operations, it utilizes node replacements, and it can simultaneously modify multiple parts of the program tree. Unlike the local, additive or reductive changes of GSGP and SLIM, TSGP performs more holistic, context-aware modifications of the programs, explaining its ability to generate semantically similar and small solutions.

Section 2 reviews semantic GP, including GSGP, SLIM, and TSGP. Section 3 describes the experimental setup, including the datasets and the configuration parameters of the chosen methods. Section 4 presents brief results on solution quality (Sect. 4.1), while the core of the analysis is presented in subsection Population-Level Dynamics (Sect. 4.2). Section 5 concludes with remarks and future work.

2 Background: Semantic Variation Operators

There exists a wide variety of semantic GP methods, differing in several key aspects [24]. These range from whether they apply semantic criteria indirectly after offspring generation [3,4,11,16] to methods that directly integrate semantics into the variation process [2,15,22]. Some studies introduce mutation-based operators [2,4,22], while others rely on crossover-based operators [3,10,11,16,18] or a combination of both [15]. In this paper, we focus on approaches that use influential direct-based mutation operators. In particular, we study Geometric Semantic Mutation of GSGP, which was the first direct semantic mutation approach [15], as well as the recently published variant SLIM [22], which successfully mitigates the excessive growth inherent in GSGP. Additionally, we explore the transformer-based mutation operator of Transformer Semantic GP (TSGP) [2], which is a fundamentally new approach of generating offspring with similar semantics using a pre-trained transformer model instead of GSOs. While these three methods all generate offspring with similar semantics, they differ in how they achieve this semantic variation at the syntactic level.

2.1 Geometric Semantic Mutation (GSGP)

The Geometric Semantic Mutation operator (GSM) [15] modifies a parent program tree T by adding a randomly generated sub-expression with minor expected semantic effect. This effect is achieved by constructing the sub-expression as the difference between two independently generated random program trees, T_{R1} and T_{R2}. In the semantic space, this corresponds to a ball mutation, where the semantics of the offspring lie within a radius of m_s around the parent's semantics. With the size m_s of the mutation step, a mutation is defined as

$$\mathrm{GSM}(T) = T + m_s \cdot (T_{R1} - T_{R2}).$$

Although such mutations guarantee semantically similar offspring and induce a unimodal error landscape [21], they inevitably increase program size and syntactic bloat.

Repeated applications of GSM result in nested linear combinations. For the example case of three iterated mutations, we obtain

$$\mathrm{GSM}^3(T) = T + m_s \cdot (T_{R1} - T_{R2}) + m_s \cdot (T_{R3} - T_{R4}) + m_s \cdot (T_{R5} - T_{R6}).$$

2.2 Inflate & Deflate Mutation (SLIM_GSGP)

Based on GSM, SLIM [22] introduces a new deflate operator alongside standard GSM (referred to as inflate mutation). The key idea is that semantic similarity can be established not only by adding new sub-expressions (inflate), but also by removing previously added ones (deflate) that were introduced in prior mutation steps. Thus, the deflation operator returns a semantically similar offspring with reduced program size.

For the example from above with three iterated GSM operations, a deflate operation can remove the term $m_s \cdot (T_{R3} - T_{R4})$, which was introduced in a previous generation:

$$\text{Deflate}\big(\text{GSM}^3(T)\big) = T + m_s \cdot (T_{R1} - T_{R2}) + \cancel{m_s \cdot (T_{R3} - T_{R4})} + m_s \cdot (T_{R5} - T_{R6})$$
$$= T + m_s \cdot (T_{R1} - T_{R2}) + m_s \cdot (T_{R5} - T_{R6}).$$

The resulting offspring becomes a potentially novel and semantically similar program of reduced size. SLIM balances the inflate and deflate operations using a hyperparameter p_{xo}, which determines the probability of applying a deflate over the inflation operation during variation.

2.3 Transformer Semantic Mutation (TSGP)

Unlike rule-based semantic mutation operators, TSGP's semantic mutation operator is a transformer model that has learned to generate syntactic variations that result in similar semantics [2]. The model is trained once during a single offline *Model Building* step and is then applied to various problems in a process called *Model Inference.* In the first step of *Model Building*, millions of diverse programs (referred to as functions f) are generated, and their semantics $s(f)$ are approximated. The semantics are calculated using a fixed evaluation dataset consisting of randomly generated input values that comprehensively cover the possible input space, which allows the program semantics to be accurately approximated.

In the second step, programs are paired with semantically similar ones to create training examples representing semantic mutations. To achieve this, a k-nearest-neighbor search is performed in the semantic space. For each program f_i the search identifies k programs f_j with minimal semantic distance $\text{SD}(s(f_i), s(f_j)) = \|s(f_i) - s(f_j)\|_2$, resulting in k training examples (f_i, f_j).

In the final step, the transformer is trained on all semantic pairings (f_i, f_j) in a sequence-to-sequence task, where the transformer model receives one of the programs f_i as the input sequence and aims to generate the other f_j as the target output sequence. Additionally, this process is conditioned on their semantic distance SD, allowing the model to learn the degree of semantic similarity.

During evolutionary search, the pre-trained model acts as a semantic mutation operator. A parent program is encoded as a token sequence and passed to the pre-trained transformer model along with a target semantic distance SD_t that controls the degree of semantic similarity of the offspring (similar to the mutation step of GSOs). The offspring are generated auto-regressively: the transformer samples tokens one at a time, conditioned on the parent sequence, previously sampled tokens, and the hyperparameter SD_t. This process is inherently probabilistic, with the model outputting a probability distribution over the entirety of available tokens at each step, from which the next token is then sampled. Recently, [1] proposed a single transformer model applicable to problems of varying input dimensionality, by also conditioning the model with the problem dimensionality.

3   Experimental Settings

Datasets. We perform experiments on symbolic regression (SR) problems, where the goal is to evolve mathematical expressions that best fit a given input-output mapping. To evaluate the performance of the compared methods, we select four datasets from Penn Machine Learning Benchmarks (PMLB) [19], a widely adopted benchmark suite also used in SRBench [12]. We randomly sample two real-world datasets (Galaxy, Rabe), ensuring each contains more than 100 samples, as well as two synthetic datasets (Feynman I 27 6, Feynman II 34 2). All datasets have an input dimensionality between 2 and 5, matching the dimensional range of the transformer model [1]. Both Feynman datasets are evaluated using subsamples of 10,000 records each to ensure computational feasibility and reliable performance estimation. For each dataset, 75% of the data is used for training, and 25% is held out for testing to assess the prediction quality.

All experiments are conducted in a standardized search space, where both input features and target values are standardized to have zero mean and unit variance. Standardization has been shown to improve GP performance by reducing dependence on constant scaling [6, 17] and also enables the transformer model to operate within its learned semantic space.

Table 1. Configuration parameters and values used for the evaluated methods.

Parameter	Value
Initialization	Ramped Half-and-Half
Primitive Set	$\{V, \mathrm{ERC}, +, -, \times, \%, \mathrm{pow}\}$
ERC Range	[-0.5, 0.5], stepsize 0.1
Population Size	100
Generations	100
Selection	Tournament selection of size 5
Evaluation Metric	Root Mean Squared Error (RMSE)
Runs	30

Parameters. The experimental setup is implemented using two frameworks: stdGP and TSGP are based on DEAP [7], while GSGP and SLIM are implemented using the SLIM framework [20]. The methods are configured according to the parameters specified in Table 1, including a population size of 100, 100 generations, tournament selection of size 5 (without elitism), and fitness evaluation based on Root Mean Squared Error (RMSE). Each experiment is repeated 30 times. Initialization is conducted with Ramped Half-and-Half with a depth range of 2 to 5. Due to implementation differences between frameworks, we retain the default initialization settings for GSO-based methods (GSGP and SLIM), resulting in slightly smaller initial programs compared to the DEAP-based setup.

The primitive set consists of terminals and functions. Terminals include input variables V up to d dimensions and ephemeral random constants (ERCs) uniformly sampled from $[-0.5, 0.5]$ with step size 0.1, ensuring compatibility with the standardized space. Functions include addition, subtraction, multiplication, and protected division (%). Additionally, we include the power operator (pow) to align with TSGP's training data. Note that pow is currently only supported in DEAP and is therefore omitted for GSO-based methods.

For standard GP (stdGP), we apply 90% crossover (with a 10% terminal bias) and 10% subtree mutation, where the mutation depth ranges from 0 to 2 [9]. For GSGP and SLIM, we use the default mutation operator provided by the SLIM framework (SLIM+2), which mutates by adding a substructure consisting of two randomly generated trees. Since our analysis focuses exclusively on direct semantic mutation operators, no crossover is performed. In SLIM, the inflate/deflate mutation rate is set to its default value of 0.2, meaning that 20% of mutations are inflationary and 80% are deflationary. For TSGP, we use the pre-trained transformer model from Anthes et al. (2025) [1], trained on 20 million functions across dimensions $d \in \{2, 3, 4, 5\}$. The target semantic distance is set to $\mathrm{SD}_t = 1$, following [1], and the input dimensionality d matches that of the given problem.

4   Experiments

We split the experimental analysis into two parts. First, we provide a brief overview of the best solutions found by each method, reporting their predictive quality and size. We then present the core of our analysis, where we examine the syntax of programs at population level, focusing on how different operators modify the program structure during variation.

4.1   Best Solutions

We analyze the best solutions found during a run, which are defined as the programs with the lowest training RMSE among all programs generated by the methods. The final predictive quality is evaluated on the held-out test set.

Table 2. Median test RMSE of the best solution per run, across 30 independent runs, for stdGP, GSGP, SLIM, and TSGP. Bold values indicate the best prediction quality and label symbols indicate their significant superiority to other methods.

Data set	$_a$stdGP	$_b$GSGP	$_c$SLIM	$_d$TSGP
Feynman I 27 6	0.3052	$_{ac}$**0.1933**	0.2330	0.2256
Feynman II 34 2	0.2253	0.1725	0.2014	$_{abc}$**0.0881**
Galaxy	0.3145	0.3213	0.3435	$_c$**0.2824**
Rabe	0.1427	0.2228	0.2243	$_{abc}$**0.0929**

Prediction Quality. Table 2 reports the median test RMSE over 30 independent runs for each dataset. Best results are highlighted in bold. Statistical significance of the best performing method is assessed using the Mann–Whitney U test with $\alpha = 0.05$, with a Bonferroni correction applied for multiple comparisons. Method labels (a, b, c, d) indicate statistically significant differences compared to worse-performing methods.

The results show that TSGP most frequently produces the best solutions with the lowest test error. This difference is statistically significant compared to all other baselines on half of the datasets. The remaining methods show comparable performance. These findings are consistent with previous work [1,2].

Table 3. Median program size (nodes) of the best solution identified per run, across 30 independent runs, for stdGP, GSGP, SLIM, and TSGP. Values in bold indicate the smallest solutions and label symbols indicate their significant superiority to other methods.

Data set	$_a$**stdGP**	$_b$**GSGP**	$_c$**SLIM**	$_d$**TSGP**
Feynman I 27 6	$_{bc}$**87**	1202	122	$_{bc}$**87**
Feynman II 34 2	96	980	121	$_{bc}$**78**
Galaxy	92	1181	98	$_{bc}$**86**
Rabe	$_b$**63**	811	80	83

Program Size. We analyze the program size of the best solutions, measured as the number of nodes in their program trees. Smaller solutions are preferred because they are easier to evaluate and interpret [8]. Table 3 reports the median program size over 30 runs with the smallest solutions highlighted in bold. Statistical significance is again assessed using the Mann–Whitney U test with $\alpha = 0.05$, and significant differences of the smallest solution over others are marked with label symbols (a, b, c, d).

As expected, we find that the smallest solutions are produced by TSGP and stdGP. Both methods significantly outperform GSGP in all cases. On three of the four datasets, they also significantly outperform SLIM, the bloat-mitigating variant of GSGP. Notably, TSGP not only produces the best solutions in terms of predictive quality but also (together with stdGP) generates the most compact ones.

A comparison of the GSO-based methods reveals that GSGP produces by far the largest programs that are between 8 and 12 times larger than those generated by SLIM. This shows that the deflate operator in SLIM has a strong effect on reducing the program size and successfully mitigates bloat. However, despite this improvement, the best solutions produced by SLIM are still significantly larger than those of TSGP in most cases.

4.2   Population-Level Dynamics

We want to better understand the observed differences in size between the different semantic methods. Thus, we analyze how semantic operators affect the syntax of the evolved programs. In contrast to the previous subsection, which focuses on the best solution per run, we now analyze the entire population of programs generated across all generations. This allows us to examine the variation behavior of each method and to draw conclusions about their syntactic effects during evolution.

Mean Program Size. Figure 1 plots the mean program size of the entire population over generations for stdGP, GSGP, SLIM and TSGP. The results average 30 runs, with standard deviation in shaded regions.

For GSGP, we observe a monotonic growth in program size caused by its purely additive mutation operator, which systematically and continuously inflates programs. TSGP and SLIM also show initial growth, but their mean program size stabilizes after a few generations. Notably, TSGP suppresses growth more effectively on the Feynman datasets, while on Galaxy, the program sizes are comparable to stdGP and SLIM. On Rabe, SLIM generates the smallest programs on average (although at the cost of significantly worse predictive performance of the best solutions, as shown in Table 2).

For all datasets, the mean program sizes generated by SLIM (Fig. 1) are much smaller than the size of the best solutions found (Table 3). This suggests that while deflation reduces the size of programs, the best solutions tend to be larger, possibly because they retain more complex structures needed for higher accuracy.

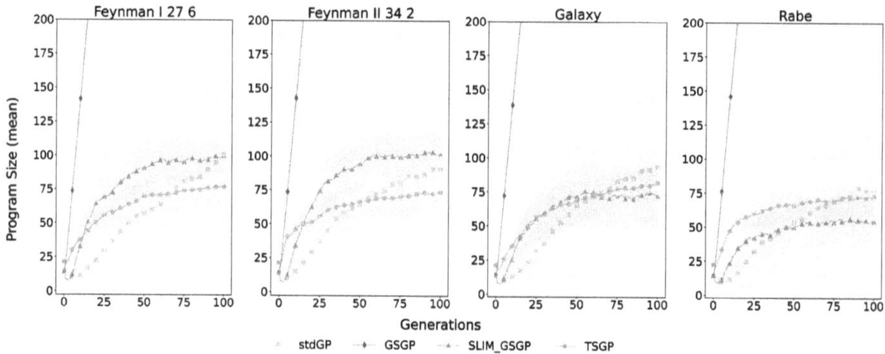

Fig. 1. Mean program size ($\pm$ standard deviation), over 100 generations, averaged across 30 independent runs. The y-axis is capped at 200 to highlight differences among non-bloating methods.

Impact of Variation on Program Size. All subsequent experiments focus exclusively on the variation step of each method. To capture subtle yet systematic effects, such as the occasional application of SLIM's inflate operator (which occurs at a low frequency but is an important part of its variation), we report mean variation dynamics.

Thus, Fig. 2 plots the mean difference in program size (nodes) between parents and offspring over generations across all runs, with standard deviation shown as shaded regions. This metric reveals whether variation tends to increase or decrease program size on average, providing insight into the growth or reduction tendencies inherent in each operator.

GSGP consistently generates offspring that are larger than their parents, reflecting its insert-only mutation operator, which systematically inflates program structure. In contrast, SLIM generates offspring that are, on average, smaller than their parents, which is a direct consequence of its deflation operator and the high probability of applying deflate (set to 0.8) instead of inflate. Although the variation operator on average reduces program size, the average program size increases over generations (compare Fig. 1). This implies that selection prefers larger programs, which is consistent with the observation that the best solutions in SLIM are often larger than the average program size (Table 3).

Both TSGP and stdGP produce offspring that are, on average, equal or slightly smaller than their parents. Thus, unlike SLIM, the ability of TSGP to find compact solutions is not due to a stronger variation bias in favor of smaller programs, but must be based on a different mechanism. Additionally, the resulting offspring size varies significantly more than with GSO-based methods. In stdGP, the standard deviation increases over time, aligning with the observation that the programs grow larger during evolution, as shown in Fig. 1, which causes crossover and mutation to have a more variable impact on the size of the offspring.

In TSGP, the variance in size deviation remains consistently high across all generations. This reflects the probabilistic nature of the transformer's generation process. Trained on diverse program structures, the model produces offspring that can be significantly different from the parent depending on the given context (parent program). This behavior reflects the model's learned ability to perform syntactically flexible mutations.

Impact of Variation on Program Structure. We want to better understand why TSGP produces compact solutions. Thus, we measure the Levenshtein distance of the syntactic modifications applied during variation. This metric quantifies the minimum number of edit operations (insertions, deletions, or substitutions of nodes) required to transform one sequence into another [13]. We compute the Levenshtein distance over the prefix traversal of each program, treating each node as a token in a sequence. This enables us to compare the syntactic dissimilarity between parents and their offspring across all methods and generations.

Figure 3 plots the mean Levenshtein distance between parent and offspring over the number of generations across all runs for stdGP, GSGP, SLIM, and TSGP. On average, TSGP and semantic methods perform stronger modifications compared to stdGP. TSGP exhibits the highest degree of structural modification among them, particularly in the early to mid stages of the search.

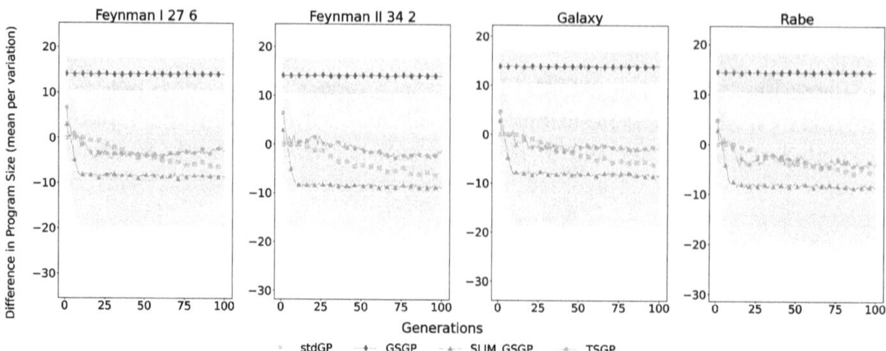

Fig. 2. Mean difference in program size between parent and offspring ($\pm$ standard deviation), over 100 generations, averaged across 30 independent runs.

The behavior of TSGP and stdGP varies across generations, reflecting their dependence on the structure of the parent program. TSGP performs major structural changes during the early search stages (generations 1–25), despite the relatively small size of programs at this stage. This indicates an exploration phase in which the model generates highly diverse syntactic variants. As evolution continues, the change becomes smaller but remains high compared to stdGP. In contrast, GSO-based methods (GSGP and SLIM) exhibit a stable and consistently high level of structural change across all generations, as each mutation adds or removes a randomly generated substructure, regardless of the structure or size of the parent.

The standard deviations of the Levenshtein distance are in line with the differences in variation behavior across methods. GSO-based methods show low and stable variance, reflecting their deterministic and uniform variation. In contrast, TSGP and stdGP exhibit high variance, suggesting that their structural modifications fluctuate considerably among different mutations. This is expected for stdGP because its operators rely on probabilistic choices that can lead to minimal changes, such as small subtree mutations, or drastic changes, such as large subtree exchanges during crossover. For TSGP, the high variance indicates that the model is able to generate programs with completely different syntax (but similar semantics). This demonstrates the structural flexibility of the transformer-based approach to express variations through a variety of syntactic forms and adapt its modifications to the context of the parent program. Despite this high flexibility and strong syntactic changes, all offspring generated by TSGP are semantically similar to their parents (significantly more than those produced by stdGP [1,2]).

Edit Operations. We now analyze the Levenshtein distance in detail by examining the different edit operations (insertion, deletion and replacement) that are applied during variation. We consider three complementary perspectives: (1) the distribution of edit operation types, (2) the number of distinct edit types applied

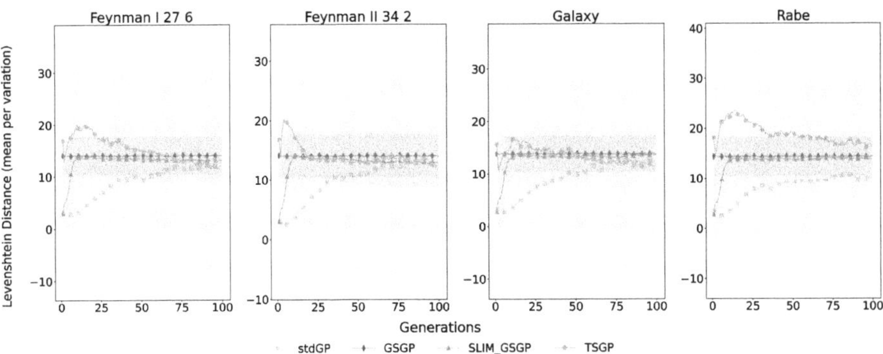

Fig. 3. Mean Levenshtein distance between parent and offspring ($\pm$ standard deviation), over 100 generations, averaged across 30 independent runs.

simultaneously per variation, and (3) the spatial distribution of edits across the program structure.

Distribution of Edit Operations. Figure 4 plots the average number of inserted, deleted, and replaced nodes per variation step, measured as the Levenshtein distance, and averaged over each generation and all runs.

For GSGP and SLIM, we observe a behavior consistent with their design: GSGP performs only insertions, resulting in a consistently positive change in offspring length. In contrast, SLIM applies both insertions and deletions, with deletions dominating due to its high deflate probability. Thus, SLIM reduces on average the size of an individual. Neither GSGP nor SLIM supports replacements, so the number of replacements is always zero.

For stdGP, we observe mainly insertions and deletions, which are the result of subtree exchanges during crossover or mutation. Although stdGP does not explicitly perform node replacements, some subtree exchanges are classified as replacements, especially those involving syntactically similar subtrees. Furthermore, the number of deleted nodes increases over the number of generations as programs grow and larger subtrees are replaced by smaller ones.

For TSGP, we observe a high number of node replacements, a behavior not observed in stdGP or GSO-based methods. On average, TSGP performs up to nine times more replacements than stdGP on the Rabe dataset, and 5.5 times more on Feynman I 27 6. This suggests that, unlike GSO-based methods, which make coarse, discrete decisions (inflate/deflate), TSGP can perform fine-grained syntactic modifications to produce syntactically different offspring with similar semantics. We also find a high number of insertions and deletions, reflecting TSGP's ability to perform diverse mutations. Notably, the standard deviation of all edit types is higher for TSGP, particularly for replacements. This suggests that the model adapts its modifications to the context of the parent, sometimes making minimal changes (e.g. replacing a single operator) and sometimes performing large-scale restructuring (e.g. rewriting entire sub-expressions).

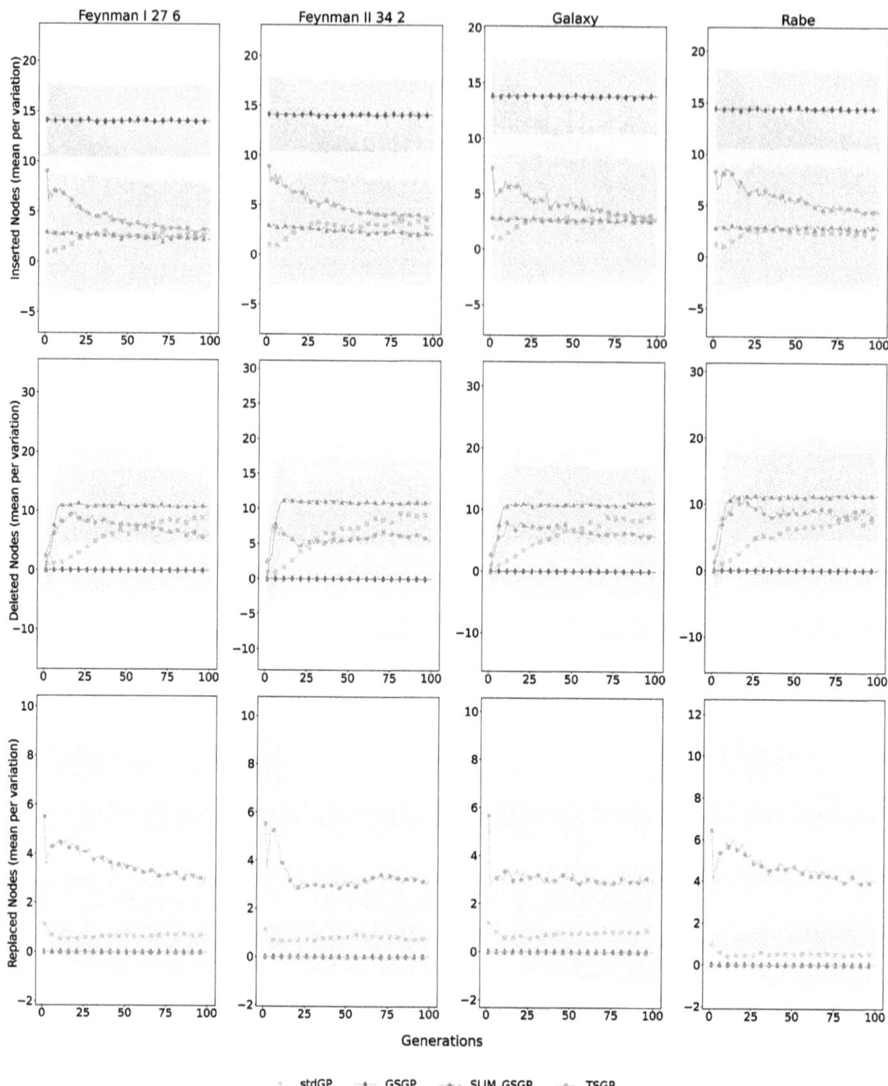

Fig. 4. Mean number of inserted, deleted, and replaced nodes per variation step ($\pm$ standard deviation), over 100 generations, averaged across 30 independent runs.

Number of Simultaneous Edit. Table 4 summarizes the results and reports the mean and standard deviation of the number of distinct edit types applied per variation step, averaged over all generations and runs. GSO-based methods apply exactly one edit type per variation: GSGP always inserts, and SLIM always either inserts or deletes, depending on the chosen operator.

In contrast, both stdGP and TSGP apply multiple edit types simultaneously per variation. On average, TSGP applies between 2.18 and 2.33 distinct edit types per variation, which is significantly more than the 1.40–1.46 applied by stdGP. This suggests that TSGP produces more complex, multimodal mutations that involve not only more numerous edits but also more diverse ones, indicating greater syntactic expressivity.

Table 4. Mean number of distinct edit types per variation step ($\pm$ standard deviation), across 30 independent runs, for stdGP, GSGP, SLIM, and TSGP.

Data set	stdGP	GSGP	SLIM	TSGP
Feynman I 27 6	1.43 ± 0.60	1.00 ± 0.00	1.00 ± 0.00	2.22 ± 0.73
Feynman II 34 2	1.45 ± 0.61	1.00 ± 0.00	1.00 ± 0.00	2.20 ± 0.75
Galaxy	1.46 ± 0.62	1.00 ± 0.00	1.00 ± 0.00	2.18 ± 0.75
Rabe	1.40 ± 0.60	1.00 ± 0.00	1.00 ± 0.00	2.33 ± 0.71

Spatial Distribution of Edits. In order to assess whether edits are localized or distributed throughout the program, we group edit operations into clusters. Edit operations that occur within a window of at most 2 consecutive positions in the prefix traversal sequence are counted to be in the same edit cluster. Consequently, Table 5 reports the mean and standard deviation of the number of edit clusters per variation step averaged over all generations and runs.

GSGP and SLIM consistently produce exactly one edit cluster per variation, reflecting their local, deterministic nature. The standard deviation of zero confirms that their modifications are confined to a single contiguous region of the program. StdGP produces one or two edit clusters per variation as a result of the combined effects of crossover and mutation.

TSGP shows the highest level of distributed changes, with an average of 3.1 to 3.6 edit clusters per variation, which is substantially more than any other method. A standard deviation of roughly 2 indicates high variation. This confirms that TSGP does not restrict edits to localized regions but performs multiple spatially separated edits simultaneously. Consequently, TSGP is not a local mutation operator, but a holistic, context-aware variation operator capable of restructuring programs across multiple, non-contiguous regions while still keeping the semantics of solutions similar.

Table 5. Mean number of edit clusters per variation step ($\pm$ standard deviation), across 30 independent runs, for stdGP, GSGP, SLIM, and TSGP.

Data set	stdGP	GSGP	SLIM	TSGP
Feynman I 27 6	1.62 ± 1.54	1.00 ± 0.00	1.00 ± 0.00	3.67 ± 2.18
Feynman II 34 2	1.64 ± 1.56	1.00 ± 0.00	1.00 ± 0.00	3.27 ± 2.09
Galaxy	1.65 ± 1.69	1.00 ± 0.00	1.00 ± 0.00	3.10 ± 2.12
Rabe	1.66 ± 1.63	1.00 ± 0.00	1.00 ± 0.00	3.62 ± 2.23

5 Conclusion and Future Work

In this paper, we analyzed how semantic variation operators express semantic similarity through syntactic transformations, and why Transformer Semantic Genetic Programming (TSGP) consistently produces significantly more compact and high-quality solutions than GSO-based methods such as GSGP and SLIM. Experiments across four symbolic regression problems reveal that TSGP's solution compactness is not due to a variational bias toward smaller programs, as in SLIM, but rather from its inherent syntactic flexibility.

The transformer model of TSGP, being trained on a set of syntactically diverse yet semantically similar programs, generates offspring through a probabilistic, context-aware process. Unlike GSO-based operators, which are constrained to fixed, local, and linear transformations, TSGP performs a wide variety of syntactic modifications, ranging from single node replacements to large-scale subtree rewrites, and operates holistically across the entire program structure. This enables TSGP to adapt its mutation behavior to the context of the parent program, exploring various syntactic forms while preserving semantic similarity. Consequently, TSGP discovers solutions that are both highly fit and syntactically compact.

Future work should focus on the TSGP training process to better understand how the diversity, distribution, and structural properties of the training data influence the model's variation behavior. Additionally, developing a semantic-aware crossover operator for TSGP could further enhance its search efficiency and solution quality, extending its advantages beyond mutation-based evolution.

References

1. Anthes, P., Sobania, D., Rothlauf, F.: Transformer semantic genetic programming for d-dimensional symbolic regression problems (2025). https://arxiv.org/abs/2511.09416
2. Anthes, P., Sobania, D., Rothlauf, F.: Transformer semantic genetic programming for symbolic regression. In: Proceedings of the Genetic and Evolutionary Computation Conference, pp. 952–960 (2025)
3. Beadle, L., Johnson, C.G.: Semantically driven crossover in genetic programming. In: 2008 IEEE Congress on Evolutionary Computation (IEEE World Congress on Computational Intelligence), pp. 111–116. IEEE (2008)

4. Beadle, L., Johnson, C.G.: Semantically driven mutation in genetic programming. In: 2009 IEEE Congress on Evolutionary Computation, pp. 1336–1342. IEEE (2009)
5. Castelli, M., Vanneschi, L., Silva, S.: Prediction of high performance concrete strength using genetic programming with geometric semantic genetic operators. Expert Syst. Appl. **40**(17), 6856–6862 (2013)
6. Dick, G., Owen, C.A., Whigham, P.A.: Feature standardisation and coefficient optimisation for effective symbolic regression. In: Proceedings of the 2020 Genetic and Evolutionary Computation Conference, pp. 306–314 (2020)
7. Fortin, F.A., De Rainville, F.M., Gardner, M.A.G., Parizeau, M., Gagné, C.: DEAP: evolutionary algorithms made easy. J. Mach. Learn. Res. **13**(1), 2171–2175 (2012)
8. Javed, N., Gobet, F., Lane, P.: Simplification of genetic programs: a literature survey. Data Min. Knowl. Disc. **36**(4), 1279–1300 (2022)
9. Koza, J.R.: Genetic programming as a means for programming computers by natural selection. Stat. Comput. **4**(2), 87–112 (1994)
10. Krawiec, K.: Medial crossovers for genetic programming. In: European Conference on Genetic Programming, pp. 61–72. Springer (2012)
11. Krawiec, K., Lichocki, P.: Approximating geometric crossover in semantic space. In: Proceedings of the 11th Annual conference on Genetic and Evolutionary Computation, pp. 987–994 (2009)
12. La Cava, W., et al.: Contemporary symbolic regression methods and their relative performance. Adv. Neural. Inf. Process. Syst. **2021**(DB1), 1 (2021)
13. Lcvenshtcin, V.: Binary codes capable or 'correcting deletions, insertions, and reversals'. In: Soviet Physics-Doklady, vol. 10 (1966)
14. McPhee, N.F., Ohs, B., Hutchison, T.: Semantic building blocks in genetic programming. In: European Conference on Genetic Programming, pp. 134–145. Springer (2008)
15. Moraglio, A., Krawiec, K., Johnson, C.G.: Geometric semantic genetic programming. In: International Conference on Parallel Problem Solving from Nature, pp. 21–31. Springer (2012)
16. Nguyen, Q.U., Nguyen, X.H., O'Neill, M.: Semantic aware crossover for genetic programming: the case for real-valued function regression. In: European Conference on Genetic Programming, pp. 292–302. Springer (2009)
17. Owen, C.A., Dick, G., Whigham, P.A.: Feature standardisation in symbolic regression. In: Australasian Joint Conference on Artificial Intelligence, pp. 565–576. Springer (2018)
18. Pietropolli, G., Farinati, D., Manzoni, L., Castelli, M., Silva, S., Vanneschi, L.: Introducing crossover in slim-GSGP. In: European Conference on Genetic Programming (Part of EvoStar), pp. 103–119. Springer (2025)
19. Romano, J.D., et al.: PMLB V1.0: an open source dataset collection for benchmarking machine learning methods. arXiv preprint arXiv:2012.00058v2 (2021)
20. Rosenfeld, L., et al.: Slim_GSGP: a Python library for non-bloating GSGP. In: Proceedings of the Genetic and Evolutionary Computation Conference, pp. 1026–1034 (2025)
21. Vanneschi, L.: An introduction to geometric semantic genetic programming. In: NEO 2015: Results of the Numerical and Evolutionary Optimization Workshop NEO 2015, 23–25 September 2015 in Tijuana, Mexico, pp. 3–42. Springer (2016)
22. Vanneschi, L.: Slim_GSGP: the non-bloating geometric semantic genetic programming. In: Giacobini, M., Xue, B., Manzoni, L. (eds.) Genetic Programming, pp. 125–141. Springer, Cham (2024)

23. Vanneschi, L., Castelli, M., Manzoni, L., Silva, S.: A new implementation of geometric semantic GP and its application to problems in pharmacokinetics. In: European Conference on Genetic Programming, pp. 205–216. Springer (2013)
24. Vanneschi, L., Castelli, M., Silva, S.: A survey of semantic methods in genetic programming. Genet. Program Evolvable Mach. **15**(2), 195–214 (2014)

Node Preservation and Its Effect
on Crossover in Cartesian Genetic
Programming

Mark Kocherovsky[(✉)] [iD], Illya Bakurov [iD], and Wolfgang Banzhaf [iD]

Department of Computer Science and BEACON Center for the Study of Evolution in
Action, Michigan State University, East Lansing, MI 48824, USA
`kocherov@msu.edu`
`https://banzhaf-lab.github.io/`

Abstract. While crossover is a critical and often indispensable component in other forms of Genetic Programming, such as Linear- and Tree-
based, it has consistently been claimed that it deteriorates search performance in CGP. As a result, a mutation-alone $(1+\lambda)$ evolutionary strategy
has become the canonical approach for CGP. Although several operators
have been developed that demonstrate an increased performance over
the canonical method, a general solution to the problem is still lacking.
In this paper, we compare basic crossover methods, namely one-point
and uniform, to variants in which nodes are "preserved," including the
subgraph crossover developed by Roman Kalkreuth, the difference being
that when "node preservation" is active, crossover is not allowed to break
apart instructions. We also compare a node mutation operator to the traditional point mutation; the former simply replaces an entire node with a
new one. We find that node preservation in both mutation and crossover
improves search using symbolic regression benchmark problems, offering
a general solution to CGP crossover.

Keywords: Cartesian Genetic Programming · Crossover · New
Directions

1   Introduction

Genetic Programming (GP) has seen several decades of research as a subfield of
Evolutionary Computation (EC). Established in the late 1980s / early 1990s by
John Koza [26,27], GP extends traditional Genetic Algorithmic (GA) methods
by evolving *programs*, ranging from basic Boolean and symbolic expressions to
complex system control algorithms. Programs are traditionally represented as
trees (TGP), where leaves contain terminals (input variables and constants)
and parent nodes contain operators. As the tree is traversed, an expression is
constructed. Crossover occurs by swapping subtrees between individuals, and
mutation changes the tree structure. A notable feature of GP is the presence
of *introns*, also referred to as *ineffective instructions*. These are parts of the

© The Author(s), under exclusive license to Springer Nature Switzerland AG 2026
L. Manzoni et al. (Eds.): EuroGP 2026, LNCS 16521, pp. 67–83, 2026.
`https://doi.org/10.1007/978-3-032-23005-8_5`

program that do not influence the final output behavior, yet they are believed to play a beneficial role by providing additional genetic material for variation operators to act upon. In contrast, *exons* are parts of the code that directly contribute to the output behavior of programs [8,38].

A consistent problem with GP is *bloat*, a phenomenon where individuals become increasingly large, thus more complex, leading to slower run-times and deteriorating interpretability. Early studies established that the majority of the genome in GP consists of introns [37,38]. More recent findings have shown that this proportion further increases when also considering genes that only marginally affect behavioral semantics [4]. Additionally, even under exploratory evolutionary dynamics, the genotypic population is often dominated by a small number of distinct phenotypes, i.e., tree structures composed primarily of exonic (active) components.

Bloat in TGP has been extensively studied, with several theories proposed to explain its emergence [3,6,7,16,29–31,43,44,47]. One prominent theory derives parallels from intron research in genetics [50], and suggests that bloat arises as a protective mechanism to preserve useful substructures, or building blocks, which are otherwise vulnerable to disruption by mutation and crossover [3,7,31,37,38, 44]. This vulnerability arises from the lack of explicit built-in mechanisms for preservation and modular reuse of building blocks in basic TGP.

As a response to TGP's limitations, numerous alternative GP paradigms have been developed. These primarily differ in the data structures used to represent candidate solutions and in the design of initialization and variation operators tailored to those representations [5,9]. One of these alternatives is Linear GP (LGP), where programs are represented as variable-length sequences of instructions executed in order [5,9,33,35,40,45]. Each instruction operates on a set of registers, which serve as a steady-state memory module that stores intermediate results. This register-based architecture facilitates the reuse of partial computations and reduces the risk of structural disruption during variation [24].

Cartesian Genetic Programming (CGP) is another form of GP; CGP uses a fixed-length linear genome to encode programs represented as directed acyclic graphs (DAGs), where each gene specifies a functional component and its connections in a cartesian grid-like structure [35]. Each node in the genome applies a function to a fixed number of inputs, selected from earlier nodes or input variables, thereby supporting explicit reuse of intermediate results. Unlike LGP, however, CGP does not use registers and instead relies on positional references within the genome [24].

The fixed-length representation in CGP helps to control bloat by forcing the evolutionary search to operate within a bounded genomic space. However, this structural rigidity can also limit evolvability, particularly in problems that benefit from larger or more modular solutions. On the other hand, the fixed-length encoding enables direct crossover at the genomic level. Thus, it became tempting for researchers and practitioners to use traditional GA-style crossover operators in CGP. Yet, empirical studies have consistently shown that such operators are often detrimental to CGP's search performance [11,24], and the scientific community mostly avoids their usage [34]. As such, a $(1 + \lambda)$ configuration is typically used in practice.

In response, researchers have proposed systematic explanations for CGP's crossover limitations and introduced alternative operators specifically tailored to its representation [10–12,14,15,19,22,23,42]. In particular, Kalkreuth [20–22] and Cui [12–15], have advanced a deeper understanding of CGP's structural and positional constraints, leading to more principled crossover operators. These efforts demonstrate that effective crossover is possible when grounded in a clear rationale, yet the field still lacks a unified theory or widely accepted best practices for recombination in CGP. In 2024, Kocherovsky and Banzhaf hypothesized that CGP lacks means to "anchor" good substructures, as the steady-state memory does in LGP [24]. Building on this, their 2025 work [25] reaffirmed the positional and structural insights from Cui [14], arguing that the failure of some CGP crossover strategies stems from implicit assumptions about positional equivalence and structural interchangeability. From this basis, we follow Kalkreuth [20] and Kocherovsky et al. [24,25] in testing one of the features of Subgraph crossover in the *general* case: crossover in Subgraph crossover *necessarily operates at node-level, instead of at the genomic level*, because otherwise its behavioral constraints would be broken. We call this technique **Node Preservation**, because the node is structurally preserved throughout the operation.

Here we thus argue that by ensuring to perform node preservation—which can be accomplished as easily as changing the representation of a program from a one-dimensional vector to a two-dimensional matrix—both crossover and mutation are *generally* improved. We test one-point and uniform crossover methods, which are classical operators in the wider field of evolutionary computation both with and without this constraint, each in the typical CGP representation and in a more LGP-like two-dimensional representation, where we find that node preservation significantly improves the likelihood of producing better solutions. We also compare these methods to the Subgraph crossover, which performs better than traditional operators, likely due to the care it takes in making sure subgraphs of different parents manage to link together.

The structure of this paper is as follows. In Sect. 2 we discuss the previous literature on CGP crossover. In Sect. 3 we explain the structure of our CGP models and introduce our Node Preservation Method. Section 4 explains our experimental conditions, Sect. 5 introduces and discusses our results, while Sect. 6 concludes the paper with a list of contributions and potential for future work.

We also make our code and supplementary material available on our github here: https://github.com/MarkKocherovsky/cgp_crossover.

2   Background

2.1   GA-Like Crossover in CGP

The use of GA-like fixed-length encoding in CGP facilitates direct crossover at the genomic level, which could justify why CGP researchers and practitioners started to use traditional GA-style recombination methods such as n-point and uniform crossover [39]. In n-point crossover, a small number of crossover points results in the exchange of contiguous segments between parents. This preserves the relative ordering of neighboring genes, under the assumption that genes located near one another are functionally related and should be inherited together to maintain their combined effect. As the number of crossover points increases, this assumption weakens. In the extreme case (uniform crossover) each gene is swapped with a probability of 0.5 (typically independently), entirely disregarding positional relationships. Such an approach treats the genome as an unordered set of genes, maximizing diversity at the cost of disrupting structural coherence.

While these assumptions often hold in many GA applications, such as the Traveling Salesman Problem [46], they break down in the context of CGP [10]. CGP encodes programs as directed acyclic graphs (DAGs), where each node can connect to any precedent node in the genome, rather than just to adjacent ones. Consequently, functional relationships between genes are not determined by proximity in the genome, and assuming otherwise can lead to the disruption of essential dependencies during crossover. Moreover, ignoring positional information altogether, as uniform crossover does, can exacerbate the problem by destroying useful structures, particularly since the functionality of a node at index i can vary significantly between parents. This issue was first raised by Cai *et al.* [10] and termed as *positional dependence*: *"The effect or meaning of a component in the evolved or resulting program is determined by its absolute or relative position in the program representation"*. Later, Goldman and Punch [17] deepened the understanding by introducing the concept of *positional bias* to describe the fact that a node's activation probability varies significantly by position, with a bias toward nodes near the input, which receive more connections due to the DAG structure in CGP, where each node can only connect to preceding nodes to avoid cycles. The presence of positional bias was further confirmed in [14,25].

To our knowledge, the first attempt to design a better crossover operator for CGP was presented in 2007 by Clegg *et al.* [11]. The authors, however, essentially adapted the real-valued crossover often used in GA literature for solving continuous optimization problems. This crossover performs a randomly-weighted average of the two parent genes. Given that traditional CGP uses an integer-based representation, the authors proposed an algorithm for converting it into a $[0, 1]$ range of floating-points. The real-valued CGP crossover was assessed on two symbolic regression problems (Koza2 and Koza3) and was found to outperform the mutation-alone strategy only on one problem. Nevertheless, it was found to achieve faster convergence in earlier iterations, which led the authors to adapt the crossover rate over generations. The adaptive variant turned out to outperform the mutation-alone strategy. However, later studies examined real-valued crossover on more problems and found no advantage in using it. The paper's methodology has also been called into question given a relatively small amount of data presented [25,48]. While Clegg et al. also used indivisible node blocks as crossover units, the authors of Cui et al. (2024) were unable to replicate these results [12].

2.2 Representation-Aware Crossover in CGP

Early efforts using GA-style crossover in CGP often failed to respect the inherent structural dependencies encoded in CGP's DAG. Consequently, a new wave of crossover research has emerged that explicitly accounts for positional dependence and positional bias, while respecting the underlying graph-based representation of CGP. In 2017, Kalkreuth *et al.* introduced the Subgraph Crossover, which acts as a one-point crossover where the crossover point is bounded by the range of active nodes in the program [22]. This was found to outperform traditional crossover methods on Boolean and symbolic regression problems [20–22]. Husa and Kalkreuth introduced block crossover, which swaps coherent blocks of instructions between parents [19]. An operator that combined well-performing parent subgraphs into a single chromosome was introduced by da Silva and Bernardino [42].

Goldman and Punch [17] observed that the distribution of active nodes across the CGP genome tends to be biased towards earlier indices and proposed two strategies to mitigate this bias. The first, called DAG, mutates connections to every other node in the genome, as long as the new connection does not create a cycle; the second, called REORDER, shuffles active nodes in the genome in such a way that the phenotype does not change. Both strategies led to an increased the amount of active nodes in evolved graphs and achieved faster convergence towards a solution with smaller genotypes. Following Goldman and Punch, Cui *et al.* demonstrated that applying the REORDER operator could mitigate positional bias and enable crossover to contribute positively to CGP search [14]. This work led to further enhancements, such as the Equidistant-REORDER operator [15], which more uniformly distributes exons across the genome. Performing Equidistant-REORDER was found to improve CGP's problem-solving capabilities across Boolean and regression benchmarks [12].

3 Methods

In Cartesian Genetic Programming, a program is defined as a set of nodes, of which there are three types: *input*, including variable and constant terminals; *function*, which take in input and carry out instructions; and *output*, which take in one argument each for the program to return. Typically, a CGP genome is represented as a linear genome in the form of a one-dimensional vector of function and output nodes:

$$p_i \in P = f_0|a_{0,1}|a_{0,2}|f_1|a_{1,1}|a_{1,2}|...|f_n|a_{n,1}|a_{n,2}|o_0|o_1|...|o_m \tag{1}$$

assuming a program $p \in P$ contains n function nodes, each with one operator $f \in F$ and an arity $a = 2$, and m output nodes. Each node in Eq. 1 is separated by a | character for the reader's convenience. In this work, although we follow [24, 25], we use a slightly different representation: each program is represented as a *matrix*, with each row representing a node. Technically, the matrix would look as shown in Table 1. Note the recording of which function nodes are active, which

Table 1. Example CGP program p in this work. The full representation has a length of $n + |I| + |O|$, where there are n function nodes, $|I|$ input and constant nodes, and $|O|$ output nodes. A unified index registry allows for easy reference to and from nodes of different types.

Index	NodeType	Value	Operator	Operand0	Operand1	Active
0	InputNode	x	NaN	NaN	NaN	NaN
1	ConstantNode	1	NaN	NaN	NaN	NaN
2	FunctionNode	$x+1$	Addition	0	1	1
3	FunctionNode	0	Multiplication	2	2	0
...	...	...	...	...	...	...
$n+2$	FunctionNode	0	Analytic Quotient	i	j	0
o_0	OutputNode	$p(x)$	NaN	2	NaN	NaN

is important for crossover algorithms and data analysis; and the separation of input and constant nodes, which is done for clarity.

This decision has four motivations: first, it is easier to program and manage on a technical level; second, it is more readable and writable on the software design level; third, this representation offers the user more information as to the internal dynamics of the model; and finally, it makes it easier to swap nodes as all a practitioner needs to is to swap rows between parents, thus following Kalkreuth [21].

3.1   Crossover

In addition to the canonical $(1 + 4)$ strategy, we also implement the traditional one-point and uniform crossovers [39]. The latter is simply a form of n-point crossover, where n is equal to half the length of the genome. While in all replicates, the matrix form in Table 1 is used generally, in runs *without* node preservation, functions are used to convert to and from the more typical form given in Eq. 1 specifically to perform crossover. For all our methods that involve crossover, a $(40 + 40)$ strategy is implemented, following [24,25].

The Subgraph crossover, developed by Kalkreuth [20–22], works in three main steps, assuming an otherwise one-point crossover. First, a crossover point is chosen *within* the range of active nodes. The operator must swap *sub*graphs of the program, hence the range constraint *and* the need to prevent crossover points from breaking apart functions. If the nodes are not preserved, then the subgraph will change during crossover. Second, the parent programs swap *nodes* and are recombined. Finally, the first active node after the crossover point has connections re-wired to ensure that both subgraphs remain active.

3.2   Mutation

We test two mutation operators. The first is point mutation, which is standard for CGP. If a model is chosen to be mutated, a single output or function node is chosen at random, and either has its operator or operands changed to a random legal value. The second is *node* mutation, where a randomly chosen node is simply replaced with an entirely new randomly generated node. We test all operators with both mutation types separately to provide a comprehensive comparison.

3.3 Selection and Fitness

For selection, we primarily use elite tournament selection, except for canonical $(1 + \lambda)$ experiments, which necessarily use elitism. Neutral drift is allowed in all cases. Following [24,25], our fitness function is based on the Pearson correlation r: $f = 1 - r^2$ [18].

4 Experiments

Following [24,25], we test our methods on ten symbolic regression problems. These are three Koza problems, four Nguyen problems, the Ackley, Rastrigin, and Levy problems. These are shown in detail in Table 2. For the simpler problems, training sets consisted of 20 points and test sets consisted of 10 points; for the complex problems, training sets consisted of 40 points, with test sets of 20 points. In all cases, the training-testing split was 0.33. All models were allowed to use the four arithmetic functions: addition $(+)$, subtraction $(-)$, multiplication $(*)$, and the Analytic Quotient $(/)$ [36].

Table 2. Problems tested; each model was given a set of random points within the given domain. Mostly reproduced from [25].

Problem	Function	Domain	Points
Koza-1	$x^4 + x^3 + x^2 + x$	$[-1, 1]$	30
Koza-2	$x^5 - 2x^3 + x$	$[-1, 1]$	30
Koza-3	$x^6 - 2x^4 + x^2$	$[-1, 1]$	30
Nguyen-4	$x^6 + x^5 + x^4 + x^3 + x^2 + x$	$[-1, 1]$	30
Nguyen-5	$\sin(x^2)\cos(x) - 1$	$[-1, 1]$	30
Nguyen-6	$\sin(x) + \sin(x + x^2)$	$[-1, 1]$	30
Nguyen-7	$\ln(x + 1) + \ln(x^2 + 1)$	$[0, 2]$	30
Ackley [2]	$-20\exp(-0.2x^2) - \exp(\cos 2\pi x) + 20 + \exp 1$	$[-32.768, 32.768]$	60
Rastrigin [41]	$10 + x^2 - 10\cos 2\pi x$	$[-5.12, 5.12]$	60
Levy [28]	See Reference	$[-10, 10]$	60

Our full list of shorthand used to refer to the various crossover methods is shown in Table 3. Each method has fifty replicates for each problem, with each replicate run for six thousand generations. Each model is given a maximum size (i.e. number of active nodes) of 64. We collect a series of metrics in each generation: best fitness, median fitness, best model size, median model size, and semantic diversity between each model in the population.

All experiments were run on Michigan State University's High-Performance Computing Cluster [1].

Table 3. Evolutionary parameters to demonstrate the effects of various crossover and mutation strategies. Mostly reproduced from [25].

Notation	Crossover	Mutation
Canonical-PM	None	Point (100%)
One-Point PM	One-Point Nodes Preserved (50%)	Point (50%)
One-Point 1D PM	One-Point (50%)	Point (50%)
Uniform PM	Uniform Nodes Preserved (50%)	Point (50%)
Uniform 1D PM	Uniform (50%)	Point (50%)
Subgraph PM	Subgraph (50%)	Point (50%)
Canonical-NM	None	Node (100%)
One-Point NM	One-Point (50%)	Node (50%)
One-Point 1D NM	One-Point Flattened (50%)	Node (50%)
Uniform NM	Uniform (50%)	Node (50%)
Uniform 1D NM	Uniform Flattened (50%)	Node (50%)
Subgraph NM	Subgraph (50%)	Node (50%)

4.1 Plackett-Luce Analysis

Following [15], we use the Plackett-Luce model to compare our results at the end of evolution and provide a value indicating the probability that a given crossover operator will produce the best model compared to the other operators for a given problem. In this study, we use the `PlackettLuce` library in R provided by Turner et al. [49]. Because it is based on rankings, it is important to remember that the Plackett-Luce model does not give us the probability that a model will be objectively "good", but only the probability that an operator will produce a better model than its competitors. This allows us to make more definite statements when comparing the results of the various operators without having to go back-and-forth between the fitness table and matrices of p-values.

5 Results

5.1 Fitness

The median value of the best fitness at the end of each run per crossover operator and problem is shown in Table 4. Additionally, box plots showing the best fitness at the end of each run are shown in Fig. 1.

Table 4. Median Fitness for the best individual in all 50 replicates. The best operator's fitness in each problem is highlighted in deep green and written in bold, and the worst fitness is highlighted in deep red.

	Canonical		One-Point		One-Point 1D	
	Point Mutation	Node Mutation	Point Mutation	Node Mutation	Point Mutation	Node Mutation
Koza 1	1.323E-02	3.288E-03	**3.389E-04**	5.052E-04	8.183E-04	9.353E-04
Koza 2	2.368E-02	2.444E-02	5.897E-03	3.126E-03	1.720E-02	1.480E-02
Koza 3	7.156E-02	3.323E-02	1.474E-02	1.166E-02	2.952E-02	2.155E-02
Nguyen 4	4.207E-03	3.803E-03	8.881E-04	5.530E-04	1.782E-03	8.516E-04
Nguyen 5	1.469E-03	5.610E-04	1.849E-04	**3.732E-05**	6.693E-04	3.764E-04
Nguyen 6	2.376E-03	1.106E-03	**2.297E-04**	2.706E-04	4.852E-04	6.026E-04
Nguyen 7	1.345E-04	1.920E-05	5.111E-06	7.018E-06	4.012E-04	2.060E-05
Ackley	4.920E-02	5.113E-02	3.520E-02	3.678E-02	3.349E-02	3.161E-02
Levy	1.383E-01	1.225E-01	8.726E-02	9.271E-02	9.657E-02	9.611E-02
Rastrigin	5.315E-01	5.648E-01	4.688E-01	4.801E-01	4.900E-01	4.758E-01
	Uniform		Uniform 1D		Subgraph	
	Point Mutation	Node Mutation	Point Mutation	Node Mutation	Point Mutation	Node Mutation
Koza 1	4.894E-04	7.112E-04	1.562E-03	1.075E-03	3.512E-04	3.614E-04
Koza 2	5.107E-03	5.990E-03	1.005E-02	1.001E-02	**2.139E-03**	2.841E-03
Koza 3	1.597E-02	**5.382E-03**	5.750E-02	2.445E-02	7.979E-03	1.115E-02
Nguyen 4	**5.400E-04**	5.920E-04	1.444E-03	9.458E-04	1.294E-03	7.360E-04
Nguyen 5	1.133E-04	9.046E-05	4.491E-04	1.778E-04	5.235E-05	5.637E-05
Nguyen 6	2.499E-04	3.928E-04	5.926E-04	7.864E-04	3.071E-04	2.903E-04
Nguyen 7	8.678E-06	7.140E-06	1.820E-05	1.450E-05	**3.169E-06**	5.500E-06
Ackley	3.466E-02	3.076E-02	3.197E-02	3.220E-02	3.202E-02	**3.015E-02**
Levy	**8.370E-02**	9.839E-02	9.829E-02	9.720E-02	9.196E-02	1.002E-01
Rastrigin	4.512E-01	4.911E-01	4.834E-01	4.953E-01	4.649E-01	**4.383E-01**

It is clear from the data that Canonical CGP performs significantly worse than other methods, even traditional one-point and uniform methods when the opposite is expected from the reports in most of the literature. This has also been reported in [12]. Without forming any definite conjectures, we regard this as a point of future investigation. Comparing the mutation methods (column-wise), it is visible that by preserving nodes in crossover and in mutation, we tend to achieve better results than if instructions were allowed to be broken apart – or, more accurately, *partially altered* by either operator.

Table 5 shows the p_best values for each operator both for each problem and averaged over all problems. This latter table is incredibly useful, as we can see that on average, Subgraph Crossover (which preserves nodes by default) is the most likely to produce the best model across all operators. Some basic statistical operations, if $\langle p(x) \rangle$ is the average probability that operator x produces the best model, demonstrate the following:

1. $\langle p(\text{Any SGX}) \rangle = 0.246$
2. $\langle p(\text{Any Full Node Mutation}) \rangle = 0.520$
3. conversely, $\langle p(\text{Any Point Mutation}) \rangle = 0.480$

It is therefore more likely that by fully replacing nodes during mutation we can produce better models than by using standard point mutation. Given that on average, the only case where this does not hold is Subgraph Crossover (and even in this case is very close to a 49%–51% split), we conjecture that our proposed mutation operator approaches a general solution. It is also clear that node preservation in crossover also seems to present a near-general solution. There are a couple of specific cases where node preservation is *less* likely to produce a better result, but the differences in these rare cases are very small.

Table 5. Probability that a crossover operator would result in the best model across all operators both per-problem and averaged over all problems.

| | Canonical | | One-Point | | One-Point 1D | |
	Point Mutation	Node Mutation	Point Mutation	Node Mutation	Point Mutation	Node Mutation
Koza 1	0.020	0.039	0.115	0.097	0.063	0.073
Koza 2	0.035	0.040	0.086	0.144	0.050	0.042
Koza 3	0.042	0.052	0.076	0.101	0.057	0.082
Nguyen 4	0.037	0.043	0.096	**0.140**	0.056	0.075
Nguyen 5	0.023	0.041	0.088	**0.159**	0.049	0.045
Nguyen 6	0.032	0.039	0.107	**0.118**	0.075	0.080
Nguyen 7	0.026	0.050	0.124	0.112	0.026	0.034
Ackley	0.028	0.025	0.090	0.081	0.085	0.094
Levy	0.030	0.048	0.108	0.095	0.086	0.072
Rastrigin	0.033	0.036	0.111	0.064	0.080	0.089
Average	0.030	0.041	0.100	0.111	0.063	0.069
Std. Dev.	0.007	0.008	0.015	0.030	0.019	0.021
	Uniform		Uniform 1D		Subgraph	
	Point Mutation	Node Mutation	Point Mutation	Node Mutation	Point Mutation	Node Mutation
Koza 1	0.088	0.095	0.059	0.072	**0.158**	0.121
Koza 2	0.103	0.080	0.066	0.060	**0.159**	0.136
Koza 3	0.080	**0.147**	0.047	0.077	0.117	0.121
Nguyen 4	0.097	0.116	0.066	0.083	0.084	0.107
Nguyen 5	0.089	0.108	0.046	0.079	0.144	0.131
Nguyen 6	0.105	0.106	0.064	0.056	0.115	0.105
Nguyen 7	0.103	0.143	0.044	0.066	**0.157**	0.116
Ackley	0.074	0.102	0.100	0.093	**0.115**	0.111
Levy	0.108	0.087	0.072	0.082	**0.113**	0.101
Rastrigin	0.107	0.081	0.083	0.072	0.106	0.139
Average	0.095	0.106	0.065	0.074	**0.127**	0.119
Std. Dev.	0.012	0.024	0.017	0.011	0.026	0.013

5.2   Complexity

In Table 6, we show that in most cases the Canonical CGP method is more likely to produce models with fewer active nodes, with one-dimensional crossover methods in distant second. However, this should not be conflated with the probability of finding an objectively better or less complex model, but only in comparison to the other tested methods. The complexity of the best models (corresponding to those in Fig. 1) is shown in Fig. 2.

Table 6. Probability that a given crossover operator will produce the smallest model across all operators.

	Canonical		One-Point		One-Point 1D	
	Point Mutation	Node Mutation	Point Mutation	Node Mutation	Point Mutation	Node Mutation
Koza 1	**0.150**	0.096	0.031	0.065	0.069	0.107
Koza 2	0.105	0.096	0.025	0.058	0.132	0.106
Koza 3	0.080	0.111	0.037	0.077	0.090	**0.114**
Nguyen 4	**0.143**	0.099	0.036	0.053	0.102	0.086
Nguyen 5	**0.133**	0.119	0.040	0.047	0.120	0.094
Nguyen 6	0.124	**0.146**	0.034	0.067	0.085	0.088
Nguyen 7	0.100	0.098	0.033	0.048	0.103	**0.180**
Ackley	0.119	0.119	0.042	0.047	0.077	0.119
Levy	0.148	**0.180**	0.031	0.037	0.099	0.106
Rastrigin	0.197	0.160	0.025	0.058	0.090	0.087
Average	**0.130**	0.122	0.033	0.056	0.097	0.109
Std. Dev.	0.033	0.030	0.006	0.012	0.019	0.028
	Uniform		Uniform 1D		Subgraph	
	Point Mutation	Node Mutation	Point Mutation	Node Mutation	Point Mutation	Node Mutation
Koza 1	0.039	0.060	0.113	0.110	0.077	0.082
Koza 2	0.032	0.052	0.068	**0.145**	0.086	0.096
Koza 3	0.054	0.071	0.067	0.106	0.081	0.112
Nguyen 4	0.039	0.065	0.114	0.089	0.050	0.125
Nguyen 5	0.038	0.046	0.098	0.085	0.078	0.102
Nguyen 6	0.045	0.059	0.082	0.103	0.062	0.105
Nguyen 7	0.041	0.053	0.106	0.116	0.047	0.074
Ackley	0.040	0.053	**0.117**	0.115	0.059	0.093
Levy	0.034	0.066	0.094	0.093	0.053	0.060
Rastrigin	0.022	0.045	0.082	0.111	0.041	0.080
Average	0.038	0.057	0.094	0.107	0.063	0.093
Std. Dev.	0.008	0.009	0.019	0.017	0.016	0.019

6   Discussion

In this work, we have demonstrated that forcing nodes to remain intact through-out the crossover process following [20] is a beneficial approach for search on symbolic regression problems. We also propose that practitioners perform "full" node mutations by entirely replacing nodes rather than altering single genes, which we find to be beneficial for search in most cases.

What does this mean for how we see the history – and future – of CGP crossover research? It tells us that *from the start*, the field has been subject to misconceptions. First, as Kocherovsky et al. demonstrated, the assumption that genes close to one another are functionally related is not applicable to CGP [25]. Second, the original design of CGP chromosomes as one-dimensional strings, an obvious inheritance from EC as a whole [32,33] will evolve over generations, but treating each part of the instructions as their own genes does not result in productive operators.

The subgraph crossover avoids this problem because its design *requires* nodes to remain intact during crossover. Kalkreuth correctly assumed that an edge of the graph is not separable from the node it is directed at, and designed the operator to take this into account. Rewiring the first active node after the

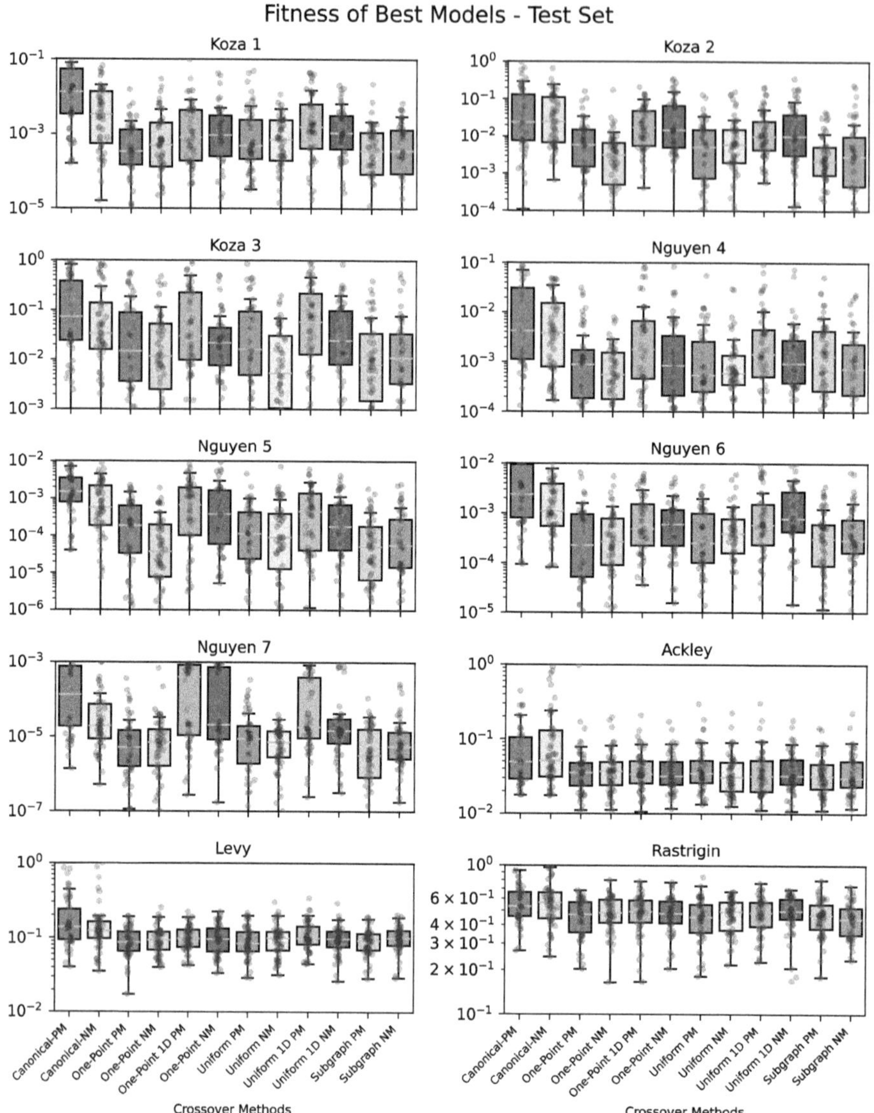

Fig. 1. Box plots of best fitness at the end of evolution for each run.

crossover point further serves to preserve good structures by partially repairing the destruction caused by one-point crossover.

Future research into CGP crossover should explicitly consider how their genomes are structured and why the researchers chose to structure them in that way. When new operators are developed, one should be cognizant of the fact that

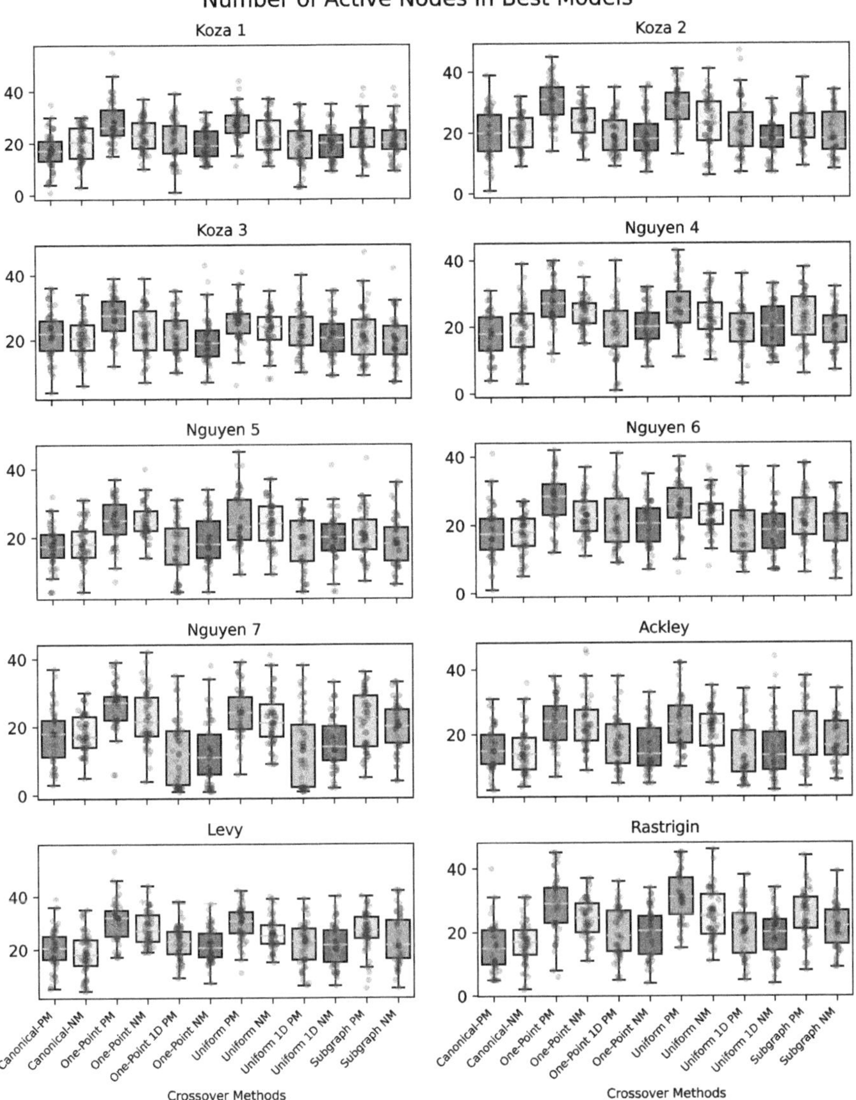

Fig. 2. Box plots of the amount of active nodes in the best models at the end of evolution for each run.

- There is a positional bias in node activity; it is more likely that nodes closer to the front will be active.
- Crossover is a destructive operation and it may be worth taking the time to repair part of the damage.
- Instructions close to each other in the genome are not necessarily – and in fact are not likely to be – related functionally. The very last node (node n)

is as likely to take the first node as input as it is its immediate predecessor (node $n - 1$).

Unless counter-measures are created for such tendencies, operators will suffer from the negative effects such tendencies can bring to an algorithm. Future research should also expand on the domain of CGP applications. It is common for Boolean problems to be considered for CGP benchmarking, and classification problems are increasingly prevalent in the literature [34], thus it would be useful to expand our work to these as well. We do not believe that the final word has been spoken on the efficiency of CGP operators, but think that node preservation holds a key role in designing them for productive application.

References

1. Hardware | Institute for Cyber-Enabled Research. https://icer.msu.edu/hpcc/hardware (2024)
2. Ackley, D.H.: A Connectionist Machine for Genetic Hillclimbing, 1st edn. Kluwer Academic Publishers (1987). https://doi.org/10.1007/978-1-4613-1997-9
3. Altenberg, L.: The evolution of evolvability in genetic programming. In: Kinnear, K. (ed.) Advances in Genetic Programming, vol. 1, pp. 47–74. MIT Press (1994). https://doi.org/10.5555/185984.185991
4. Banzhaf, W., Bakurov, I.: On the nature of the phenotype in tree genetic programming. In: Proceedings of the Genetic and Evolutionary Computation Conference, GECCO 2024, pp. 868–877. Association for Computing Machinery, New York, NY, USA (2024). https://doi.org/10.1145/3638529.3654129
5. Banzhaf, W., Francone, F.D., Keller, R.E., Nordin, P.: Genetic Programming: An Introduction: On the Automatic Evolution of Computer Programs and Its Applications. Morgan Kaufmann Publishers Inc., San Francisco, CA, USA (1998). https://doi.org/10.5555/280485
6. Banzhaf, W., Langdon, W.B.: Some considerations on the reason for bloat. Genet. Programm. Evol. Mach. **3**, 81–91 (2002). https://doi.org/10.1023/A:1014548204452
7. Blickle, T., Thiele, L.: Genetic programming and redundancy. In: Hopf, J. (ed.) Genetic Algorithms within the Framework of Evolutionary Computation (Workshop at KI-94, Saarbrücken), pp. 33–38. Max-Planck-Institut für Informatik (MPI-I-94-241) (1994)
8. Brameier, M., Banzhaf, W.: Evolving teams of predictors with linear genetic programming. Genet. Programm. Evol. Mach. **2**, 381–407 (2001). https://doi.org/10.1023/A:1012978805372
9. Brameier, M.F., Banzhaf, W.: Linear genetic programming. In: Genetic and Evolutionary Computation, 1st edn., pp. 36–37. Springer (2007). https://doi.org/10.1007/978-0-387-31030-5
10. Cai, X., Smith, S.L., Tyrrell, A.M.: Positional independence and recombination in cartesian genetic programming. In: Collet, P., Tomassini, M., Ebner, M., Gustafson, S., Ekárt, A. (eds.) Genetic Programming, pp. 351–360. Springer, Heidelberg (2006). https://doi.org/10.1007/11729976_32
11. Clegg, J., Walker, J.A., Miller, J.F.: A new crossover technique for cartesian genetic programming. In: Proceedings of the 9th Annual Conference on Genetic and Evolutionary Computation, GECCO 2007, pp. 1580–1587. Association for Computing Machinery, New York, NY, USA (2007). https://doi.org/10.1145/1276958.1277276

12. Cui, H., Heider, M., Hähner, J.: Positional bias does not influence cartesian genetic programming with crossover. In: Affenzeller, M., et al. (eds.) Parallel Problem Solving from Nature – PPSN XVIII, pp. 151–167. Springer, Switzerland, Cham (2024). https://doi.org/10.1007/978-3-031-70055-2_10

13. Cui, H., Margraf, A., Hähner, J.: Analysing the influence of reorder strategies for cartesian genetic programming. SN Comput. Sci. **6**(7), 754 (2025). https://doi.org/10.1007/s42979-025-04296-4

14. Cui, H., Margraf, A., Heider, M., Hähner, J.: Towards understanding crossover for cartesian genetic programming. In: Towards Understanding Crossover for Cartesian Genetic Programming, pp. 308–314, January 2023. https://doi.org/10.5220/0012231400003595

15. Cui, H., Margraf, A., Hähner, J.: Equidistant reorder operator for cartesian genetic programming. In: 15th International Conference on Evolutionary Computation Theory and Applications, pp. 64–74, January 2023. https://doi.org/10.5220/0012174100003595

16. Dignum, S., Poli, R.: Generalisation of the limiting distribution of program sizes in tree-based genetic programming and analysis of its effects on bloat. In: Proceedings of the 9th Annual Conference on Genetic and Evolutionary Computation, GECCO 2007, pp. 1588–1595. Association for Computing Machinery, New York, NY, USA (2007). https://doi.org/10.1145/1276958.1277277

17. Goldman, B.W., Punch, W.F.: Length bias and search limitations in cartesian genetic programming. In: Proceedings of the 15th Annual Conference on Genetic and Evolutionary Computation, GECCO 2013, pp. 933–940. Association for Computing Machinery, New York, NY, USA, July 2013. https://doi.org/10.1145/2463372.2463482

18. Haut, N., Banzhaf, W., Punch, B.: Correlation versus RMSE loss functions in symbolic regression tasks. In: Trujillo, L., Winkler, S.M., Silva, S., Banzhaf, W. (eds.) Genetic Programming Theory and Practice XIX, pp. 31–55. Springer (2023). https://doi.org/10.1007/978-981-19-8460-0_2

19. Husa, J., Kalkreuth, R.: A comparative study on crossover in cartesian genetic programming. In: Castelli, M., Sekanina, L., Zhang, M., Cagnoni, S., García-Sánchez, P. (eds.) Genetic Programming, pp. 203–219. Springer International Publishing, Cham (2018). https://doi.org/10.1007/978-3-319-77553-1_13

20. Kalkreuth, R.: A comprehensive study on subgraph crossover in cartesian genetic programming. In: 12th International Joint Conference on Computational Intelligence, November 2020. https://doi.org/10.5220/0010110700590070

21. Kalkreuth, R.: Reconsideration and extension of cartesian genetic programming. Ph.D. thesis, Department of Computer Science, University of Dortmund, December 2021. https://doi.org/10.17877/DE290R-22504

22. Kalkreuth, R., Rudolph, G., Droschinsky, A.: A new subgraph crossover for cartesian genetic programming. In: McDermott, J., Castelli, M., Sekanina, L., Haasdijk, E., García-Sánchez, P. (eds.) European Conference on Genetic Programming, pp. 294–310. Springer International Publishing, Cham (2017). https://doi.org/10.1007/978-3-319-55696-3_19

23. Kaufmann, P., Platzner, M.: Advanced techniques for the creation and propagation of modules in cartesian genetic programming. In: Proceedings of the 10th Annual Conference on Genetic and Evolutionary Computation, GECCO 2008, pp. 1219–1226. Association for Computing Machinery, New York, NY, USA (2008). https://doi.org/10.1145/1389095.1389334

24. Kocherovsky, M., Banzhaf, W.: Crossover destructiveness in cartesian versus linear genetic programming. In: ALIFE 2024: Proceedings of the 2024 Artificial Life Conference, p. 20. ALIFE 2022: The 2022 Conference on Artificial Life, July 2024. https://doi.org/10.1162/isal_a_00735
25. Kocherovsky, M., Kianinejad, M., Bakurov, I., Banzhaf, W.: On the effectiveness of crossover operators in cartesian genetic programming. In: Xue, B., Manzoni, L., Bakurov, I. (eds.) Genetic Programming, pp. 68–84. Springer Nature Switzerland, Cham (2025). https://doi.org/10.1007/978-3-031-89991-1_5
26. Koza, J.R.: Genetic programming: a paradigm for genetically breeding populations of computer programs to solve problems. Technical report, Stanford University, Stanford, CA, USA (1990)
27. Koza, J.R.: Genetic programming as a means for programming computers by natural selection. Stat. Comput. **4**, 87–112 (1994). https://doi.org/10.1007/BF00175355
28. Laguna, M., Marti, R.: Experimental testing of advanced scatter search designs for global optimization of multimodal functions. J. Glob. Optim. **33**, 235–255 (2005). https://doi.org/10.1007/s10898-004-1936-z
29. Langdon, W.B., Poli, R.: Fitness causes bloat. In: Chawdhry, P.K., Roy, R., Pant, R.K. (eds.) Soft Computing in Engineering Design and Manufacturing, pp. 13–22. Springer, London (1998)
30. Langdon, W.B., Poli, R.: Fitness causes bloat: mutation. In: Banzhaf, W., Poli, R., Schoenauer, M., Fogarty, T.C. (eds.) European Conference on Genetic Programming, pp. 37–48. Springer, Heidelberg (1998). https://doi.org/10.1007/BFb0055926
31. McPhee, N.F., Miller, J.D.: Accurate replication in genetic programming. In: Proceedings of the 6th International Conference on Genetic Algorithms, pp. 303–309. Morgan Kaufmann Publishers Inc., San Francisco, CA, USA (1995). https://doi.org/10.5555/645514.657773
32. Miller, J., Thomson, P., Fogarty, T., Ntroduction, I.: Designing electronic circuits using evolutionary algorithms. Arithmetic circuits: a case study. In: Genetic Algorithms and Evolution Strategies in Engineering and Computer Science (1999)
33. Miller, J.F.: An empirical study of the efficiency of learning Boolean functions using a cartesian genetic programming approach. In: Proceedings of the 1st Annual Conference on Genetic and Evolutionary Computation, GECCO'99, vol. 2. pp. 1135–1142. Morgan Kaufmann Publishers Inc., San Francisco, CA, USA (1999). https://doi.org/10.5555/2934046.2934074
34. Miller, J.F.: Cartesian genetic programming: its status and future. Genet. Program Evolvable Mach. **21**(1–2), 129–168 (2020). https://doi.org/10.1007/s10710-019-09360-6
35. Miller, J.F., Harding, S.L.: Cartesian genetic programming. In: Proceedings of the 10th Annual Conference Companion on Genetic and Evolutionary Computation, GECCO 2008, pp. 2701–2726. Association for Computing Machinery, New York, NY, USA (2008). https://doi.org/10.1145/1388969.1389075
36. Ni, J., Drieberg, R.H., Rockett, P.I.: The use of an analytic quotient operator in genetic programming. IEEE Trans. Evol. Comput. **17**(1), 146–152 (2013). https://doi.org/10.1109/TEVC.2012.2195319
37. Nordin, P., Banzhaf, W.: Complexity compression and evolution. in: proceedings of the 6th international conference on genetic algorithms, pp. 310–317. Morgan Kaufmann Publishers Inc., San Francisco, CA, USA (1995). https://doi.org/10.5555/645514.658069

38. Nordin, P., Francone, F., Banzhaf, W.: Explicitly defined introns and destructive crossover in genetic programming. In: Advances in Genetic Programming, vol. 2. The MIT Press, October 1996. https://doi.org/10.7551/mitpress/1109.003.0010

39. Oltean, M., Groşan, C., Oltean, M.: Encoding multiple solutions in a linear genetic programming chromosome. In: Computational Science - ICCS 2004, pp. 1281–1288. Springer, Heidelberg (2004). https://doi.org/10.1007/978-3-540-24688-6_165

40. O'Neill, M., Ryan, C.: Grammatical evolution. IEEE Trans. Evol. Comput. **5**(4), 349–358 (2001). https://doi.org/10.1109/4235.942529

41. Rudolph, G.: Globale Optimierung mit parallelen Evolutionsstrategien. Diploma thesis, Department of Computer Science, University of Dortmund (1990)

42. da Silva, J.E., Bernardino, H.S.: Cartesian genetic programming with crossover for designing combinational logic circuits. In: 2018 7th Brazilian Conference on Intelligent Systems (BRACIS), pp. 145–150 (2018). https://doi.org/10.1109/BRACIS.2018.00033

43. Silva, S., Costa, E.: Dynamic limits for bloat control in genetic programming and a review of past and current bloat theories. Genet. Programm. Evol. Mach. **10**, 141–179 (2009). https://doi.org/10.1007/s10710-008-9075-9

44. Soule, T., Heckendorn, R.B.: An analysis of the causes of code growth in genetic programming. Genet. Program Evolvable Mach. **3**(3), 283–309 (2002). https://doi.org/10.1023/A:1020115409250

45. Spector, L., Robinson, A.: Genetic programming and autoconstructive evolution with the push programming language. Genet. Program Evolvable Mach. **3**(1), 7–40 (2002). https://doi.org/10.1023/A:1014538503543

46. Sun, C.: A study of solving traveling salesman problem with genetic algorithm. In: 2020 9th International Conference on Industrial Technology and Management (ICITM), pp. 307–311 (2020). https://doi.org/10.1109/ICITM48982.2020.9080397

47. Tackett, W.A.: Recombination, selection, and the genetic construction of computer programs. Ph.D. thesis, University of Southern California, USA (1994). Not available from University Microfilms, Incorporated

48. Turner, A.J.: Improving crossover techniques in a genetic program. Master's thesis, Department of Electronics, University of York, April 2012

49. Turner, H.L., van Etten, J., Firth, D., Kosmidis, I.: Modelling Rankings in R: The PlackettLuce Package. Comput. Stat. **35**, 1027–1057 (February 2020). https://doi.org/10.1007/s00180-020-00959-3

50. Wu, A.S., Lindsay, R.K.: A survey of intron research in genetics. In: Voigt, H.M., Ebeling, W., Rechenberg, I., Schwefel, H.P. (eds.) Parallel Problem Solving from Nature — PPSN IV, pp. 101–110. Springer, Heidelberg (1996). https://doi.org/10.1007/3-540-61723-X_974

New Perspectives on Cartesian Genetic Programming: A Survey

Mark Kocherovsky[1(✉)] , Henning Cui[2] , Illya Bakurov[1] ,
Michael Heider[2] , Roman Kalkreuth[3] , and Wolfgang Banzhaf[1]

[1] Department of Computer Science and BEACON Center for the Study of Evolution
in Action, Michigan State University, East Lansing, MI 48824, USA
`kocherov@msu.edu`
[2] Organic Computing Group, University of Augsburg, 86159 Augsburg, Germany
`michael.heider@uni-a.de`
[3] Chair for AI Methodology (AIM), Faculty of Computer Science, RWTH Aachen
University, 52062 Aachen, Germany
`kalkreuth@aim.rwth-aachen.de`
`https://banzhaf-lab.github.io/` , `https://aim.rwth-aachen.de/`

Abstract. Over the past twenty-five years, certain practices and
assumptions in Cartesian Genetic Programming (CGP) have become
conventional wisdom, yet recent research challenges their validity. This
position paper critically examines these long-standing beliefs and proposes evidence-based alternatives for the CGP community. We address
four misconceptions: The purported ineffectiveness of crossover operators; the overlooked impact of positional bias; problems with tournament selection on Boolean benchmarks; and the limitations of single-
domain analysis. Through a review of recent literature, we identify a key
principle underlying successful CGP operators—the preservation of node
structural integrity during genetic operations. We discuss current best
practices including rigorous hyperparameter tuning, cross-domain benchmarking, node-preserving operators, and modern fitness function design.
Our analysis reveals that many accepted CGP practices, including the
ubiquitous $(1 + 4)$ evolution strategy, lack generalizability across problem domains. We believe that by reconsidering these assumptions and
adopting the recommendations presented here, researchers and practitioners can develop more effective and robust CGP implementations.

Keywords: Cartesian Genetic Programming · Crossover · Positional
Bias · Node Preservation · Best Practices · Future Perspectives

1 Introduction

Cartesian Genetic Programming (CGP) is a branch of Genetic Programming
(GP) introduced by Julian Miller and his collaborators in the late 1990s [62,64].
The basic idea behind CGP is that a program is represented as a direct acyclic
graph (DAG) consisting of a series of nodes, each taking inputs, performing

an operation, and providing an output. Since then, there has been a wealth of investigation into CGP, its various dynamics, and towards the development of more effective operators.

Over the past twenty-five years, certain assumptions about CGP have become the default mode of operation for researchers in the field. These include the assumption that crossover is generally deleterious in CGP search, the lack of mitigation for positional bias, and the CGP benchmarking methodology. This work aims to discuss these assumptions holistically in order to make recommendations for potential best practices for CGP implementations. We base our advice on the last twenty-five years of work in and around CGP. With this contribution, we also want to propose avenues for future work.

Our position covers four related topics. After a brief introduction to CGP (Sect. 2), we first cover misconceptions in the older literature that are beginning to be unveiled by newer work in Sect. 3. Second, in Sect. 4 we holistically cover operators shown to be well-performing in the recent literature. In Sect. 5, we discuss best practices in CGP for users regarding traversing the search space and establishing strong generalized operators. Finally, in Sect. 6, we discuss new directions of research based on the overview given in the other sections.

2 Brief Overview of Cartesian Genetic Programming

CGP individuals are feed-forward directed acyclic graphs, and thus consist of nodes and edges. A node is either an input node, which takes in data samples; an output node, which redirects the output of previous nodes without any additional modification and forms part of the program output; or a function node, which performs calculation(s) on inputs and returns an output. It is important to note that the input for function nodes may be from any node closer to the start of the genome, meaning that function node at position m, can take input from any input node i or a subset of function nodes at positions $[0, m)$. The edges of the graph therefore carry values from one node to another for operations or output.

CGP individuals are normally represented by a one-dimensional string of values. A CGP node has to have a given number of possible operator inputs called an *arity*. An arity of two ($a = 2$) means that the operations will be either unary or binary. A CGP individual p can be written as

$$p = i_0|i_1|\ldots|i_\alpha|g_0|a_{01}|a_{02}|g_1|a_{11}|a_{12}|\ldots|g_\beta|a_{\beta 1}|a_{\beta 2}|o_0|o_1\ldots o_\gamma \qquad (1)$$

where, given α-dimensional data, at most β function nodes, and a target output-dimensionality of γ, we will thus far have α genes representing input nodes, $\beta \cdot a$ genes for function nodes (with an operator gene g_m and a genes indicating which nodes to use as input), and γ genes representing which indices to return as output. Each gene is separated by the | character to make it easier to read. Input nodes are excluded from genomic operations since their value is a given data sample rather than dependent on other nodes' calculations [54].

Simple point mutation, for example, can be executed by picking a gene at random and exchanging it for another legal value. One-point crossover can be implemented by choosing an index at random, and performing recombination as in any other genetic algorithm. Typically, a $(1+\lambda)$ evolution strategy (ES) is used in the evolutionary loop, as it is generally thought that basic crossover methods are not effective for CGP [13], although we discuss this as a misconception in Sect. 3.1.

3 Misconceptions in the CGP Community

Over the years, standard practices to CGP's usage have been established. However, they are also oftentimes adopted without question in theoretical studies or use-case specific applications, even though they might impair CGP's evolutionary processes. In this section, we want to draw attention to selected major misconceptions in CGP to aid researchers and practitioners in the future.

3.1 Effects of Crossover

Early experiments in CGP operators leaned towards standard genetic algorithmic operators rather than specifically GP-oriented operators. Clegg et al. [13] from 2007 is often cited as the original source that stated the CGP crossover problem, in which crossover is generally regarded as a universally-deleterious operator in CGP. In the paper, the authors implement a new crossover based on real-valued crossovers found in the genetic algorithm literature [1] and describe it as more effective than a crossover operator matching a one-point crossover from standard GAs where the nodes are kept intact (more about this kind of method will be described in Sect. 5.2). However, Turner [83] and Kocherovsky et al. [54] found that the real-valued operator was not effective compared to other crossover operators.

Recent results by Kalkreuth [44,45,47], Kocherovsky et al. [52], and Cui et al. [14] demonstrate an alternate possibility. In these works, crossover operators are not generally deleterious, and can even improve performance compared to the canonical $(1+\lambda)$ method. We thus postulate that the negative attitude towards crossover in CGP may be misplaced.

3.2 Tournament Selection Negatively Affects CGP on Boolean Benchmarks

According to Cui et al.'s [14] findings, *standard tournament selection* greatly worsens the performance of CGP on *Boolean benchmarks*[1]. For example, on all tested Boolean benchmarks, all CGP versions using tournament selection in a (μ, λ)-ES with elitism had worse fitness values than CGP with random selection in a $(\mu+\lambda)$-ES. This outcome was independent of the CGP version and crossover used. It is also important to note that such poor performances occurred even after an extensive hyperparameter-tuning process.

---

[1] Other benchmark types, e.g. symbolic regression, were not affected.

A possible reason for this aforementioned problem might be due to a specific characteristic of Boolean benchmarks: They have a *deceptive fitness landscape* [86]. Due to the Boolean nature of this benchmark type, various different solutions lead to the same fitness value and numerous solutions are also treated as equally good. Thus, CGP with tournament selection might not be suitable for this specific kind of benchmark. This might be due to the properties of this selection method: Blickle and Thiele [10] proved that tournament selection has the lowest loss of population diversity. This is paired with the highest expected variance of the fitness distribution of the population. The combination of both characteristics might lead to CGP not being able to properly explore the search space (Fitness landscapes are discussed more in-depth in Sect. 6.4).

To the best of our knowledge, this finding has not been mentioned yet by other publications, and might have substantial consequences for the CGP research community. Some foundational work (e.g. [42,62], or [88]) used CGP with a tournament selection. *If* other researchers are able to replicate and validate this outcome, parts of the literature's empirical work and conclusions should then be subject to revision.

3.3 Positional Bias

CGP enforces a feed-forward grid. This leads to *positional bias* (also called *length bias*), as was found by Goldman and Punch [30]. This bias describes a non-uniform distribution of active nodes in an individual. Computational nodes near input nodes have a *higher chance* of being active compared to the nodes near output nodes. This is due to the feed-forward nature of CGP. Function nodes near input nodes can be used by the majority of other nodes in the CGP graph. However, function nodes near the output nodes have far fewer opportunities to be used. As a result, there are also less nodes that could mutate their connections to nodes on the output side, decreasing their probability to become active.

Positional bias negatively impacts CGP's evolutionary search process because it increases the difficulty of solving certain tasks or evolving specific structures, leading to decreased performance [31,73]. To counteract positional bias, Goldman and Punch [30,32] and Cui et al. [15,17] proposed extensions to CGP. By *reordering active nodes*, this issue can be mitigated and fitness improved.

3.4 Cross-Domain Analysis

In the past, two problem domains in particular have been used to better understand the search properties of CGP. These are symbolic regression (SR) and logic synthesis (LS), the latter of which can be considered a significant pillar in the application domain since the early days of CGP. Various studies focusing on hyperparameter parameterization optimization of CGP revealed significant differences when benchmarks from SR or LS were used with CGP. In the framework of the first study, Kaufmann and Kalkreuth [49] tuned essential parameters of a $(1 + \lambda)$-ES, such as offspring population size (λ), genome length, and mutation rate by using a SOTA hyperparameter optimizer called `irace` [58]. The outcome of this first study indicated a rather contradictory situation between settings that work effectively in SR and LS. For SR, it was found that large populations can be effectively used, while in LS configurations with minimal populations performed best. Moreover, the study clearly demonstrated that the $(1 + 4)$-ES cannot be considered a general effective choice for configuring CGP, which had been postulated by Miller [63] based on his experiments with Boolean functions.

Later work by Kalkreuth [45] further pursued this topic by including recombination operators such as subgraph crossover and block crossover [42]. Mainly, the study confirmed the results reported in [49], but it also demonstrated that recombination-based algorithms can work to some extent in SR. In the LS domain, on the other hand, it was found that a mutation-only $(1 + \lambda)$-ES performed best for the Boolean functions considered. The effectiveness of the minimal setting of the λ parameter was later confirmed on a larger scale by Kalkreuth et al. [48] who introduced a benchmark suite for logic synthesis with respective baseline results obtained with CGP.

In [45], it was highlighted that SR and LS as tackled within the GP paradigm markedly differ by the nature of the fitness functions. Practitioners typically use a continuous fitness function, such as mean squared error for symbolic regression problems, but in logical synthesis problems, the Hamming distance is used predominantly. Fitness functions are described further in Sect. 5.4.

Overall, cross-domain studies in CGP broke with the dogma that the $(1+4)$-ES can be considered a generally effective choice since there is more evidence that mutation-only algorithms with a very small population size perform quite well in LS, but the success and effectiveness of algorithms that use larger populations and recombination has been demonstrated in SR to some extent.

4 Successful Genetic Operators in CGP - A Holistic View

A holistic analysis of successful crossover and mutation strategies in CGP reveals a consistent pattern: the structural integrity of *nodes*—as self-contained computational units—must be preserved to ensure effective variation. This conclusion stems from two major contributions.

First, Kalkreuth [44] demonstrated the benefits of explicitly constraining one-point crossover to act only on active nodes of CGP's DAG, thus performing recombination of active subgraphs rather than arbitrary genome fragments (cf. Sect. 4.1). In this way, intra-node dependencies were kept intact and crossover's destructiveness was reduced.

Second, more recently, Kocherovsky et al. [52] generalized the principles behind subgraph crossover beyond one-point crossover. Specifically, it was shown that explicitly preserving node boundaries—regardless of whether the nodes are active or inactive—substantially improves the likelihood of producing fitter offspring. This method resulted in classical one-point and uniform crossover operators achieving performance close to that of subgraph crossover, and in some symbolic regression benchmarks, even outperforming it. Beyond crossover, node boundaries preservation was also extended to mutation. Kocherovsky et al. found that using node-level mutation (replacing the entire instruction), instead of traditional gene-level mutation (modifying a single gene within a node), consistently led to better search outcomes [52]. Moreover, as proposed by Kalkreuth [45], these findings allow to question the strict reliance on phenotypically-active substructures. In fact, enforcing node preservation regardless of the node's activity status may offer additional benefits by promoting neutrality, which was found to be advantageous for evolutionary search due to its role in enabling a broad exploration and contributing to simplicity bias [5–7, 38–40]. These aspects, however, still need to be investigated further.

These results point to a general principle: node-preserving operators in CGP contribute to the evolvability. Moreover, we argue that treating nodes as atomic units of computation can facilitate the modularity, emergence, retention, and reuse of useful building blocks, ultimately leading to more robust and quicker evolutionary search. In the following subsections we discuss subgraph crossover and node connectivity in more detail.

4.1 Subgraph Crossover

The subgraph crossover technique for CGP draws inspiration from the subtree crossover used in tree-based GP. This method performs recombination while taking the CGP phenotype into careful consideration. Simply exchanging parts of the genotype, as noted by Clegg et al. [13], can lead to poor results.

In CGP, an individual's phenotype is defined by its active graph trace identified during evaluation, which ultimately determines its semantic output. Therefore, subgraph crossover focuses solely on recombining the genetic material found within these active paths. This approach aims to minimize the disruption typically caused by traditional genotypic single-point crossover in CGP, offering a more meaningful recombination of subgraphs.

To describe the procedure of subgraph crossover, let n_i represent the predefined number of input nodes and n_f the predefined number of function nodes. In typical CGP representations, inputs are indexed from 0 to $n_i - 1$ and the function nodes of each graph are indexed from n_i to $n_i + n_f - 1$. The nodes which lie between the input and output nodes are denoted as function nodes. Crossover is performed with two pre-selected parents that are denoted as P_1 and P_2. For the crossover procedure, the number of nodes of the active function nodes is necessary. The node numbers of the active nodes of P_1 and P_2 are stored in two lists M_1 and M_2. The active nodes are determined by a backward search in the evaluation procedure. To define one suitable crossover point, we define two possible crossover points C_{P1} and C_{P2} of the two parents. Given this information on active nodes and the length of a path, we can choose two possible crossover points. The possible crossover points C_{P1} and C_{P2} are chosen by chance in the range of the active function nodes which are stored in M_1 and M_2. Input or output nodes are not allowed as possible crossover points. A general crossover point CP is defined by choosing the smaller crossover point from C_{P1} and C_{P2}. The reason for this is that the subgraphs of the parents, which will be placed in front of or behind the crossover point of an offspring's genome should be balanced. The representation of CGP allows active paths of an individual, possibly starting in the middle or back of the graph. The subgraph which will be placed in front of the crossover point has to start at more leading active nodes.

4.2   Node Connectivity

In subgraph crossover, covered in Sect. 4.1, it is required to perform crossover in such a way that individual instructions remain intact until *after* recombination, at which point a node can be rewired for continuity [44]. In a similar vein, Kocherovsky et al. [53] postulated that, whereas linear genetic programming makes use of memory external to the program as a way to anchor substructures, CGP has no such method to prevent destructive operations, and thus when a gene is altered, a CGP program is more likely to suffer catastrophic effects on the graph trace. Based on their experimental results, [54] suggested that "traditional" methods of crossover failed to take into account that instructions are *interdependent* because instructions take other instructions as inputs. The authors also remarked that nodes adjacent to each other in the genotype are functionally related, as each node can take input from any other node towards the front of the genome.

In 2025, Kocherovsky et al. demonstrated that Kalkreuth's subgraph operator dramatically improves crossover operations [52]. Subgraph crossover explicitly takes into account the structure of the DAG; it includes a step where the first active node immediately after the crossover point has its connections rewired to ensure that both subgraphs remain active. The attention to node connectivity is likely one source of the literature's observations of subgraph being effective.

Following these works, we stress that *properly understanding and accounting for the nuances of node connectivity is vital for developing more effective crossover operators in CGP*. Without taking into account these nuances, it is less likely that a beneficial pattern will emerge. This limits the ability of evolving programs to exploit beneficial substructures by opening these substructures to large amounts of destructive forces. Making sure node integrity is maintained decreases the chances of a substructure being destroyed by an errant crossover or mutation in the wrong place.

5   CGP Best Practices

In the following sections, we highlight some best practices which we deem vital for using CGP and advancing its state-of-the-art productively.

5.1 Hyperparameter Tuning and Experimental Design

The Evolutionary Computation community agrees on the importance of good experimental design—which also includes *tuning hyperparameters* if necessary. However, most CGP related researchers still use an elitist $(\mu + \lambda)$-ES with $\mu = 1$ and $\lambda = 4$. While this does indeed create some comparability to previous works, it may also lead to major problems. For example, in the context of Boolean and symbolic regression benchmarks, Kaufmann and Kalkreuth [49,50] and Trautwein et al. [82] showed that the $(1 + 4)$-ES is seldom the best choice. In almost all benchmarks, different combinations of μ and λ lead to better performance values. Without tuning hyperparameters, experimental results could lead to fallacies and wrong conclusions.

Another characteristic of good experimental design is a *broad spectrum of up-to-date benchmarks.* But many research papers still use out-dated and/or too simple benchmarks—even though they might be disapproved by the GP community [61,87]. By using state-of-the-art benchmarks or benchmark suites [28,48,72], this problem can easily be eliminated.

Both problems also play a vital role in *crossover's negative reputation in CGP.* Clegg et al.'s [13] work is often cited as an argument against the use of standard crossover techniques. However, given today's standards, the authors' experimental design could be called into question. They only based their results on a single, simple quadratic function without any hyperparameter tuning. Both aspects might greatly impair CGP performance, and their conclusion of crossover's negative influence on CGP should be faced with skepticism from a modern perspective.

Lastly, good benchmarking practices and experimental design also include testing the same configuration multiple times with independent random seeds. In some cases, a train-test split is also necessary.

We also want to highlight the importance of (state-of-the-art) statistical analysis [11,56]. Oftentimes, using a Student's t-test (or its variants) might lead to faulty conclusions as wrong statistical prerequisites are assumed. Recently, Cui et al. [14–17] or Kocherovsky [52] et al. used the Plackett-Luce model [11] for operator comparisons. In short, the model uses the rankings of found models across operators to determine the probability that a given operator will produce a model better than the other tested operators. This probability can show differences in performance that are perhaps not obvious from looking at the distribution or quartiles alone. In addition, it doubles as a relative significance measurement. For a more in-depth guide on best practices, we refer to Bartz-Beielstein et al. [8].

5.2 Enforcing Node Integrity

In Sect. 4.2, assumptions on the functional relationships of nodes were discussed, where it was put forward that Subgraph Crossover owes its effectivity in part to explicitly taking into account the structure of the DAG. Another reason, as put forth in Kocherovsky et al. [52], is that the subgraph crossover *requires function nodes to remain intact* throughout the operation. In other words, a crossover point could not be chosen that would separate $a_{n1}a_{n2}g_n$ (see Eq. (1))—*as is already standard in tree-based GP and LGP*. The authors tested generalizing this strategy to ten symbolic regression problems using n-point and uniform crossover and found that forcing node integrity improves the evolutionary algorithm's ability to search in most cases. The authors called this "enforcing node integrity" and "node preservation" since the node is preserved throughout. We want to point out that a similar method was presented by Clegg et al. in 2007 [13] for one-point crossover, but as covered in Sect. 5.1, there is reason to be skeptical of their methodology.

The authors of [52] also found that the same effect occurs when the same restriction is applied to mutation. Instead of traditional point mutation, a "macro" mutation was performed by replacing whole nodes with new randomly-generated operations. Even comparing canonical $(1 + \lambda)$ CGP with the different mutation operators shows that explicit node preservation during mutation also improves the search, and forcing the restriction on both crossover and mutation together provides even more benefits.

Despite these results being very new, we can comfortably recommend enforcing node integrity as an effective practice for CGP users. We also highly recommend further investigation into the topic to measuring the different degrees of phenotypic destruction brought about by node preservation compared to a more typical operator.

5.3   Cross-Domain Benchmarking

Based on the results of the hyperparameter studies surveyed in Sect. 3.1, we emphasize that cross-domain benchmarking can lead to a more holistic perspective on the success of recombination-based or mutation-only algorithms in CGP. We stress that merely focusing on one problem already led to premature generalization of dogmas such as the assumption that the use of $(1 + 4)$-ES is a good choice in general. We therefore recommend evaluating new methodologies on at least two SOTA benchmarks that differ remarkably in terms of the optimization objective and application scope. In the continuous domain, we suggest SRBench [72], while in the discrete domain General Boolean Function Benchmark Suite (GBFS) [48], and General Program Synthesis Benchmark Suite I and II (PSB) [36,37] can be considered adequate choices. For classification problems, Penn Machine Learning Benchmark (PMLB) [71] is a contemporary choice.

5.4   Fitness Functions and Evaluation Strategies

The CGP literature features various fitness functions depending on the type of problems used for benchmarking. As mentioned in Sect. 3.4, logical synthesis benchmarks typically use the Hamming distance as measurement [44] or an alternative metric such as proportion of correct mappings [14,15,17]. In symbolic regression benchmarks, a continuous function is used. For example, the sum of absolute differences (SAD) [44–46], mean absolute error [14,15], and root mean squared error (RMSE) [34]. Clegg et al. used sum of square error in 2007 [13]. In addition, recent literature includes *iterations to solution*; how many fitness evaluations are necessary until a solution is found. Note that while in logical synthesis an exact solution is expected, a solution in symbolic regression problems can be defined as a fitness under a given threshold.

In 2023, Haut et al. [34] found that using the Pearson Correlation r in the fitness function tends to significantly improve search. Specifically, using $f(x) = 1 - r^2$ found models that, had much smaller RMSE values than those models found using RMSE itself. This is because the correlation function takes into account the global behavior of models to match its overall shape to the given data, in contrast to local error functions that only take into account the distance between individual points at a given x value. Correlation has been used in lieu of RMSE in several recent papers [3,4,52–54]. In the Boolean domain it was found that fitness functions that effectively translate the symmetry of solutions into neutrality can lead to more efficient fitness functions [40].

6 New Directions for Research

After the introduction of best practices, as well as common mistakes, we now want to highlight future research directions we believe are of high relevance.

6.1 Foundational Research

Over the last several years, new ideas have emerged and have been developed in the literature. The idea of standard crossover operators having a negative influence on CGP was challenged by some researchers as covered in Sects. 3.1, 4.1, and 4.2. Furthermore, CGP using a standard tournament selection (with or without elitism) as selection/replacement strategy performs exceptionally bad on Boolean benchmark compared to CGP using its standard $(1+4)$-ES, at least with the traditional Hamming distance fitness function (cf. Sect. 3.2).

Based on the previously mentioned findings, we want to draw attention to more CGP related foundational research:

1. Why does tournament selection worsen CGP's performance on Boolean benchmarks?
2. In general, how do different evolutionary operators and their combination affect CGP?
3. Can the positive effects of standard crossover operators (1-point, 2-point, uniform crossover, . . .) be validated by other researchers?

6.2 Analyzing Search Spaces at the Semantic Level

Based on the result of various cross-domain benchmarking that has been surveyed in Sect. 5, a step that should be considered in future work would be the analysis of the respective search spaces and the search behavior of the algorithms tested to ultimately better understand what constitutes the success of a CGP algorithm in the respective problem domain. Initial work in this direction has been presented by Kalkreuth [45] by obtaining the distribution of the semantic spaces in SR and LS that could explain the difference in the search behavior between two problem domains. However, this has been merely done on a case-by-case basis by considering a relatively small set of problems. Therefore, we think that a more quantitative study is missing in the field that allows more general conclusions.

6.3   Homologous Methods in CGP

Molecular biologists have long noted that in real-world reproductive processes, genetic recombination tends to occur where genomes are rather similar to one another. This phenomenon is termed *Homology.* and is thought to emerge from base-pair bonding mechanisms [29, 41]. Hence, crossover operators inspired by homologous operations have been developed for genetic algorithms [43, 65] but have received particular attention in the genetic programming space.

While some early work was considered part of schema theory [76], Francone et al. [29] introduced a method which they call "sticky crossover", which swaps instructions in linear genetic programs by forcing the crossover point to be at the same index in each program, arguing that homology—i.e. similarity between instructions at a given index—will emerge on its own. They reported that sticky crossover improved fitness and reduced bloat, but also reported that the emergence of homology relies on sticky crossover making up an overwhelming majority of crossover operations. A few years later, Platel et al. [74] introduced what they called "maximum homologous crossover" for LGP, which computes the best alignment for nodes. "Alignment" in this case refers to using string edit distance metrics to transform program x into program y using the minimal amount of operations.

However, to our knowledge homologous methods have not been consciously introduced into CGP until 2023. Torabi et al. [81] implemented a homologous crossover with CGP because they were specifically assembling convolutional neural networks (CNNs), and thus represent instructions as strings, where each character maps to a CNN component. From a technical perspective, one-to-one mapping enabled the authors to explicitly take homology into account. Therefore, in-line with our recommendations on accounting for node connectivity and enforcing node integrity (Sects. 4.2 and 5.2), we recommend introducing homologous crossovers to CGP by representing instructions as whole units rather than a collection of disparate genes.

6.4   Fitness Landscape in GP and Search Trajectories

In GP, the fitness landscape (FL) defines the mapping between each candidate program and its corresponding fitness value. The geometry of this landscape governs how evolution navigates towards optimality [7].

In continuous optimization the FL can be understood as measurable fitness gradients that can be visualized and analyzed. In two dimensions, for example, the FL can be visualized as a 3D surface. Smooth, convex landscapes typically lead to efficient optimization, whereas rugged landscapes with multiple local optima make convergence more difficult [59,60]. However, when a search space is a high-dimensional mix between mostly discrete terms and real-valued constants, and there is no a consistent way to measure *distance* between solutions that reflects both structure and semantics [85], the FL becomes much more difficult to understand. Small structural mutations may produce large and unpredictable behavioral changes, making it difficult to define meaningful notions of distance, continuity, or neighborhood. Moreover, the mapping between genotype, phenotype, and behavior (the program's output) introduces neutrality, further complicating visualization and interpretation [5,6,20].

To better understand the behavior of metaheuristics, Search Trajectory Networks (STNs) have been proposed as data-driven tools to visualize and analyze how evolutionary algorithms explore FLs [67,68]. In STNs, nodes represent distinct search states (e.g., genotypes, phenotypes, or behaviors), and edges represent transitions between them across generations. By capturing evolutionary trajectories, neutral drifts, and stepping-stone transitions, STNs reveal patterns of exploration and exploitation and the connectivity of promising regions. Initially, this tool was proposed in the context of continuous problem solving, but later it was applied on Boolean functions with linear GP [40], navigation tasks with tree GP [66], and symbolic regression with CGP [20]. In particular, De La Torre et al. [20] found that both the genotype and the phenotype space suffer from a high degree of structural and semantic redundancy on symbolic regression tasks, making it difficult to interpret search trajectories of CGP. To provide a coarser and more understandable look into population dynamics, they proposed using STNs on the behavioral level, which dramatically reduced the number of unique locations, revealing to which extent different configurations and runs genuinely explored the search space.

By bridging FL understanding and STNs, it becomes possible to investigate why and how different hyper-parameters and/or evolutionary operators succeed, fail, or interact. This allows advancing our understanding of GP's search dynamics beyond simple population-level statistics. Despite the existence of tools like STNs, FL analysis in GP and CGP in particular is still poorly covered in the literature, although it can reveal very important features about an algorithm's suitability for solving a given problem.

6.5 Self-adaptive CGP

The effects of hyperparameterisation on evolutionary algorithms have been studied since the early days of the field [19,23]. A more direct approach consists of sampling a range of different hyperparameters for a given problem using cross-validation (aka hyperparameter tuning), and then using the optimal configuration to solve it. At time of writing, this is the established best practice as described in Sect. 5.1, but it carries several drawbacks:

1. The test values are typically predefined, therefore a limited range of hyper-parameters is explored and the optimal choice can be eventually overlooked;
2. A significant amount of computational resources must be allocated during the testing phase to befittingly estimate the optimal hyperparameters;
3. Since the hyperparameters often interact in a complex manner, if the values are tuned one at a time, one may end with a suboptimal choice. Alternatively, if the values are explored simultaneously, the search space of the hyperparameters can be unbearably large;
4. Different stages of the evolutionary process may require different optimal hyperparameter values;
5. Optimal values for a given algorithm tendentiously differ from one problem to another, which implies repeating the tuning for each new problem.

An alternative to static tuning is parameter control, where hyperparameters change dynamically during the evolutionary run [23]. Deterministic control relies on some time-dependent function to update hyperparameters [80], while adaptive control adjusts hyperparameter values dynamically based on information extracted from the evolutionary process. A significant amount of work suggests the superiority of hyperparameter control over tuning [21,23,57,70].

Traditionally, researchers focused on the control of specific hyperparameters related to genetic operators [18,23,33,57] or selection methods [2,21]. Several works found that population size was one of the most influential control variables in evolutionary search [9,22,23,25] as it can have large impacts on the behavior (final performance, population diversity, search speed, and others) of the optimization process. For traditional GP, considerable research has focused on understanding the impact of population size on its search dynamics and on crafting strategies to adapt or control it [12,24,26,27,55,77,79]. Additionally, multiple studies have explored adaptive probabilities for crossover and mutation [51,69,89], as well as broader forms of control including elitism, crossover, mutation, and replacement parameters [78].

Hyperparameter tuning remains a best-practice at time of publication because until recently, (self-)adaptive optimizers have received little research attention in CGP. Although Kalkreuth et al. [47] had some success in 2015, recently, optimizers and especially modern (self-)adaptive versions were not the core focus of the field. However, both in GP and especially the wider EC community, (self-)adaptivity of evolutionary algorithms has shown great advantages over the non-adaptive counterparts [21,23,33,35,55,57] Having (self-)adaptive operators as part of a CGP optimizer could provide two essential benefits. First, the optimization might become more efficient and more effective. Second, self-adaptive operators often reduce the effect of specific hyperparameters on the performance. Ideally, this leads to hyperparameter-free training.

While this can come at a cost regarding the overall computational effort needed for a single training instance, it can lead to less hyperparameter tuning (or better guided tuning) and, crucially, assist those practitioners without in-depth knowledge on beneficial configurations of CGP parameters, which in turn leads to wider adoption of CGP. We thus encourage further study of and a shift towards self-adaptive methods once the dynamics and effects are better understood.

6.6 Hybridization of Reorder and Subgraph

To this day, *other influences* of positional bias or reordering nodes are *unclear*. In particular, its effect on *crossover* strategies are not fully understood yet, although first steps were accomplished by Cui et al. [14]. Further investigating the effects of reordering nodes with crossover might lead to new insights. Thus, we further propose the investigation of a *hybridization between reordering nodes and the subgraph crossover mechanism* (cf. Sect. 4.1).

Kalkreuth's [46] recombination method only swaps one subgraph between individuals. As a result, it can also be seen as a "highly guided and specialized one-point crossover" (with one-point crossover showing favorable properties in GP [75]). However, this operator must choose one crossover point within the range of active nodes. Due to positional bias, this crossover point is highly likely to be close to the input nodes—as almost all active nodes cluster near input nodes. This was demonstrated by Kocherovsky et al. in 2025 [54]. By eliminating positional bias, we believe that the crossover point for the subgraph crossover is moved towards the middle of CGP's graph. This leads to a more uniform distribution of swapped genetic material. In turn, CGP's genotype changes are more substantial. This increases *neutral genetic drift*, a desirable trait in CGP which leads to performance boosts [39, 84, 90].

7 Conclusion

Cartesian Genetic Programming was proposed nearly thirty years ago, and its study and refinement has progressed significantly since then. However, as with any design paradigm, certain practices and assumptions become frozen accidents.

In this work, we have sought to discuss misconceptions prevalent in the field, particularly around:

1. Crossover, positional bias, and benchmarking;
2. A holistic review successful genetic operators, such as subgraph and node-preserving crossover;
3. Promoting *new* best practices based on the recent literature, such as hyperparameter tuning, cross-domain benchmarking, and certain methods of evaluation;
4. Introducing new directions for research around selection methods, crossover fundamentals, homologous operators, fitness landscapes, adaptive methods, and practice hybridization.

It is our hope that this work contributes to the general field of CGP by encouraging both researchers and general practitioners to approach CGP from new angles in academic, educational, and industrial contexts. It is our position that a fresh look at CGP from new points of view—indebted to older viewpoints, of course—could be vital to overcoming existing issues and discovering both new dynamics to be observed and new problems to be solved, enhancing the abilities robustness not just of CGP models themselves, but the field as a whole.

Acknowledgements. Partial support was received by the Bavarian State Ministry of Science and the Arts through the Augsburg AI Production Network as part of the High-Tech Agenda Plus. This work was supported by an Alexander von Humboldt Professorship in AI awarded to Holger H. Hoos, with funding provided by the German Federal Ministry of Education and Research.

References

1. Adewuya, A.: New methods in genetic search with real-valued chromosomes. Bachelor's thesis, Mississippi State University (1996)
2. Affenzeller, M., Wagner, S.: Offspring selection: a new self-adaptive selection scheme for genetic algorithms. In: Ribeiro, B., Albrecht, R.F., Dobnikar, A., Pearson, D.W., Steele, N.C. (eds.) Adaptive and Natural Computing Algorithms, pp. 218–221. Springer, Vienna (2005). https://doi.org/10.1007/3-211-27389-1_52
3. Bakurov, I., Haut, N., Banzhaf, W.: Sharpness-Aware Minimization in Genetic Programming, pp. 151–175. Springer, Singapore (2025). https://doi.org/10.1007/978-981-96-0077-9_8
4. Bakurov, I., Murphy, A., Ofria, C., Banzhaf, W.: A comparison of tournament and lexicase selection paradigms in regression problems: error-based fitness versus correlation fitness. In: Proceedings of the Genetic and Evolutionary Computation Conference, GECCO 2025, pp. 970–979. Association for Computing Machinery, New York, NY, USA (2025). https://doi.org/10.1145/3712256.3726448
5. Banzhaf, W., Bakurov, I.: On the nature of the phenotype in tree genetic programming. In: Proceedings of the Genetic and Evolutionary Computation Conference, GECCO 2024, pp. 868–877. Association for Computing Machinery, New York, NY, USA (2024). https://doi.org/10.1145/3638529.3654129
6. Banzhaf, W., Hu, T., Ochoa, G.: How the combinatorics of neutral spaces leads genetic programming to discover simple solutions. In: Winkler, S., Trujillo, L., Ofria, C., Hu, T. (eds.) Genetic Programming Theory and Practice XX, pp. 65–86. Genetic and Evolutionary Computation, Springer, Michigan State University, USA (2023). https://doi.org/10.1007/978-981-99-8413-8_4
7. Banzhaf, W., Leier, A.: Evolution on Neutral Networks in Genetic Programming, pp. 207–221. Springer US, Boston, MA (2006). https://doi.org/10.1007/0-387-28111-8_14
8. Bartz-Beielstein, T., et al.: Benchmarking in optimization: best practice and open issues (2020). https://arxiv.org/abs/2007.03488
9. Bingchuan, W., Shui, Z.Y., Feng, Y., Ma, Z.: Evolutionary algorithm with dynamic population size for constrained multiobjective optimization. Swarm Evol. Comput. **73**, 101104 (2022). https://doi.org/10.1016/j.swevo.2022.101104

10. Blickle, T., Thiele, L.: A mathematical analysis of tournament selection. In: Proceedings of the 6th International Conference on Genetic Algorithms, pp. 9–16. Morgan Kaufmann Publishers Inc., San Francisco, CA, USA (1995). https://doi.org/10.5555/645514.658088

11. Calvo, B., Ceberio, J., Lozano, J.A.: Bayesian inference for algorithm ranking analysis. In: Proceedings of the Genetic and Evolutionary Computation Conference Companion, GECCO 2018, pp. 324–325. Association for Computing Machinery, New York, NY, USA (2018). https://doi.org/10.1145/3205651.3205658

12. Castelli, M., Manzoni, L., Silva, S., Vanneschi, L., Popovič, A.: The influence of population size in geometric semantic GP. Swarm Evol. Comput. **32**, 110–120 (2017). https://doi.org/10.1016/j.swevo.2016.05.004

13. Clegg, J., Walker, J.A., Miller, J.F.: A new crossover technique for cartesian genetic programming. In: Proceedings of the 9th Annual Conference on Genetic and Evolutionary Computation, GECCO 2007, pp. 1580–1587. Association for Computing Machinery, New York, NY, USA (2007). https://doi.org/10.1145/1276958.1277276

14. Cui, H., Heider, M., Hähner, J.: Positional bias does not influence cartesian genetic programming with crossover. In: Affenzeller, M., et al. (eds.) Parallel Problem Solving from Nature – PPSN XVIII, pp. 151–167. Springer Nature, Switzerland, Cham (2024). https://doi.org/10.1007/978-3-031-70055-2_10

15. Cui, H., Margraf, A., Hähner, J.: Analysing the influence of reorder strategies for cartesian genetic programming. SN Comput. Sci. **6**(7), 754 (2025). https://doi.org/10.1007/s42979-025-04296-4

16. Cui, H., Margraf, A., Heider, M., Hähner, J.: Towards understanding crossover for cartesian genetic programming. In: Towards Understanding Crossover for Cartesian Genetic Programming, pp. 308–314 (2023). https://doi.org/10.5220/0012231400003595

17. Cui, H., Margraf, A., Hähner, J.: Equidistant reorder operator for cartesian genetic programming. In: van Stein, N., Marcelloni, F., Lam, H.K., Cottrell, M., Filipe, J. (eds.) Proceedings of the 15th International Joint Conference on Computational Intelligence - ECTA, 13–15 November 2023, in Rome, Italy, pp. 64 – 74 (2023). https://doi.org/10.5220/0012174100003595

18. Das, S., Abraham, A., Chakraborty, U.K., Konar, A.: Differential evolution using a neighborhood-based mutation operator. IEEE Trans. Evol. Comput. **13**(3), 526–553 (2009). https://doi.org/10.1109/TEVC.2008.2009457

19. De Jong, K.: Parameter setting in EAs: a 30 year perspective, pp. 1–18. Springer, Heidelberg (2007). https://doi.org/10.1007/978-3-540-69432-8_1

20. De La Torre, C., Cussat-Blanc, S., Wilson, D., Lavinas, Y.: On search trajectory networks for graph genetic programming. In: Proceedings of the Genetic and Evolutionary Computation Conference Companion, GECCO 2024 Companion, pp. 1681–1685. Association for Computing Machinery, New York, NY, USA (2024). https://doi.org/10.1145/3638530.3664169

21. Droste, S., Jansen, T., Wegener, I.: Dynamic parameter control in simple evolutionary algorithms. In: Martin, W.N., Spears, W.M. (eds.) Foundations of Genetic Algorithms, vol. 6, pp. 275–294. Morgan Kaufmann, San Francisco (2001). https://doi.org/10.1016/B978-155860734-7/50098-6. https://www.sciencedirect.com/science/article/pii/B9781558607347500986

22. Eiben, A.E., Marchiori, E., Valkó, V.A.: Evolutionary algorithms with on-the-fly population size adjustment. In: Yao, X., et al. (eds.) Parallel Problem Solving from Nature - PPSN VIII, pp. 41–50. Springer, Heidelberg (2004). https://doi.org/10.1007/978-3-540-30217-9_5

23. Eiben, A., Hinterding, R., Michalewicz, Z.: Parameter control in evolutionary algorithms. IEEE Trans. Evol. Comput. **3**(2), 124–141 (1999). https://doi.org/10.1109/4235.771166
24. Farinati, D., Bakurov, I., Vanneschi, L.: A study of dynamic populations in geometric semantic genetic programming. Inf. Sci. **648**, 119513 (2023). https://doi.org/10.1016/j.ins.2023.119513.   https://www.sciencedirect.com/science/article/pii/S0020025523010988
25. Farinati, D., Vanneschi, L.: A survey on dynamic populations in bio-inspired algorithms. Genet. Program Evolvable Mach. **25**(2), 19 (2024). https://doi.org/10.1007/s10710-024-09492-4
26. Fernandez, F., Tomassini, M., Vanneschi, L.: Saving computational effort in genetic programming by means of plagues. In: The 2003 Congress on Evolutionary Computation, CEC 2003, vol. 3, pp. 2042–2049 (2003). https://doi.org/10.1109/CEC.2003.1299924
27. Fernandez, F., Vanneschi, L., Tomassini, M.: The effect of plagues in genetic programming: a study of variable-size populations. In: Ryan, C., Soule, T., Keijzer, M., Tsang, E., Poli, R., Costa, E. (eds.) Genetic Programming, pp. 317–326. Springer, Heidelberg (2003). https://doi.org/10.1007/3-540-36599-0_29
28. de Franca, F.O., et al.: SRBench++: principled benchmarking of symbolic regression with domain-expert interpretation. IEEE Trans. Evol. Comput. **29**(4), 1127–1137 (2025). https://doi.org/10.1109/TEVC.2024.3423681
29. Francone, F.D., Conrads, M., Banzhaf, W., Nordin, P.: Homologous crossover in genetic programming. In: Proceedings of the 1st Annual Conference on Genetic and Evolutionary Computation, GECCO'99, vol. 2, pp. 1021–1026. Morgan Kaufmann Publishers Inc., San Francisco, CA, USA (1999). https://doi.org/10.5555/2934046.2934059
30. Goldman, B.W., Punch, W.F.: Length bias and search limitations in cartesian genetic programming. In: Proceedings of the 15th Annual Conference on Genetic and Evolutionary Computation, GECCO 2013, pp. 933–940. Association for Computing Machinery, New York, NY, USA (2013). https://doi.org/10.1145/2463372.2463482
31. Goldman, B.W., Punch, W.F.: Reducing wasted evaluations in cartesian genetic programming. In: Proceedings of the 16th European Conference on Genetic Programming, EuroGP 2013, pp. 61–72. Springer, Heidelberg (2013). https://doi.org/10.1007/978-3-642-37207-0_6
32. Goldman, B.W., Punch, W.F.: Analysis of cartesian genetic programming's evolutionary mechanisms **19**(3), 359–373 (2015). https://doi.org/10.1109/TEVC.2014.2324539
33. Hansen, N., Ostermeier, A.: Completely derandomized self-adaptation in evolution strategies. Evol. Comput. **9**(2), 159–195 (2001). https://doi.org/10.1162/106365601750190398
34. Haut, N., Banzhaf, W., Punch, B.: Correlation versus RMSE loss functions in symbolic regression tasks. In: Trujillo, L., Winkler, S.M., Silva, S., Banzhaf, W. (eds.) Genetic Programming Theory and Practice XIX, pp. 31–55. Springer (2023). https://doi.org/10.1007/978-981-19-8460-0_2
35. Heider, M., Krischan, M., Sraj, R., Hähner, J.: Exploring self-adaptive genetic algorithms to combine compact sets of rules. In: 2024 IEEE Congress on Evolutionary Computation (CEC), pp. 1–8 (2024). https://doi.org/10.1109/CEC60901.2024.10612101

36. Helmuth, T., Kelly, P.: PSB2: the second program synthesis benchmark suite. In: Proceedings of the Genetic and Evolutionary Computation Conference, GECCO 2021, pp. 785–794. Association for Computing Machinery, New York, NY, USA (2021). https://doi.org/10.1145/3449639.3459285
37. Helmuth, T., Spector, L.: General program synthesis benchmark suite. In: Proceedings of the 2015 Annual Conference on Genetic and Evolutionary Computation, GECCO 2015, pp. 1039–1046. Association for Computing Machinery, New York, NY, USA (2015). https://doi.org/10.1145/2739480.2754769
38. Hu, T., Banzhaf, W.: Neutrality and variability: two sides of evolvability in linear genetic programming. In: Proceedings of the 11th Annual Conference on Genetic and Evolutionary Computation, GECCO 2009, pp. 963–970. Association for Computing Machinery, New York, NY, USA (2009). https://doi.org/10.1145/1569901.1570033
39. Hu, T., Banzhaf, W.: Neutrality, robustness, and evolvability in genetic programming. In: Riolo, R., Worzel, B., Goldman, B., Tozier, B. (eds.) Genetic Programming Theory and Practice XIV, pp. 101–117. Springer International Publishing, Cham (2018). https://doi.org/10.1007/978-3-319-97088-2_7
40. Hu, T., Banzhaf, W., Ochoa, G.: How neutrality shapes evolution: simplicity bias and search. In: Proceedings of the Genetic and Evolutionary Computation Conference, GECCO 2025, pp. 1008–1016. Association for Computing Machinery, New York, NY, USA (2025). https://doi.org/10.1145/3712256.3726454
41. Hunter, N.: Meiotic recombination: the essence of heredity. Cold Spring Harbor Perspect. Biol. **7 12** (2015). https://doi.org/10.1101/cshperspect.a016618. https://api.semanticscholar.org/CorpusID:33542130
42. Husa, J., Kalkreuth, R.: A comparative study on crossover in cartesian genetic programming. In: Castelli, M., Sekanina, L., Zhang, M., Cagnoni, S., García-Sánchez, P. (eds.) Genetic Programming, pp. 203–219. Springer International Publishing, Cham (2018). https://doi.org/10.1007/978-3-319-77553-1_13
43. Iqbal, S., Hoque, M.T.: A homologous gene replacement based genetic algorithm. In: Proceedings of the 2016 on Genetic and Evolutionary Computation Conference Companion, pp. 91–92 (2016)
44. Kalkreuth, R.: A comprehensive study on subgraph crossover in cartesian genetic programming. In: 12th International Joint Conference on Computational Intelligence, pp. 59—70 (2020). https://doi.org/10.5220/0010110700590070
45. Kalkreuth, R.: Reconsideration and extension of cartesian genetic programming. Ph.D. thesis, Department of Computer Science, University of Dortmund (2021). https://doi.org/10.17877/DE290R-22504
46. Kalkreuth, R., Rudolph, G., Droschinsky, A.: A new subgraph crossover for cartesian genetic programming. In: McDermott, J., Castelli, M., Sekanina, L., Haasdijk, E., García-Sánchez, P. (eds.) European Conference on Genetic Programming, pp. 294–310. Springer International Publishing, Cham (2017). https://doi.org/10.1007/978-3-319-55696-3_19
47. Kalkreuth, R., Rudolph, G., Krone, J.: Improving convergence in cartesian genetic programming using adaptive crossover, mutation and selection. In: 2015 IEEE Symposium Series on Computational Intelligence, pp. 1415–1422 (2015). https://doi.org/10.1109/SSCI.2015.201
48. Kalkreuth, R., Vašíček, Z., Husa, J., Vermetten, D., Ye, F., Bäck, T.: General Boolean function benchmark suite. In: Proceedings of the 17th ACM/SIGEVO Conference on Foundations of Genetic Algorithms, FOGA 2023, pp. 84–95. Association for Computing Machinery, New York, NY, USA (2023). https://doi.org/10.1145/3594805.3607131

49. Kaufmann, P., Kalkreuth, R.: Parametrizing cartesian genetic programming: an empirical study. In: Kern-Isberner, G., Fürnkranz, J., Thimm, M. (eds.) KI 2017: Advances in Artificial Intelligence, pp. 316–322. Springer International Publishing, Cham (2017). https://doi.org/10.1007/978-3-319-67190-1_26

50. Kaufmann, P., Kalkreuth, R.: On the parameterization of cartesian genetic programming. In: 2020 IEEE Congress on Evolutionary Computation (CEC), pp. 1–8 (2020). https://doi.org/10.1109/CEC48606.2020.9185492

51. Kefer, K., Affenzeller, M., Winkler, S.: Adaptive operators for genetic programming to identify optimal energy flow controllers. Procedia Comput. Sci. **253**, 1991–2002 (2025). https://doi.org/10.1016/j.procs.2025.01.261. https://www.sciencedirect.com/science/article/pii/S1877050925002698. 6th International Conference on Industry 4.0 and Smart Manufacturing

52. Kocherovsky, M., Bakurov, I., Banzhaf, W.: Node preservation and its effect on crossover in cartesian genetic programming (2025). https://arxiv.org/abs/2511.00634

53. Kocherovsky, M., Banzhaf, W.: Crossover destructiveness in cartesian versus linear genetic programming. In: ALIFE 2024: Proceedings of the 2024 Artificial Life Conference, p. 20. Artificial Life Conference Proceedings, International Society for Artificial Life (2024). https://doi.org/10.1162/isal_a_00735

54. Kocherovsky, M., Kianinejad, M., Bakurov, I., Banzhaf, W.: On the effectiveness of crossover operators in cartesian genetic programming. In: Xue, B., Manzoni, L., Bakurov, I. (eds.) Genetic Programming, pp. 68–84. Springer Nature Switzerland, Cham (2025). https://doi.org/10.1007/978-3-031-89991-1_5

55. Kouchakpour, P., Zaknich, A., Braunl, T.: Dynamic population variation in genetic programming. Inf. Sci. **179**, 1078–1091 (2009). https://doi.org/10.1016/j.ins.2008.12.009

56. Kruschke, J.K.: Bayesian estimation supersedes the T test. J. Exp. Psychol. Gen. **142**(2), 573 (2013). https://doi.org/10.1037/a0029146

57. Lima, C.F., Lobo, F.G.: Parameter-less optimization with the extended compact genetic algorithm and iterated local search. In: Deb, K. (ed.) Genetic and Evolutionary Computation – GECCO 2004, pp. 1328–1339. Springer, Heidelberg (2004). https://doi.org/10.1007/978-3-540-24854-5_127

58. López-Ibáñez, M., Dubois-Lacoste, J., Pérez Cáceres, L., Birattari, M., Stützle, T.: The irace package: iterated racing for automatic algorithm configuration. Oper. Res. Perspect. **3**, 43–58 (2016). https://doi.org/10.1016/j.orp.2016.09.002. https://www.sciencedirect.com/science/article/pii/S2214716015300270

59. Malan, K.M., Engelbrecht, A.P.: Quantifying ruggedness of continuous landscapes using entropy. In: 2009 IEEE Congress on Evolutionary Computation, pp. 1440–1447 (2009). https://doi.org/10.1109/CEC.2009.4983112

60. Malan, K.M., Engelbrecht, A.P.: A survey of techniques for characterising fitness landscapes and some possible ways forward. Inf. Sci. **241**, 148–163 (2013). https://doi.org/10.1016/j.ins.2013.04.015. https://www.sciencedirect.com/science/article/pii/S0020025513003125

61. McDermott, J., et al.: Genetic programming benchmarks: looking back and looking forward. SIGEVOlution **15**(3) (2022). https://doi.org/10.1145/3578482.3578483

62. Miller, J.F.: An empirical study of the efficiency of learning Boolean functions using a cartesian genetic programming approach. In: Proceedings of the 1st Annual Conference on Genetic and Evolutionary Computation, GECCO 1999, vol. 2, pp. 1135–1142. Morgan Kaufmann Publishers Inc., San Francisco, CA, USA (1999). https://doi.org/10.5555/2934046.2934074. https://dl.acm.org/doi/10.5555/2934046.2934074

63. Miller, J.F., Smith, S.L.: Redundancy and computational efficiency in cartesian genetic programming. IEEE Trans. Evol. Comput. **10**(2), 167–174 (2006). https://doi.org/10.1109/TEVC.2006.871253

64. Miller, J.F., Harding, S.L.: Cartesian genetic programming. In: Proceedings of the 10th Annual Conference Companion on Genetic and Evolutionary Computation, GECCO 2008, pp. 2701–2726. Association for Computing Machinery, New York, NY, USA (2008). https://doi.org/10.1145/1388969.1389075

65. Mühlenbein, H., Schomisch, M., Born, J.: The parallel genetic algorithm as function optimizer. Parallel Comput. **17**, 619–632 (1991). https://doi.org/10.1016/S0167-8191(05)80052-3. https://api.semanticscholar.org/CorpusID:2485505

66. Nadizar, G., Rusin, F., Medvet, E., Ochoa, G.: The role of stepping stones in map-elites: insights from search trajectory networks. In: Genetic Programming: 28th European Conference, EuroGP 2025, Held as Part of EvoStar 2025, Trieste, Italy, 23–25 April 2025, Proceedings, pp. 224–239. Springer, Heidelberg (2025). https://doi.org/10.1007/978-3-031-89991-1_14

67. Ochoa, G., Malan, K.M., Blum, C.: Search trajectory networks of population-based algorithms in continuous spaces. In: Applications of Evolutionary Computation: 23rd European Conference, EvoApplications 2020, Held as Part of EvoStar 2020, Seville, Spain, 15–17 April 2020, Proceedings, pp. 70–85. Springer, Heidelberg (2020). https://doi.org/10.1007/978-3-030-43722-0_5

68. Ochoa, G., Malan, K.M., Blum, C.: Search trajectory networks: a tool for analysing and visualising the behaviour of metaheuristics. Appl. Soft Comput. **109**, 107492 (2021). https://doi.org/10.1016/j.asoc.2021.107492. https://www.sciencedirect.com/science/article/pii/S1568494621004154

69. Oh, S., Suh, W.H., Ahn, C.W.: Self-adaptive genetic programming for manufacturing big data analysis. Symmetry **13**(4) (2021). https://doi.org/10.3390/sym13040709. https://www.mdpi.com/2073-8994/13/4/709

70. Ojha, V., Timmis, J., Nicosia, G.: Assessing ranking and effectiveness of evolutionary algorithm hyperparameters using global sensitivity analysis methodologies. Swarm Evol. Comput. **74**, 101130 (2022). https://doi.org/10.1016/j.swevo.2022.101130. https://www.sciencedirect.com/science/article/pii/S2210650222001006

71. Olson, R.S., La Cava, W., Orzechowski, P., Urbanowicz, R.J., Moore, J.H.: PMLB: a large benchmark suite for machine learning evaluation and comparison. BioData Mining **10**(1), 36 (2017). https://doi.org/10.1186/s13040-017-0154-4

72. Orzechowski, P., La Cava, W., Moore, J.H.: Where are we now? A large benchmark study of recent symbolic regression methods. In: Proceedings of the Genetic and Evolutionary Computation Conference, GECCO 2018, pp. 1183–1190. Association for Computing Machinery, New York, NY, USA (2018). https://doi.org/10.1145/3205455.3205539

73. Payne, A.J., Stepney, S.: Representation and structural biases in CGP. In: 2009 IEEE Congress on Evolutionary Computation, pp. 1064–1071 (2009). https://doi.org/10.1109/CEC.2009.4983064

74. Platel, M.D., Clergue, M., Collard, P.: Maximum homologous crossover for linear genetic programming. In: Proceedings of the 6th European Conference on Genetic Programming, EuroGP 2003, pp. 194–203. Springer, Heidelberg (2003). https://doi.org/10.5555/1762668.1762687

75. Poli, R., Langdon, W.B.: Genetic programming with one-point crossover. In: Chawdhry, P.K., Roy, R., Pant, R.K. (eds.) Soft Computing in Engineering Design and Manufacturing, pp. 180–189. Springer, London (1998). https://doi.org/10.1007/978-1-4471-0427-8_20

76. Poli, R., Langdon, W.B.: Schema theory for genetic programming with one-point crossover and point mutation. Evol. Comput. **6**(3), 231–252 (1998). https://doi.org/10.1162/evco.1998.6.3.231
77. Poli, R., McPhee, N.F., Vanneschi, L.: The impact of population size on code growth in GP: analysis and empirical validation. In: Proceedings of the 10th Annual Conference on Genetic and Evolutionary Computation, GECCO 2008, pp. 1275–1282. Association for Computing Machinery, New York, NY, USA (2008). https://doi.org/10.1145/1389095.1389341
78. Riekert, M., Malan, K.M., Engelbrect, A.P.: Adaptive genetic programming for dynamic classification problems. In: Proceedings of the Eleventh Conference on Congress on Evolutionary Computation, CEC 2009, pp. 674–681. IEEE Press (2009). https://doi.org/10.1109/CEC.2009.4983010
79. Rochat, D., Tomassini, M., Vanneschi, L.: Dynamic size populations in distributed genetic programming. In: Keijzer, M., Tettamanzi, A., Collet, P., van Hemert, J., Tomassini, M. (eds.) Genetic Programming, pp. 50–61. Springer, Heidelberg (2005). https://doi.org/10.1007/978-3-540-31989-4_5
80. Tanabe, R., Fukunaga, A.S.: Improving the search performance of shade using linear population size reduction. In: 2014 IEEE Congress on Evolutionary Computation (CEC), pp. 1658–1665 (2014). https://doi.org/10.1109/CEC.2014.6900380
81. Torabi, A., Sharifi, A., Teshnehlab, M.: Using cartesian genetic programming approach with new crossover technique to design convolutional neural networks. Neural Process. Lett. **55**(5), 5451–5471 (2023). https://doi.org/10.1007/s11063-022-11093-0
82. Trautwein, J., Heider, M., Cui, H., Hähner, J.: Ant-based metaheuristics struggle to solve the cartesian genetic programming learning task. In: Xue, B., Manzoni, L., Bakurov, I. (eds.) Genetic Programming, pp. 139–155. Springer Nature Switzerland, Cham (2025)
83. Turner, A.J.: Improving crossover techniques in a genetic program. Master's thesis, Department of Electronics, University of York (2012)
84. Turner, A.J., Miller, J.F.: Neutral genetic drift: an investigation using cartesian genetic programming **16**(4), 531–558 (2015). https://doi.org/10.1007/s10710-015-9244-6
85. Vanneschi, L.: Fitness landscapes and problem hardness in genetic programming. In: Proceedings of the 12th Annual Conference Companion on Genetic and Evolutionary Computation, GECCO 2010, pp. 2711–2738. Association for Computing Machinery, New York, NY, USA (2010). https://doi.org/10.1145/1830761.1830916
86. Vašíček, Z.: Bridging the gap between evolvable hardware and industry using cartesian genetic programming. In: Emergence, Complexity and Computation, vol. 28, pp. 39–55. Springer International Publishing (2018). https://doi.org/10.1007/978-3-319-67997-6_2
87. White, D., et al.: Better GP benchmarks: community survey results and proposals **14**, 3–29 (2013)
88. Wilson, G., Banzhaf, W.: A comparison of cartesian genetic programming and linear genetic programming. In: O'Neill, M., et al. (eds.) Genetic Programming, pp. 182–193. Springer, Heidelberg (2008). https://doi.org/10.1007/978-3-540-78671-9_16

89. Yin, Z., Brabazon, A., O'Sullivan, C.: Adaptive genetic programming for option pricing. In: Proceedings of the 9th Annual Conference Companion on Genetic and Evolutionary Computation, GECCO 2007, pp. 2588–2594. Association for Computing Machinery, New York, NY, USA (2007). https://doi.org/10.1145/1274000.1274029
90. Yu, T., Miller, J.: Finding needles in haystacks is not hard with neutrality. In: Foster, J.A., Lutton, E., Miller, J., Ryan, C., Tettamanzi, A. (eds.) Genetic Programming, pp. 13–25. Springer, Heidelberg (2002). https://doi.org/10.1007/3-540-45984-7_2

Semantic Search Trajectory Networks for Understanding Genetic Programming

Josip Hrvatić[1]([⊠])[iD], Magda Smolić-Ročak[1][iD], Marko Đurasević[1][iD],
and Gabriela Ochoa[2][iD]

[1] Faculty of Electrical Engineering and Computing, University of Zagreb,
10000 Zagreb, Croatia
{josip.hrvatic,magda.smolic-rocak,marko.durasevic}@fer.hr
[2] University of Stirling, Stirling FK9 4LA, UK
gabriela.ochoa@stir.ac.uk

Abstract. Genetic programming has become a powerful tool for solving symbolic regression and similar problems, yet its inner workings often remain a mystery, prompting the need for techniques that can reveal and explain its evolutionary behaviour. Therefore, in this study, we adapt search trajectory networks (STNs) to analyse and visualise the search process of tree-based genetic programming (GP) for symbolic regression problems. STNs are graph-based models of the dynamics of search and evolutionary algorithms, where nodes represent search states and edges represent transitions between consecutive states. As network models, they can be analysed and visualised with the rich graph theory and complex networks toolsets. We consider three STN models: a standard search space or genotype model, where nodes are syntax trees, and two semantic or phenotype models, where nodes encode regression output vectors. The two semantic models differ in the partitioning technique used to group sets of output vectors into STN nodes: (i) standard hypercubes, and (ii) a newly proposed approach using an advanced clustering method. Standard tree-based GP is contrasted against a recent variant of geometric semantic GP. Our analysis reveals that STNs quantitatively and visually capture the essential search and performance differences between these two GP paradigms. The newly proposed semantic STNs produce compact and interpretable models, which can scale up to real-world symbolic regression problems.

Keywords: Genotype-to-phenotype mapping · Genetic programming ·
Algorithm modelling · Algorithm analysis · Search trajectories ·
Complex networks · Visualisation

1 Introduction

Genetic programming (GP) is a prolific subfield of evolutionary computation, with numerous variants, including tree-based, linear, Cartesian, gene expression, and grammatical evolution, among others. Each variant employs a distinct

L. Manzoni et al. (Eds.): EuroGP 2026, LNCS 16521, pp. 106–121, 2026.
https://doi.org/10.1007/978-3-032-23005-8_7

solution representation and/or specialised set of operators, leading to different evolutionary dynamics and search behaviours. This makes it difficult to determine why one variant (or configuration) succeeds or fails on a specific problem, as the evolutionary process encompasses complex genetic operations and mappings to behaviour or semantics. Current standard methods for comparing these variants rely primarily on aggregate metrics at the end of runs, such as best fitness, success rates, and runtime. Convergence curves are useful as they illustrate the evolutionary progress and how quickly and reliably each variant approaches an optimal or near-optimal solution over time. However, they can become cluttered and hard to interpret when contrasting many variants. Most importantly, convergence curves only show the dynamics in the fitness (or objective) space, oblivious to what is occurring in the genotype and/or phenotype spaces. Overall, these classic metrics and visualisation tools often conceal crucial differences in how variants explore and exploit the search space.

To advance GP research and enable methodical algorithm design, there is a clear need for additional visualisation and analysis tools. Search trajectory networks (STNs) [11,12] help to fill this gap. However, to the best of our knowledge, STNs have not yet been applied to tree-based genetic programming, despite their strong potential to uncover structural patterns in its search process. In an STN, solutions encountered during an optimisation run are represented as nodes, and the algorithm transitions between nodes form directed edges. By building these networks from multiple runs and different GP variants, we can visualise, analyse and contrast the overall search behaviour. This facilitates a principled analysis of search trajectories, convergence patterns, and the manifestation of phenomena such as program bloat. As graph models, STNs can be analysed with graph-theoretic tools and metrics such as centrality, clustering, and path analysis, to quantify structural properties of the search, which traditional metrics cannot achieve. Such tools would help to shift the field from ad-hoc experimentation toward data-driven understanding and selection of GP variants. This paper adapts STNs to analyse and visualise the search dynamics of tree-based genetic programming. Our main contributions are to:

- extend the application of search trajectory networks (STNs) to tree-based genetic programming,
- provide STN models on both the genotype space (tree-based encoding), and the semantic space (phenotype or behaviour) in the context of symbolic regression,
- validate STNs as a robust analysis framework by demonstrating that their structure and dynamics capture the known behavioural differences between semantic and standard tree-based genetic programming,
- scale up STNs to real-world symbolic regression tasks by using an advanced clustering method for space partitioning.

2   Search Trajectory Networks

Search trajectory networks (STNs) [12] are a graph-based tool for visualising and analysing the dynamics of heuristic optimisation algorithms. In its basic form, the search trajectories model the decision space, where nodes represent individual visited solutions (genotypes). However, for analysing and visualising large search spaces, solutions should be grouped into *locations*, representing a partition of the search space. The partition method can either use search space reduction or clustering [12] or redefine nodes in another space, such as phenotype or behaviour spaces [13]. To define a graph-based model, we need to specify its nodes and edges. The general definitions are given below.

Representative solution. A solution to the optimisation problem at a given time step that represents the state of the search algorithm (e.g. best in the population in a given iteration, incumbent solution for single-point meta-heuristics).

Location. A non-empty subset of solutions resulting from a predefined partition of the search space.

Trajectory. Given a sequence of representative solutions in the order they are encountered during the search process, a search trajectory is defined as a sequence of locations resulting from replacing each solution with its corresponding location.

Nodes (N). The set of locations in a search trajectory of the search process being modelled.

Edges (E). Directed, connecting two consecutive nodes in the search trajectory. Edges are weighted with the number of times a transition between two given nodes occurred during the process of sampling and constructing the STN.

STN. Directed graph STN $= (N, E)$, with nodes N and edges E as defined above.

2.1   Related Work

The first application of STNs to genetic programming considered linear GP searching for abstract Boolean functions [5]. The aim was to gain insight into the evolutionary progression in the context of genotype-phenotype maps with large levels of neutrality (when mutations do not lead to phenotype changes). The study quantifies phenotypic traits such as genotypic abundance and Kolmogorov complexity, linking them to evolutionary paths visualised through STNs. It finds that complex phenotypes are rare and harder to evolve, while simpler ones are common and often act as evolutionary "stepping-stones". The notion of stepping-stones is further studied in the context of designing controllers for simple robot navigation tasks [10]. Stepping-stones loosely refer to intermediate solutions that aid in reaching better outcomes in evolutionary optimisation. To explore their role and make the concept more precise, the study uses STNs and proposes characterising stepping stones as nodes with high betweenness centrality. It compares map-elites [9] and genetic algorithms across different controller representations, showing strong evidence of stepping-stones in map-elites, especially with more direct and local representations.

STNs have also been used to explore the behaviour of Cartesian genetic programming (CGP) in symbolic regression [2]. The analysis reveals interesting dynamics, including "portal" nodes (a notion related to stepping stones discussed above) that enable sudden, advantageous changes in the search process. The article analysed a single regression task and explored STNs in both genotype and phenotype (behaviour) space. The size of the genotype space makes the modelling impractical, while STNs in behaviour space offer insights into functional evolution within CGP. The study, however, only considers a standard form for partitioning the behaviour space (hypercubes), limiting the model's potential for capturing convergence regions during the search process.

2.2   Proposed STN Models

We explore two types of models, operating in different spaces: Genotype and Phenotype STNs. The Phenotype STNs enable scaling up the model to realistic problem sizes. We consider two alternative (semantic) space partitioning methods for Phenotype STNs. Therefore, three models in total are described below.

Genotype STN. Nodes are unique genotypes (GP trees) flattened into their prefix notation. Edges represent transitions between representative genotypes in consecutive generations.

Phenotype STN. Nodes are locations in the *semantic* space, which in symbolic regression is given by vectors of real numbers indicating output values of GP individuals, across all the training observations. Edges represent the aggregation of transitions between the corresponding locations in the semantic space produced by consecutive representative genotypes in the search process. We explored two space partitioning methods in semantic space.

1. *Hypercube Partitioning.* This is the standard STN partitioning method for continuous space [12], where discrete hypercubes of a fixed size (0.1 in our study), define locations. These locations (hypercubes) represent the STN nodes.
2. *Clustering Partitioning.* We propose using an advanced clustering method HDBSCAN (Hierarchical Density-Based Spatial Clustering of Applications with Noise) [1], which excels in discovering clusters of varying densities and shapes and identifying noise (outliers). This is consistent with data collected from running stochastic optimisation algorithms and allows us to scale STN models to realistic problem sizes. Locations, and thus STN nodes, are clusters of varying densities and outliers as identified by HDBSCAN.

The STN models are constructed from multiple runs (10 in our study) for each GP variant and symbolic regression problem. When constructing a model, the visited locations and their transitions are aggregated into a single graph.

Note that some locations and transitions may appear multiple times during the search process. However, the graph model retains as nodes each unique location, and as edges each unique transition between visited locations. From the perspective of model interpretability, these overlaps are desirable as they represent regions of search convergence. Counters are maintained as attributes of the graph, indicating the frequency of occurrence of each (unique) node and edge.

3 Experimental Setup

Genetic Programming Variants. Our study compares the evolutionary process of standard tree-based genetic programming (stdGP) against a recently proposed variant of Geometric Semantic GP (GSGP) [8], designed to reduce bloat, the Semantic Learning algorithm with Inflate and Deflate Mutations (slimGSGP) [15]. GSGP focuses the evolutionary search directly on the programs' semantics (the behaviour or output), rather than solely on their syntax (their structural tree representation), which allows it to induce a unimodal error surface for supervised learning problems [15]. The semantics of a program are typically defined by the vector of its outputs when evaluated over a fixed set of training inputs (fitness cases). GSGP uses specially designed geometric semantic operators for crossover and mutation that are guaranteed to have a predictable, geometric effect on this semantic space. A common challenge for GSGP is that the resulting program trees can grow excessively large (known as bloat) compared to those evolved by standard GP. slimGSGP [15] mitigates the issue of code bloat by incorporating the *deflate* mutation, a geometric semantic operator engineered to produce syntactically smaller offspring than their progenitor programs, thereby promoting the evolution of more concise solutions. Our experiments use a publicly available Python implementation of both stdGP and slimGSGP[1].

Benchmarks. Consistent with our main contributions, the purpose of our study is not to advance the state of the art in GP's performance on specific problems, but rather to extend the use of STNs as a tool for contrasting tree-based GP variants to gain insight into *why* and *when* one approach outperforms another by inspecting their search trajectories. Therefore, we deliberately selected a small set of symbolic regression problems, including (i) two classic non-linear benchmark functions [7]: **Koza-1**: $x^4 + x^3 + x^2 + x$, and **Nguyen-5**: $\sin(x^2) \times \cos(x) - 1$, both using a set of 20 equidistant points in the interval $[-1, 1]$; and (ii) two real-world symbolic regression applications often used as benchmarks in the GP literature [14,15]: **Istanbul** (7 features and 536 samples) and Residential Buildings Sales Price (**RBSP**) (107 features and 372 samples). Instances were chosen as illustrative examples where the two GP variants show performance differences. The synthetic instances serve as a proof of concept, while the real-world instances scale up the approach to realistic settings.

---

[1] https://github.com/DALabNOVA/Slim.

Table 1. Parameter settings for stdGP and slimGSGP variants

Parameter name	Value
Parameters common to both stdGP and slimGSGP	
Population size	100 individuals
Termination criterion (Generations)	500
Initial tree depth	3 (Synthetic) / 6 (Real-world)
Maximum tree depth	9 (Synthetic) / 17 (Real-world)
Initialisation	ramped half-and-half
Selection	tournament size 2; elitism
Function set	$\{+, -, \times, \text{protected} \div\}$
Fitness function	RMSE
Parameters specific to stdGP	
Crossover probability	0.8
Mutation probability	0.2
Parameters specific to slimGSGP	
SLIM variant	SLIM+sig2
Inflate mutation rate	0.3
Deflate mutation rate	0.7
Mutation step upper bound	1.0
Mutation step distribution	uniform $U(0, 1)$

Parameter Settings. Ten independent runs were executed with each algorithm on every benchmark function. Fitness was determined as the root mean squared error (RMSE) between true outputs and those generated by an individual. Addition, multiplication, subtraction, and protected division (returns 1 if the absolute value of the divisor is less than 0.001, otherwise divides as usual) are used as binary function nodes on Koza-1 and the real-world regression problems, while functions sine and cosine are added as unary function nodes in the case of Nguyen-5. The best individual from each generation is always passed to the next generation (elitism). The synthetic instances have a single variable x, while the real-world instances contains as many variables as features in the corresponding dataset. Table 1 lists the parameters used for both GP variants and instance types (synthetic/real-world).

4   Results

4.1   Synthetic Instances

The box and whisker plot in Fig. 1 (top row) shows the distribution of best fitness values (root mean squared errors) across 10 independent runs obtained by slimGSGP and stdGP on the synthetic functions. stdGP shows a higher spread on both functions, with a wider spread for Koza-1, whereas slimGSGP's fitness values are more contained. For both functions, the stdGP best run outperforms the best run of slimGP.

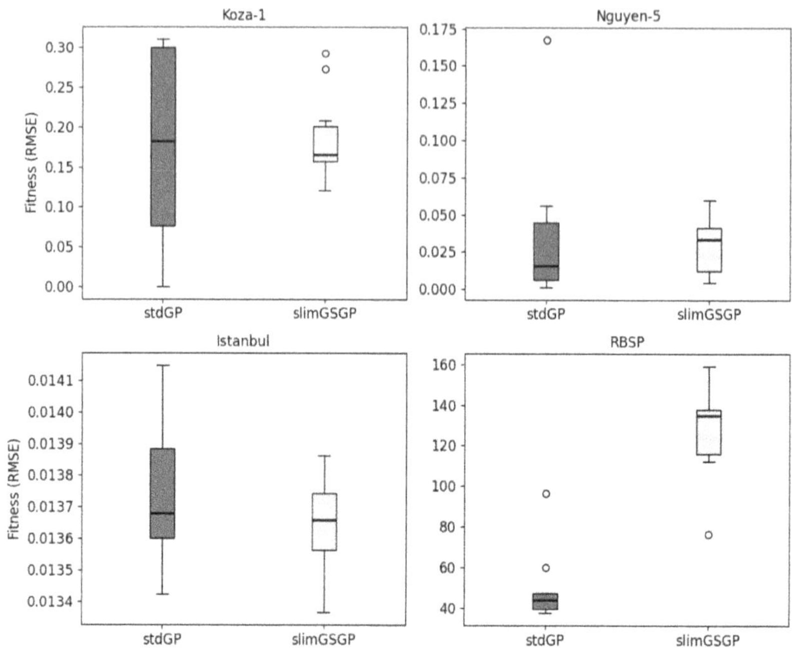

Fig. 1. Distribution of best fitness values (root mean squared errors) obtained by the two GP variants across 10 independent runs in all benchmark problems.

Network Metrics. To characterise the STN models, we report three basic network metrics that reflect their size, connectivity, and reachability of best solutions: the number of nodes, the number of connected components, and the average length of shortest paths from a start node to a best solution. Table 2 summarises these network metrics for the synthetic instances, the two GP variants and the three STN models based on genotypes and phenotypes partitioned with the hypercube and HDBSCAN clustering methods, respectively (see Sect. 2.2). The hypercube partitioning uses cubes of dimension 0.1, that is, output vectors within 0.1 precision in all dimensions belong to the same STN node. HDBSCAN clustering has a fundamental parameter controlling the minimum size of clusters; clusters of sizes below this value are considered noise (outliers). Since our data combines 10 runs to construct STNs, we set up a minimum cluster size of 5, because meaningful groupings could join locations (outputs) visited by at least half of the runs. We conducted experiments with values from 2 to 8 for this parameter, and found that minimum cluster sizes of 4 or 5 provide compact and interpretable STN models. We decided then to use a minimum cluster size of 5 in this study. In the HBDSCAN STN model, nodes are the clusters and outliers obtained after running the HDBSCAN method on the output vectors.

Table 2. Synthetic instances: network metrics across GP variants and STN models.

variant	model	#nodes	#components	path-length
		Koza-1		
slimGSGP	Genotype	193	5	20
	Hypercube	187	4	19
	HDBSCAN	50	1	2.6
stdGP	Genotype	48	1	6.12
	Hypercube	22	1	4.75
	HDBSCAN	11	1	2.75
		Nugyen-5		
slimGSGP	Genotype	206	8	21.5
	Hypercube	163	4	12
	HDBSCAN	62	1	3.78
stdGP	Genotype	124	5	12
	Hypercube	49	1	18
	HDBSCAN	53	1	9

The main observations from Table 2 are the following. For both GP variants, the Genotype STNs contain a higher number of nodes, components and path lengths than the phenotype (semantic) models with hypercube and HDBSCAN partitioning. The semantic models provide a single connected component in all cases except for the slimGP hypercube model on the Nugyen-5 instance, which has 4 components. Notice that a lower number of components is desirable as it facilitates the analysis of regions of convergence. The HDBSCAN models are generally more compact than the hypercube ones.

When observing the two GP variants, slimGP produces larger models with a higher number of nodes and path lengths for all STN models as compared with stdGP. This indicates that slimGP traverses a larger number of solutions in both the genotype and semantic spaces. Notice, however, that a larger number of visited solutions does not necessarily imply improved performance (see boxplots in Fig. 1). slimGP may be conducting smaller search "steps" on both spaces and approaching good solutions at a slower pace than stdGP. The next section will visually inspect STN models to capture features of the search process difficult to infer from network metrics.

Network Visualisation. Network models can be visualised using node-edge diagrams, where nodes are shapes, and edges are lines or curves. The placement of nodes and edges (graph layout) as well as their colours and sizes can be used to highlight relevant network features. When analysing the behaviour of algorithms, network visualisation can bring new insights into their dynamics, including connectivity patterns, convergence regions, and the reachability of best solutions.

To visualise STN models, we identify four types of nodes: (i) Start (■): nodes with solutions at the beginning of each of the independent runs, (ii) Medium (○): nodes containing intermediate solutions in the search process, (iii) End (▲): nodes with solutions at the end of trajectories with worse fitness than the best found by the GP variant, and (iv) Best (●): nodes containing solutions with the best fitness found by the GP variant across its 10 runs. These four node types are assigned different node shapes and colours (see Figs. 2 and 3). Because tree size is an important characteristic of tree-based GP, it is used to determine the size of the STN nodes. When a node represents a cluster of solutions, its size is set to the median tree size of the solutions in that cluster. Note that node size could alternatively be based on other features, such as sampling frequency, incoming weighted degree (strength), cluster (or partition) size, centrality metrics, and so on. This choice will depend on the goal of the analysis. The width of edges represents their sampling frequency, that is, how many times edges between pairs of nodes are traversed in the 10 runs used to build the networks. Two network layouts are used, a force-directed graph layout [3] called Kamada-Kawai [6], which facilitates the visual identification of separated components (see Fig. 2), and a custom layout where the y coordinate reflects fitness and the x coordinate follows a force-directed graph layout (stress majorisation [4]). Thus, while the x-axis carries no semantic meaning—serving only to highlight connectivity structure via a force-directed layout—the y-axis explicitly illustrates the search convergence toward solutions with increasingly better fitness (see Fig. 3).

Figure 2 shows the three STN models for function Koza-1 using the Kamada-Kawai graph layout. The plots show clear differences between the evolutionary processes of each GP variant, and confirm that stdGP produces smaller STN models (lower number of nodes) than slimGP. For slimGP, the Genotype and Hypercube STN models show several components and little trajectory overlap, only on some starting green squares; while the HDBSCAN model shows a single connected component, where all trajectories are attracted towards the same region of the semantic space. stdGP yields more compact models, with a single component and overlapping nodes (with high degree) in all cases. Thicker edges indicate frequently traversed regions of the syntax and semantic spaces, respectively.

The custom fitness layout shown in Fig. 3 gives us a different perspective. slimGP shows longer and gradual path lengths as compared to the shorter paths with more pronounced steps by stdGP. This observation is related to the way stdGP constructs new solutions. By using crossover and mutation operators in the syntax space, stdGP produces solutions that are noticeably different from their parents. In consequence, when a better solution is found, it generally makes a rather significant jump in terms of fitness. slimGSGP, on the other hand, produces new solutions by tweaking existing solutions rather than by introducing

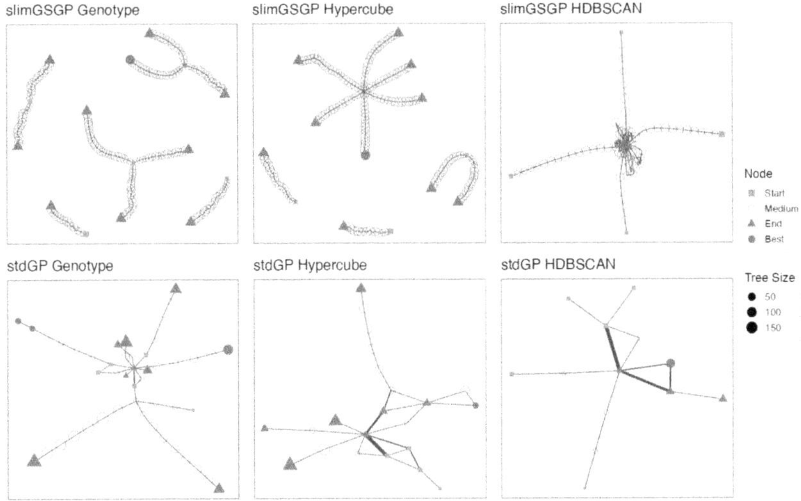

Fig. 2. Koza-1 STNs for the two GP variants using the Kamada-Kawai graph layout.

larger changes. Having passed the output of another tree through a sigmoid function and scaled it, it adds the result to the existing tree. slimGSGP also simplifies the ever-growing solutions by factoring and cancelling out opposite terms. Hence, slimGSGP converges slowly but steadily. This could be disadvantageous: slimGSGP never produced a perfect fit on the Koza-1 function, whereas stdGP managed to do so twice. As the HDBSCAN STN plot for slimGSGP shows, many trajectories are trapped in the central blue triangle below 0.25 fitness value, where several thick edges intersect, and only some can escape and reach the best node for this variant, i.e. the red circle, which is of inferior fitness than the best one found by stdGP. This is not the case for the Nguyen-5 function (Fig. 4), where both GP variants reach good solutions, but slimGSGP shows more consistent convergence toward best solutions (red circles); the convergence is clearly visually captured by the HDBSCAN STN in Fig. 4.

STNs can also be used to shed light on the problem of bloat during evolutionary search. The size of nodes in Figs. 2, 3 and 4 is proportional to the size of the representative solution's syntax tree. Since different genotypes generally map to a single phenotype, node sizes in the semantic models are calculated as the median of the genotype sizes mapped onto it. We can notice that both algorithms suffer from bloat. However, slimGSGP's nodes tend to grow much faster, whereas stdGP's nodes are smaller in the beginning. It seems that stdGP has smaller solutions in general, but during some runs, it generates solutions larger than slimGSGP. It must be stressed that slimGSGP has a built-in simplification operator (deflate mutation), and yet it still has difficulty dealing with bloat.

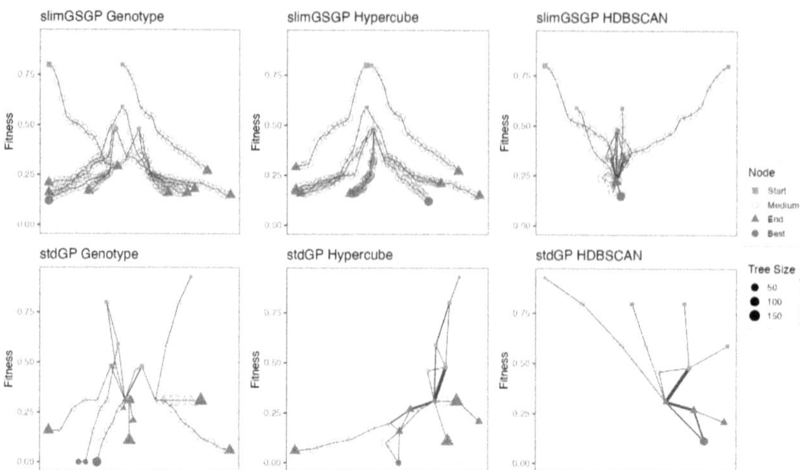

Fig. 3. Koza-1 STNs for the two GP variants using the custom fitness graph layout.

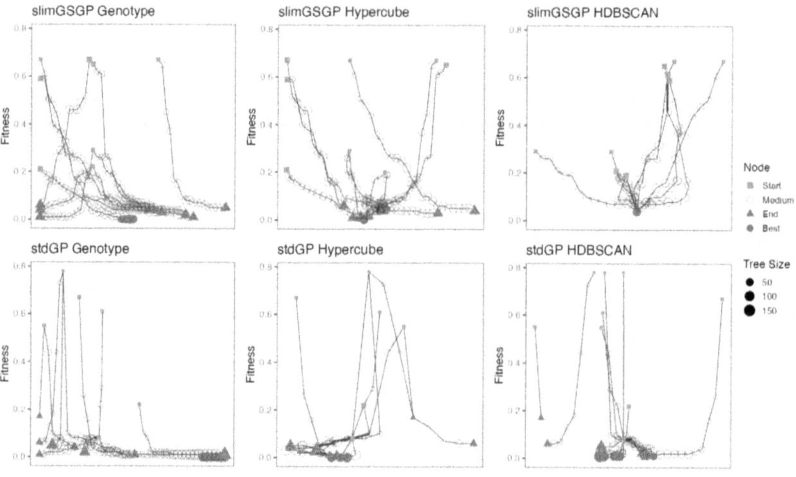

Fig. 4. Nguyen-5 STNs for the two GP variants using the custom fitness graph layout.

4.2 Real-World Instances

The box and whisker plots in Fig. 1 (bottom row) show the distribution of best fitness values (root mean squared errors) across 10 independent runs obtained by slimGSGP and stdGP on the two real-world instances. Notice the widely different range of fitness values on both datasets. The Istanbul RMSE values are all lower than 0.015, while they are 3 to 4 orders of magnitude higher for RBSP. On the Istanbul instance, both GP variants show similar performance with slimGSGP producing slightly better solutions and lower variability; the situation is reversed on the RBSP dataset, where stdGP shows significantly better performance and stability.

Network Metrics. Table 3 summarises the network metrics for the real-world instances, the two GP variants and the three STN models. The same partition parameters as for the synthetic instances are used, that is, hypercubes of dimension 0.1, and a minimum cluster size of 5. As expected, the models are larger for the real-world instance as compared to the synthetic instances. However, the same trend is observed; Genotype STNs contain a higher number of nodes, components and path lengths as compared with the hypercube and HDBSCAN semantic STNs. The semantic models have a low number of connected components in all cases except for slimGP on RBSP, where all models feature 10 components. This is the maximum possible number of components, one for each run, indicating that despite partitioning, there is no overlap in the slimGSGP search trajectories. When contrasting the models for the two GP variants, they differ across the two instances. For the Istanbul benchmark, slimGSGP models are smaller (in terms of the number of nodes) than stdGP models, while the path lengths to best solutions (red circles) and the number of components are similar for both GP variants. The RBSP models show the opposite trend, with smaller models and higher connectivity (with one or two connected components) for stdGP. This indicates that for RBSP, the trajectories of stdGP converge or overlap into some common regions of the search space, while this is not the case for slimGSGP, where independent runs traverse completely separated regions in both the syntactic and semantic spaces, featuring 10 separated components. This indicates a lack of convergence of slimGSGP in this instance. To gain a deeper insight into the contrasting behaviour of the two GP variants, the next section visualises the STN models.

Network Visualisation. Due to space constraints, we show figures using the custom fitness layout only. Figure 5 shows the STN models for the two GP variants on the Istanbul instance. The genotype STNs show search overlap only at some start nodes (green squares), indicating that search trajectories diverge in the syntactic space as expected on this large problem. The semantic models show convergence, especially the hypercube STNs. For slimGSGP, all trajectories in the hypercube model converge to the same region of the semantic space. There is overlap as well on the stdGP hypercube model, but it is more prominent at the early stages of the search process, whereas the end of trajectories (blue triangles and red circles) visit different regions. The HDBSCAN models are more divergent, showing overlaps only at early stages of the search process. Figure 5 also confirms that both GP variants reach best solutions (red circles) of similar fitness values, with slimGSGP showing more consistent convergence.

Finally, Fig. 6 shows the STNs for the hardest RBSP instance using the fitness graph layout. For slimGSGP, the three STN models show divergent search trajectories with no overlap. The HDBSCAN models for slimGSGP are smaller than the hypercube models, but aggregate solutions in the same trajectory rather

Table 3. Real-world instances: network metrics across GP variants and STN models.

variant	model	#nodes	#components	path-length
			Istanbul	
slimGSGP	Genotype	358	7	20.33
	Hypercube	89	1	8.86
	HDBSCAN	151	4	8.38
stdGP	Genotype	508	3	33.09
	Hypercube	132	1	7.36
	HDBSCAN	156	1	9.63
			RBSP	
slimGSGP	Genotype	492	10	47.5
	Hypercube	473	10	38.5
	HDBSCAN	138	10	3
stdGP	Genotype	380	2	49.17
	Hypercube	358	1	46.71
	HDBSCAN	111	1	6.17

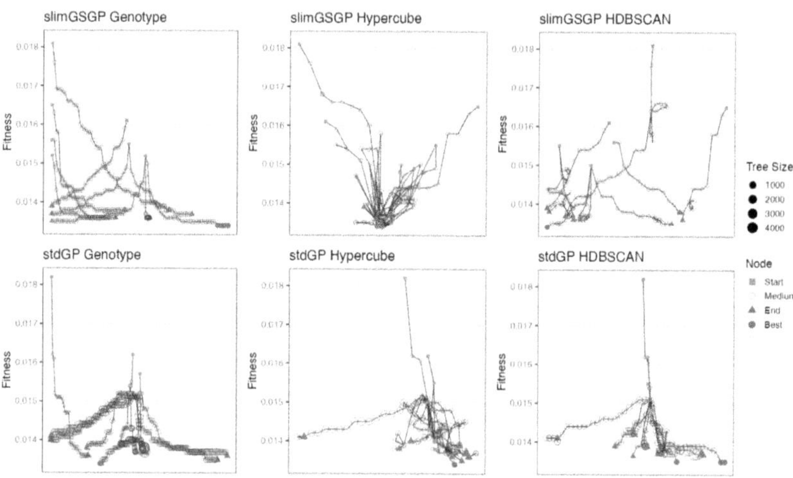

Fig. 5. Istanbul STNs for the two GP variants using the custom fitness graph layout.

than across trajectories, so the divergent behaviour is still present. Notice the little incremental fitness improvement achieved by slimGSGP in this problem; the final solutions hardly improve the initial randomly generated ones, and the trajectories look flat in their fitness values. This indicates that slimGSGP slow convergence over the desired unimodal error surface is not always advantageous. For the RBSP instances, the more drastic program changes induced by stdGP, especially early in the search process, allow it to reach much better solutions.

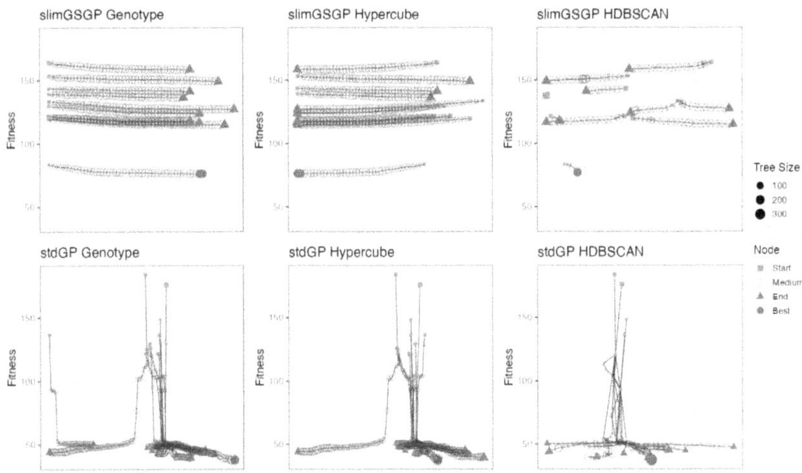

Fig. 6. RBSP STNs for the two GP variants using the custom fitness graph layout.

5   Conclusion

We adapted search trajectory networks (STNs) to analyse and visualise the evolution of tree-based genetic programming (GP). We chose synthetic and real-world symbolic regression benchmarks that show performance differences between standard GP and geometric semantic GP (GSGP). STNs on both the genotype (syntactic) and phenotype (semantic) spaces were constructed, analysed and visualised. Genotype STNs could be used to analyse small problems, but do not scale up well to realistic problem sizes due to the immensely large syntactic search spaces. However, semantic STNs showed promise for analysing and visualising the GP convergence behaviour and reachability of good solutions in real-world settings. Our analysis shows that the controlled convergence over a unimodal error surface induced by GSGP is not always desirable; sometimes, the prominent program changes of standard GP allow reaching better regions of the syntactic and semantic spaces.

The search space partitioning methods used, hypercube and clustering with HDBSCAN, show complementary advantages; one is not always preferred over the other, and more research should be devoted to partitioning methods that improve the usability and interpretability of STNs. We argue, however, that semantic STNs are a promising tool, still in development, which can assist researchers and practitioners in understanding and designing effective GP algorithms.

This study opens several avenues for further exploiting STNs in the context of GP. One research direction is to use STNs to compare the evolutionary processes of different GP variants, such as Cartesian GP, gene expression programming, and linear GP. Another is to extend this study beyond symbolic regression by employing STNs in other GP applications, such as classification tasks or problems in which GP acts as a hyper-heuristic.

Acknowledgements. This research has been supported by the Croatian Science Foundation under project MOBDOK-2023-1584, and the UK Leverhulme Trust under project RPG-2025-185.

References

1. Campello, R.J., Moulavi, D., Zimek, A., Sander, J.: Hierarchical density estimates for data clustering, visualization, and outlier detection. ACM Trans. Knowl. Disc. Data (TKDD) **10**(1), 5 (2015)
2. De La Torre, C., Cussat-Blanc, S., Wilson, D., Lavinas, Y.: On search trajectory networks for graph genetic programming. In: Genetic and Evolutionary Computation Conference Companion, pp. 1681–1685. ACM, Melbourne VIC Australia (2024)
3. Fruchterman, T.M.J., Reingold, E.M.: Graph drawing by force-directed placement. Softw. Pract. Experience **21**(11), 1129–1164 (1991)
4. Gansner, E.R., Koren, Y., North, S.: Graph drawing by stress majorization. In: Graph Drawing, pp. 239–250. Springer, Berlin, Heidelberg (2005)
5. Hu, T., Ochoa, G., Banzhaf, W.: Phenotype search trajectory networks for linear genetic programming. In: European Conference on Genetic Programming, EuroGP 2023. LNCS, vol. 13986, pp. 52–67. Springer (2023)
6. Kamada, T., Kawai, S.: An algorithm for drawing general undirected graphs. Inf. Process. Lett. **31**(1), 7–15 (1989)
7. McDermott, J., et al.: Genetic programming needs better benchmarks. In: Genetic and Evolutionary Computation Conference, GECCO, pp. 791–798. ACM, New York, NY, USA (2012). https://doi.org/10.1145/2330163.2330273
8. Moraglio, A., Krawiec, K., Johnson, C.G.: Geometric semantic genetic programming. In: Parallel Problem Solving from Nature - PPSN XII, pp. 21–31. Berlin, Heidelberg (2012)
9. Mouret, J.B., Clune, J.: Illuminating search spaces by mapping elites (2015). https://doi.org/10.48550/arXiv.1504.04909
10. Nadizar, G., Rusin, F., Medvet, E., Ochoa, G.: The role of stepping stones in map-elites: Insights from search trajectory networks. In: European Conference on Genetic Programming, EuroGP 2025, pp. 224–239. Springer Nature Switzerland, Cham (2025)
11. Ochoa, G., Malan, K.M., Blum, C.: Search trajectory networks of population-based algorithms in continuous spaces. In: European Conference on Applications of Evolutionary Computation, EvoApps, pp. 70–85. Springer, Cham (2020)
12. Ochoa, G., Malan, K.M., Blum, C.: Search trajectory networks: a tool for analysing and visualising the behaviour of metaheuristics. Appl. Soft Comput. **109**, 107492 (2021)
13. Sarti, S., Adair, J., Ochoa, G.: Neuroevolution trajectory networks of the behaviour space. In: European Conference on Applications of Evolutionary Computation, EvoApps. LNCS, vol. 13224, pp. 685–703. Springer (2022)

14. Tsanas, A., Xifara, A.: Accurate quantitative estimation of energy performance of residential buildings using statistical machine learning tools. Energy Build. **49**, 560–567 (2012)
15. Vanneschi, L.: SLIM GSGP: the non-bloating geometric semantic genetic programming. In: European Conference on Genetic Programming, EuroGP, pp. 125–141. Springer-Verlag, Berlin, Heidelberg (2024)

A Hybrid LLM-Coevolution Framework to Generate Abusive Tax Strategies

Joy Sera Bhattacharya$^{(\boxtimes)}$, Erik Hemberg , and Una-May O'Reilly

MIT CSAIL, ALFA Group, Cambridge, USA
{jbsera,ehemberg,unamay}@mit.edu
http://alfagroup.csail.mit.edu/

Abstract. We present a new framework to automatically generate abusive tax strategies. The first design of its kind, it combines a competitive coevolutionary algorithm, `CompCoevAlg`, and a large language model (`LLM`). Its `CompCoevAlg` iteratively adapts a population of strategies towards abuse and concealment by adversarially competing it against a population of audit patterns that seek to expose abusive strategies. While, per convention, the `CompCoevAlg` oversees selection, replacement, and fitness evaluation, it uses the `LLM` to employ natural-language representations for strategies and audit patterns. The `LLM` also performs similarity-based matching between audit patterns and transactions in each strategy and it mutates the transactions and pattern at the natural language level. We demonstrate the framework by using the Installment Bogus Optional Basis (iBOB) scheme. It can generate variants of a prototypical iBOB strategy. In addition to its novel integration of an `LLM`, results reveal the dual value of combining an `LLM` and Evolutionary Algorithm. Genetic operators and a search approach can make up for the `LLM`'s shallow knowledge of a problem domain, while the `LLM` can eliminate the need to laboriously translate and encode the semantic rules and behavior of the problem domain.

Keywords: coevolution · tax · evolutionary algorithm

1 Introduction

"Taxes are the lifeblood of government and no taxpayer should be permitted to escape the payment of his just share of the burden of contributing thereto." – Arthur Vanderbilt [1]

In this paper we focus on *abusive* tax avoidance strategies. In general, tax avoidance strategies exploit tax *loopholes*. A *loophole* consists of provisions or ambiguities in tax regulations that allow a tax liability to be lowered. Loopholes can be either intentionally created by legislators. [6]. Examples of intended loopholes include allowances for certain deductions such as mortgage interest and adjustments to income based on economically substantive actions. Alternatively, legislatively-unintended loopholes can be surreptitiously discovered by abusers.

L. Manzoni et al. (Eds.): EuroGP 2026, LNCS 16521, pp. 122–139, 2026.
https://doi.org/10.1007/978-3-032-23005-8_8

A tax scheme, i.e. cluster of abusive strategies, typically targets a particular set of tax regulations. For example, the Installment Bogus Optional Basis (iBOB) scheme exploits loopholes in the United States Internal Revenue Code (IRC) in §754 which allow optional basis adjustment in various circumstances [8]. A prototypical example of iBOB involves an initial scenario with a focal taxpayer, Mr. Jones, and three other financial entities related by partnership to Mr. Jones, each holding assorted assets one of which is a hotel. Mr. Brown, another actor is outside the network. Mr. Jones uses a strategy consisting of a sequence of transactions before selling the hotel to Mr. Brown, to qualify himself for an optional, but illegitimate basis adjustment that, abusively, avoids taxation.

For any set of tax regulations with loopholes, it is useful to collect a wide variety of strategies that start from the initial scenario of the scheme's prototype's and end with some taxable action, e.g. the sale of the hotel to Mr. Brown in the case of iBOB, while in between they consist of a sequence of financial transactions that lead to significant tax avoidance and, possibly, audit detection. These additional strategies assist legislators in amending tax laws to align more accurately with their intent or in issuing guidance to taxpayers, to steer them away from what is possible, but not legitimate. From an abuser's perspective, ideal strategies are *effective*, i.e. they avoid paying all or most of tax that would be expected, and covert – they are concealed from audits.

Prior work, namely the STEALTH project, has formulated the task of strategy collection as a search and optimization problem solved by a coevolutionary algorithm [9]. The project integrates a significant set of relevant tax domain semantics and actions into its evolutionary algorithm using Grammatical Evolution (GE) [17]. GE uses a grammar, a problem specific decoder, and an integer-genome. In STEALTH, GE bridges the concepts and language symbols of the tax domain (actors, actions, relationship) to a representation that could undergo key evolutionary operations such as blind variation, execution, and fitness evaluation. STEALTH's grammar expresses domain knowledge by its production rules. STEALTH's decoder converts the integer-genome, using the grammar, into a transaction sequence that constitutes a strategy. The grammar ensures a syntactically correct transaction sequence that later needs to be checked for semantic feasibility. This check rejects sequences with errors such as an asset being transferred by an actor who did not own it.

Coincidentally, a contemporary Large Language Model (LLM) can be expected to have tax knowledge implicitly encoded in its parameters due to that domain's knowledge being present in its training dataset. This domain knowledge could potentially reduce the amount of explicit domain reasoning and translational design and coding required by STEALTH. We therefore ask:

– Can an LLM integrated with a coevolutionary algorithm offer an alternative approach to collecting abusive tax strategies?

The hybrid LLM and coevolutionary algorithm framework we present here, appearing almost a decade after STEALTH, specifically investigates:

RQ 1. Could the integer-genome, grammar and decoder of GE be eliminated? Could the evolutionary algorithm directly generate natural-language based transaction sequences, continuing to execute and evaluate them as strategies? Using an LLM, how will syntactic rules be enforced without a grammar? Can semantics be enforced at the time of transaction sequence generation, eliminating the need to later check them?

RQ 2. Prior work has used an LLM as a variation operator which offers domain-informed adaptation. Will this approach work in the tax domain? Can transaction sequences (or audit information) be effectively mutated by an LLM?

Additional research questions focus on auditing. Auditing acts as a competitive force that keeps strategy search from converging. Practically, it provides insights to regulators. The STEALTH audit genome has a gene for each of a set of attributes of a transaction sequence that might be observed when the sequence is executed. The gene encodes a weight that prioritizes its attribute and the weighted sum of observed attributes composes an audit score. The objective of the audit search is to optimize the weights to identify highly abusive strategies while ignoring compliant strategies or strategies with lower abuse. A limitation of the STEALTH approach is that it restricts the genome search space to designated attributes, potentially missing unanticipated audit attributes. We ask:

RQ 3. Given the unbounded nature of natural language, is it feasible to develop similarly unrestricted, natural-language-based audit patterns to advert missing unanticipated attributes?

RQ 4. Is it feasible to adapt and compare audit patterns expressed in natural language? Can it match them to tax abuse strategies that are themselves LLM-generated?

Figure 1 presents an overview of our framework. The framework includes a tax calculator that executes a transaction sequence and uses tax regulation logic to calculate a strategy's tax liability. This component's coding has been co-piloted with an LLM. The framework includes a `CompCoevAlg` that uses language representations of tax transactions sequences (a.k.a strategies), scenarios, and patterns without grammar-based encoding and decoding. Without code, it uses an LLM to create initial populations of strategies and audit patterns, to mutate strategies and audit patterns, and to even match audit patterns to the transactions of strategies. It conventionally iteratively directs fitness evaluation, selection, and replacement.

The paper's contributions are:

(RQ 1.) The framework introduces a natural language *prompt template* that enforces syntactic constraints on any transaction or strategy, while it uses iteration and interim calculations to combine the transactions into feasible, semantically-correct sequences. The framework tasks the LLM with generating one transaction at a time, appending the new transaction to the current transaction sequence. To ensure overall feasibility, at each iteration, it executes the current transaction sequence and updates an ownership network which it feeds into the prompt on the next iteration. Simple task formulation guides the LLM to behave accurately.

(RQ 2.) It introduces a pair of natural language prompt templates for mutation. Each template describes the task in specific detail, specifies constraints, and includes positive and negative examples.

(RQ 3.) To generate audit patterns, we use a natural language prompt template that passes the current strategies to the LLM and instructs it to act as a tax auditor that analyzes transaction data in a tax network. The prompt tasks the LLM to identify general patterns and *suspicious activity* within the transaction data. *Note, the prompt template does not define a suspicious pattern.* This gives the LLM (linguistic) freedom to define the structure and content of each pattern. The LLM responds with patterns that describes attributes of the examples *eliminating any need to pre-specify them.*

(RQ 4.) To match audit patterns to a transaction, the framework exploits to the LLM's natural language input representation and the LLM's built-in similarity-based inference. A prompt template simply tasks the LLM to label a given transaction with the matching patterns, assigning at least one pattern. The accuracy of the matching arises from the clearly defined structure of a transaction and showing the LLM transactions when patterns are initially generated.

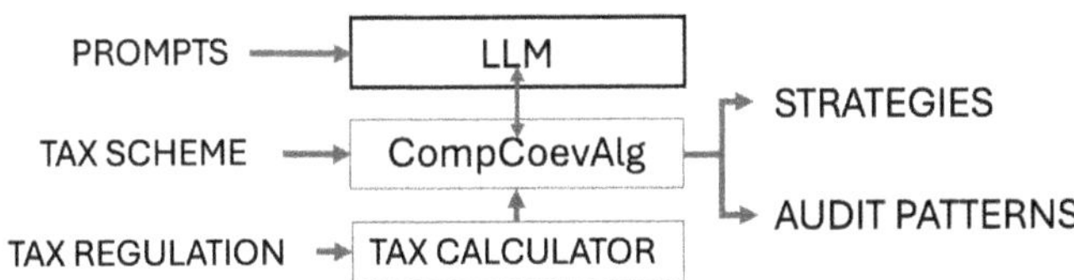

Fig. 1. Overview, given a tax scheme, regulation and prompts tax avoidance strategies and audit patterns are co-evolved with LLM operators and a tax calculator.

We proceed with background and related work (Sect. 2), methodology (Sect. 3), empirical experiments (Sect. 4), and conclusions (Sect. 5).

2 Background

2.1 The iBOB Tax Avoidance Scheme

The abusive Installment Bogus Optional Basis (iBOB) scheme, i.e. cluster of strategies, takes advantage of the IRC Section §754 election (§754 election) to artificially step up the inside basis of partnership assets to eliminate tax liability [22]. A prototypical example of an iBOB strategy is illustrated in Fig. 2.

Initially, Mr. Jones has a 99% partnership in both Jones Co. and Family Trust. Jones Co., in turn, has a 99% partnership in New Co. New Co. owns a hotel with a Fair Market Value (FMV) of $200 and an inside basis of $120 that they wish to sell to Mr. Brown (Fig. 2a).

If New Co. were to sell the hotel directly to Mr. Brown, Mr. Jones would be taxed on the difference between the hotel's FMV and his share of the inside

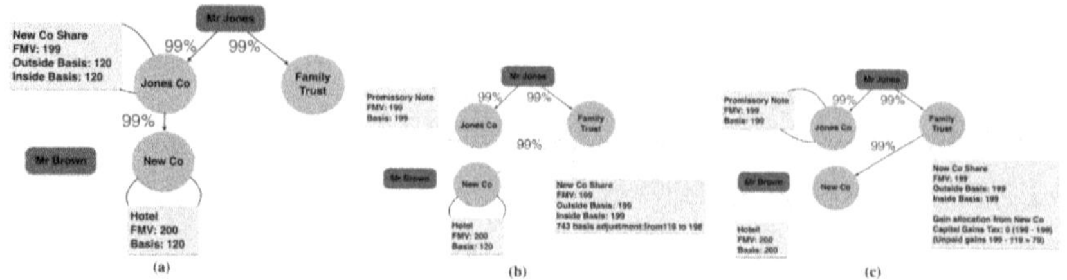

Fig. 2. Tax Network with iBOB strategy [9]. (a) Mr. Jones's tiered ownership of a hotel with \$200 FMV and \$120 basis; (b) Family Trust buys New Co.'s interest for a \$199 promissory note and makes a §754 election, stepping the hotel's inside basis to \$199; (c) New Co. sells the hotel for \$200, leaving virtually no taxable gain.

basis. So, his taxable gain would be around \$79 (\$200 FMV×0.99 share−\$120 inside basis, Fig. 2a).

To avoid paying this tax, Family Trust can purchase Jones Co.,'s partnership interest in New Co. using a promissory note with a face value of \$199 (Fig. 2b) and a §754 election. Importantly, because the promissory note is essentially a loan within Mr. Jones's network of controlled entities, it will never actually be repaid. With the §754 election, this transfer triggers a basis adjustment under IRC §743(b), stepping up the inside basis of the hotel to \$199, nearly equal to the hotel's fair market value.

Following the basis step-up, New Co. sells the hotel to Mr. Brown for \$200 (Fig. 2c). Since the inside basis has been increased to nearly match the sales price, the taxable gain is essentially eliminated for Mr. Jones.

The initial state of the iBOB scenario is grounded with actor names, assets, and appropriate types of transactions that transfer assets by selling them. The framework describes this to the **LLM** when it constructs two initial populations –strategies, and patterns. One actor is the focal point of the tax scheme, in iBOB this is Mr. Jones.

2.2   Related Work

Evolutionary computation algorithms and LLMs have been used for many different applications, e.g. program synthesis, symbolic regressions and aesthetics [13, 13, 15, 16, 23]. However, never with the combination of natural language representation, competitive coevolution and LLM variation operators in the tax domain.

Previous studies on tax planning and evolutionary modeling illustrate how taxpayers and auditors continuously adjust their strategies in response to one another. For instance, agent-based models and genetic algorithms have shown how avoidance techniques evolve [4, 9, 20, 21]. In parallel, researchers in natural language processing (NLP) have attempted to convert tax statutes into machine-readable formats, encountering numerous challenges along the way [18]. Modern large language models show considerable potential, yet they frequently misread

Table 1. Symbol Table

Symbol	Description
$s \in S$	strategy s in population S
τ	a transaction
$s = (\tau_1 \dots \tau_T)$	strategy is sequence of transactions
$d \in D$	audit pattern d in population D
$M_{d,s}$ **Algorithm 2**	set of audit patterns $d \in D$ matching some strategy s
$M_{\tau,s}$ **Algorithm 3**	set of transactions τ in strategy s that match some audit pattern d
$M_{d,S}$ **Algorithm 3**	set of strategies $s \in S$ that match some audit pattern d
Strategy Fitness	
`Taxable`(s)	tax liability of s after all transactions and elections
`AuditScore`(s)	detection score assigned by audit pattern population D to a strategy s based on matching, see Algorithm 2
Fitness(s)	C_1`Taxable`(s) $+ C_2$`AuditScore`(s)
Audit Pattern Fitness	
`MATCH?`(d,τ)	predicate: audit pattern d matches transaction τ
`Frequency`(d)	how frequently audit pattern d matches S
`TaxableScore`(d)	taxable liability identified by the audit pattern d because of its matches, see Algorithm 3
Fitness(d)	C_3`Frequency`(d) $+ C_4$`TaxableScore`(d)
`CompCoevAlg` **Parameters**	
C_M	Number of LLM mutations
C_T	Number of generations
C_t	Tournament size
C_e	Elite size
C_{s_S}	Strategy population size
C_{s_D}	Pattern population size
ρ_{G_S}	Strategy generation prompt template
ρ_{G_D}	Pattern generation prompt template
ρ_{E_D}	Pattern match prompt template (Evaluate)
ρ_{M_S}	Strategy mutation prompt template
ρ_{M_D}	Pattern mutation prompt template

statutory language and, at times, hallucinate legal provisions [2,3,5,10,24]. To curb these errors, researchers have tested role-based prompting techniques and cycles of iterative refinement [11,12,14]. Together, these approaches highlight the need for computational methods to model tax avoidance.

We extend earlier agent-based and evolutionary work on tax planning [9,21] by replacing hand-coded agents with a `CompCoevAlg` that co-adapts transaction sequences and patterns that model them being exposed. To address findings that LLMs often misread statutes [2,5,18], we developed a tax calculator ourselves. Unlike one-shot detection frameworks [7], the competition with conflicting objectives allows audit patterns to adjust as abuse strategies change. The evolutionary, generation-based co-adaptive approach is similar in spirit to iterative refinement [12,14] but it employs two populations with opposing goals.

3   Methodology

The framework's goal is to generate a cluster of abusive strategies around a set of regulations. We now describe the preparation (Fig. 2) and steps of the framework's method (Fig. 3 and Algorithm 1). We ground some descriptions of the method in terms of how they would be used to generate variants of the iBOB scheme for §754 described in Sect. 2.1. All notations used are defined in Table 1.

Algorithm 1 . CompCoevAlg (C: parameters (Tab. 1), ρ: prompt templates (Tab. 1)), LLM

1: $S = \texttt{initialize}(\rho_{G_S})$ ▷ Initialize Strategies
2: $D = \texttt{initialize}(S, \rho_{G_D})$ ▷ Initialize Audit Patterns
3: $F_S = \texttt{evaluate}(S, D, C_E)$ ▷ Evaluate initial Strategies with Tax Calculator
4: $F_D = \texttt{evaluate}(S, D, C_E, \rho_{E_D})$ ▷ Evaluate initial Audit Patterns with LLM
5: **for** $t \in [1, \ldots, C_T]$ **do**
6:     $S' = \texttt{selection}(S, C_t, C_e)$ ▷ Select Elite Strategies
7:     $D' = \texttt{selection}(D, C_t, C_e)$ ▷ Select Elite Audit Patterns
8:     $S' = \texttt{mutate}(S', \rho_{M_S}, C_M)$ ▷ Mutate Strategies with LLM
9:     $D' = \texttt{mutate}(D', \rho_{M_D}, C_M)$ ▷ Mutate Audit Patterns with LLM
10:    $F_S = \texttt{evaluate}(S', D', C_E)$ ▷ Evaluate Strategies with Tax Calculator
11:    $F_D = \texttt{evaluate}(S', D', C_E, \rho_{E_D})$ ▷ Evaluate Audit Patterns with LLM
12:    $S = \texttt{replace}(S', C_{s_S})$ ▷ Replace worst Strategies
13:    $D = \texttt{replace}(D', C_{s_D})$ ▷ Replace worst Audit Patterns
14: **Return** S, D ▷ Return Strategies and Audit Patterns

Preparations. Prior to using the framework, a set of regulations with a known tax scheme is selected. An initial scenario from a prototypical strategy of the scheme is formulated. It consists of ENTITIES, i.e. people and businesses involved in transactions, ASSETS i.e. what is bought and sold, and the initial OWNERSHIP NETWORK between ENTITIES and ASSETS. Additionally, a high-level description of scenario-appropriate TRANSACTIONS is formulated. E.g, in the initial scenario of the iBOB prototype, see Fig. 2a), the hotel (with basis $120, FMV $200) is initially owned by New Co. Each entity holds $200 in cash and a promissory note worth $200. The description is:
ENTITIES = Mr. Jones, Jones Co., New Co., Family Trust, Mr. Brown
ASSETS = Hotel, Cash, Promissory Note, Partnership

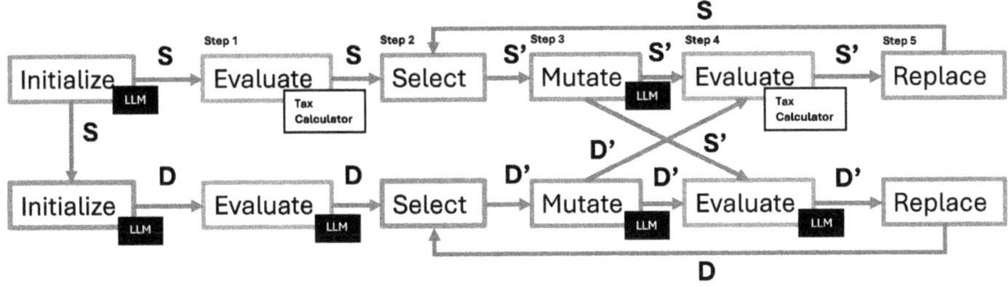

Fig. 3. Overview of framework `CompCoevAlg`. Tax avoidance strategies S (red) are co-adapted with audit patterns D (blue) with the use of LLM based operators and an evaluator (green). (Color figure online)

OWNERSHIP NETWORK = ((Mr. Jones,Jones Co.,0.99),(Mr. Jones,Family Trust,0.99), (Jones Co.,New Co.,0.99))

A transaction $\tau = (E_s, A_s, E_r, A_r, \text{ELECTION754?})$ describes the transfer of ASSET A_s from ENTITY E_s to ENTITY E_r, in exchange for ASSET A_r. A Boolean variable ELECTION754? indicates if a basis adjustment was elected.

The steps of the algorithm are:

Initialization A problem-specific calculator is implemented to calculate the tax liability of a strategy after its transactions are modeled. In the iBOB example, this calculator processes the transactions starting from the initial OWNERSHIP NETWORK. For each transaction, τ, it updates the ownership status within the network. After the sale of the taxable object, it calculates the tax liability for the focal entity according to the tax regulations.

To initialize the population of strategies S, the `CompCoevAlg` uses a creation loop that generates one strategy at a time. A sequence loop prompts the LLM using a prompt template to obtain each strategy, one transaction at a time, see **RQ 1**. This addresses the problem that the LLM could not generate a semantically feasible sequence of transactions in one shot, despite many attempts at passing it explicit rules and specific positive and negative examples. The system message of the prompt which informs the LLM that it is a tax planner that is responsible for generating valid transactions within a network of entities and assets. The LLM is told that its goal is to propose valid transactions that align with ownership rules, entity constraints, and asset guidelines that are in the prompt template. Each iteration, the calculator processes the most recently generated transaction and updates the ownership network of the scenario in the prompt template. This is the data in the prompt template that is updated every iteration. With this update, the only task the LLM has is to generate one ensuing transaction that is consistent with the ownership description. The template states the **TASK** as:

```
Be creative! Choose diverse entities, assets, and transaction
combinations while following the guidelines.
- Include a variety of entities and assets.
```

- Experiment with asset types: cash, promissory notes, hotel, partnerships.
- Use cash values between 0 and 200.
- Try both True and False for ELECTION754.

At the end of the sequence loop, if it is not present, the sale of the taxable good is appended to the end of the transaction sequence. This forces the taxable transfer in question, e.g. the sale of the hotel in iBOB from Mr. Jones to Mr. Brown and the calculator's final calculation for the strategy determines the strategy's (Mr. Jones' as focal entity) tax liability ($Taxable$).

While the prompt template is too long to include here, they are presented in Appendix B [19]. An example of a LLM generated transaction sequence is:

'new_co gives hotel to mr_brown for promissory1_mrbrown with a 754 election',

'mr_jones gives promissory1_mrjones to jones_co for $100 without a 754 election'

The CompCoevAlg next prompts the LLM to generate a population of audit patterns D to "audit" the initial population of strategies, S. In this step, we address how to flexibly generate patterns, see **RQ 3**, by employing another prompt template. The system message of the prompt template states:

You are a tax auditor that analyzes transaction data in a tax network. Your task is to identify general patterns and suspicious activity within the transaction data you observe.

The transaction data that the LLM is told to observe is filled into the prompt template using the initial strategy population. The template states the **TASK** as:

- Identify {max_patterns} number of general patterns found in the individual transactions.

The linguistic cues in the prompt template include the terms *suspicion, tax auditing,* and *analyzing.* These provide context for a very generally stated task. *Notably, the prompt template does not define a suspicious pattern.* This gives the LLM full (linguistic) freedom to define the structure and content of each pattern. The LLM responds with patterns that describes attributes of the examples *eliminating any need to pre-specify them.* Examples of patterns produced by LLM are:

pattern:1, *transaction involving 754 election*

pattern:2, *transaction with promissory notes*

pattern:3, *transaction from new_co to other entities*

pattern:4, *multiple transfers involving the same entities*

pattern:5, *multiple transactions involving promissory notes with inter-party transfers*

Step 1. Adversarial Fitness Evaluation. The CompCoevAlg next evaluates the fitness of each member s of the strategy population S and each member

d of the audit pattern population D. The fitness for each population are asymmetric.

Strategy Fitness:

The fitness assigned to each strategy sums two attributes. The strategy's (or specifically Mr. Jones') tax liability, i.e. `Taxable`, resulting from the sale of the good. The lower the liability, the more abusive the strategy. The second attribute `AuditScore`, see Algorithm 2 expresses concealment from auditing. This depends on the patterns in the current pattern population which model auditing rules, expressing the adversarial nature of objectives. Transactions are tagged with each pattern that matches them, modeling how suspicious they may appear when audited. For every strategy where at least one transaction is matched, that strategy's taxable liability is added to an audit score which is then averaged. Mathematically the fitness incentivizes abuse and concealment (or conversely disincentivizes compliance and audit detection. The coefficient C_1 tunes the value of the taxable liability, while C_2 tunes the value of the AuditScore. A strategy's objective is to minimize both its taxable liability and audit score.

Algorithm 2. `AuditScore`(s)

1: $M_{d,s} = \emptyset$
2: **for** $\forall \tau \in s$ **do**
3:     **for** $\forall d \in D$ **do**
4:         **if** `MATCH?`(d, τ) **then**
5:             $d \rightarrow M_{d,s}$
6: Return `AuditScore`$(s) = \frac{1}{|M_{d,s}|} \sum\limits_{d \in M_{d,s}} $ `Frequency`(d)

Audit Pattern Fitness: The `CompCoevAlg` also evaluates each audit pattern d of population D. The fitness of audit pattern d is the sum of two attributes: `Frequency`(d) which is how frequently it matches with strategies in S, given its matches to their transactions, and `TaxableScore` which factors in the tax liability of the strategies d matches, see Algorithm 3. Higher frequency raises audit pattern fitness, while lower abuse (among its matched strategies) decreases the fitness. The scaling factors C_3 and C_4 can be used to penalize patterns that match many transactions but that point to strategies with lower abuse.

Of note, regarding **RQ 4**, the `CompCoevAlg` again uses a prompt template to task the LLM to match each transaction sequence with patterns from the population. This is called 'labeling' and is stated in the prompt template as: `Label the transaction with the matching pattern_numbers.`

Step 2: Select The `CompCoevAlg` next generates mutations of both populations under the assumption that they may score better than existing ones. It uses tournament selection to create an updated population, it recalculates scores then ranks the new population, and finally, it selects, from each respective population, a small elite number of the best scoring strategies and patterns.

Algorithm 3. `TaxableScore`(d)

1: $M_{d,S} = \emptyset$
2: **for** $\forall s_i \in S$ **do**
3:     $M_{\tau,s} = \emptyset$
4:     **for** $\forall \tau \in s_i$ **do**
5:         **if** `MATCH?`(d, τ) **then**
6:             $\tau \to M_{\tau,s}$
7:     **if** $|M_{\tau,s}| \geq 1$ **then**
8:         $s_i \to M_{d,s}$
9: Return `TaxableScore`$(d) = \frac{1}{|M_{d,S}|} \sum_{s \in M_{d,S}}$ `Taxable`(s)

Step 3: Mutate (RQ 2) For C_M adaptations, the `CompCoevAlg` then chooses
a transaction from the elite subset and instructs the LLM to alter its entity,
asset, or §754 election? decision, while ensuring that ownership constraints
remain intact. The prompt tells the LLM to `mutate a given transaction`
`while adhering to the ownership rules`.... The prompt is extensive. It
includes specific adaptation guidelines, ownership rules, common invalid cases
(to avoid), four transaction-specific mutations options/types, and previously
mutation transactions (to avoid duplication). It stipulates that the returned
transaction must be valid in terms of ownership rules, different from prior
mutated transactions, matching the provided structure and rules, and it must
`BE CREATIVE`.

For M adaptations, the `CompCoevAlg` then chooses a pattern and tasks
the LLM to `MODIFY it in a meaningful way so that it captures a more`
`GENERAL, FREQUENT, and SUSPICIOUS pattern`
`that is meaningfully different from the existing patter and from`
`any of the previously identified patterns`.

Step 4: Re-Evaluate Fitness of Elite Adaptions The `CompCoevAlg` next
merges the new adaptations to each population and recalculates all scores.
Frequently co-occurring patterns are also composed, expanding the search
space.

Step 5: Replace Before returning to Step 1, the `CompCoevAlg` ranks each population and culls low scoring members to maintain the original size.

The framework terminates after the `CompCoevAlg` repeats Steps 2-5 for a
specified number of iterations (C_T) then returns its final populations.

Limitations Our goal is to provide a replicable methodological approach. The
implemented framework relies upon the iBOB prototype we selected and a calculator coded explicitly for iBOB. This limits it to investigating the iBOB scheme,
while we expect prototypes will exist for other schemes and their tax calculators
can be coded. This implementation uses OpenAI's Gpt-4o mini as released at
the time of our experiments. This rules out precise experimental reproducibility,
even later by the authors. Our software will be shared upon request.

4   Explorations

To form the basis for more in-depth, future experiments, we present three explorations of the framework that broadly investigate whether its behavior is reasonable.

4.1   Fitness Function Exploration

The framework has coupled asymmetric fitness functions, each with two attributes, and solves a minimax optimization problem. A strategy's first fitness attribute is its tax liability `Taxable`. If the tax liability is lower than expected by compliant conduct, it is abusive. The second attribute is `AuditScore`. This could be considered the magnitude of the suspicion a strategy draws from the audit patterns. Low suspicion is equivalent to high concealment. The strategy's objective is to maximize abuse and concealment (conversely stated, to minimize compliance and audit suspicion).

A pattern's first fitness attribute is the `Frequency` of its matches to strategies in S. Its second attribute factors in the abuse of the strategies it matches. Its objective is to maximize its ability to frequently match with highly abusive strategies.

Our first exploration is into two extremes of the coupled interactions noting that E.g., `Taxable` in a strategy's fitness is integrated into an audit pattern's `TaxableScore` and `Frequency` is integrated into the `AuditScore`. Parameter settings are: generations (T): 10, strategy population size (C_{s_S}): 10, audit pattern population size (C_{s_D}): 4, and tournament selection (C_t) of 2. We choose parameter settings that allow us to quickly discern distinct similarities or differences. The population sizes of the strategies and audit patterns are chosen to reflect an asymmetry between the diverse space of avoidance strategies and a smaller set of abstract auditing heuristics. The settings coax coevolutionary dynamics to emerge under our limited computation budget. In other settings, different parameter settings may be needed. Our experimental environment uses OpenAI GPT-4o-mini. Given our budget, we execute 6 runs.

We first set C_4 to zero. This relieves any pressure on audit patterns to match with abusive strategies. Regardless, the strategy population has auditing pressure from `AuditScore`. The runs yield an average of 57.17(14.36) abusive strategies in the iBOB family with high variance. The average pattern match `Frequency` is high – 0.96(0.02) and tight (low variance). The average transaction sequence length is 6.98(0.79). We also look for dynamics where, in a cycle, a strategy emerges and then is lost as the strategy fitness responds to auditing pressure. We call this an oscillation, but we observe none.

To investigate another extreme, we zero out C_1 and reset C_4 to one. Now there is no pressure for a strategy to reduce its tax liability. We observe that much fewer, roughly 80%, abusive strategies are found and this quantity highly varies: 12.83(15.37). We also observe much shorter transaction sequences (length $2.56(2.68) vs 6.98(0.79))$. And we observe oscillations where the same strategy emerges, disappears, then arises again. Finally, we experiment with non-zero coefficients, arriving at $(C_1 = 2, C_2 = C_3 = 1, C_4 = 100)$. All subsequent explorations go forward with this setting.

Table 2. Fitness function attribute extremes exploration. Mean(std) values from six runs. **Red cells** mark row-wise **maximums**, **blue cells** mark **minimums**

Metric	$C_4 = 0$	$C_1 = 0$	$C_1 = 2, C_2 = C_3 = 1, C_4 = 100$
# abusive strategies	54.17 (14.36)	12.83 (15.37)	38.17 (24.27)
Avg Frequency	0.96 (0.03)	0.97 (0.03)	0.96 (0.05)
Avg $|s|$	6.98 (0.79)	2.56 (2.68)	5.35 (2.68)
Oscillations	0.0 (0.0)	0.33 (0.82)	0.33 (0.82)

Across the three different variants, see Table 2, quite optimistically, we observe that, though prevalence and stability vary, abusive strategies can nonetheless be discovered. Taken together, these results suggest that tuning the coefficients of all attributes supports explorations and extended empirical investigations into a range of dynamics. In each of the 3 settings, an abusive strategy was discovered quickly, by generations 3–4. It also uncovered a novel strategy that was different from iBOB's prototype:

(a) Jones transfers `partnership2` to `JonesCo` for `promissory1_jonesco`, electing §754.
(b) Exchanges `partnership1` with Brown for `promissory1_mrbrown`, electing §754.
(c) `NewCo` swaps `promissory1_newco` for `partnership2`, electing §754.
(d) `NewCo` sells the `hotel` to Brown for $200 *without* a §754 election.

The three elections inflate inside basis to $200 while outside basis remains near $198, producing a tax liability $2 less than the iBOB prototypical strategy. Relative to the prototypical iBOB strategy, the evolved variant distributes basis adjustments across multiple transactions. While financially similar, this diffuses suspicious activity, increases concealment, and protects more against detection.

Audit Patterns The framework also reliably flagged suspicious behavior characteristic of the iBOB scheme, with high-fitness patterns such as:

(a) Repeated promissory-note transfers with §754 elections followed by hotel sales.
(b) Chains of partner-to-partner swaps with elections, culminating in negative tax.
(c) Frequent use of promissory notes in high-turnover transactions.

These patterns correspond to common audit attributes, such as repeated intra-network debt instruments and basis elections preceding asset sales. Remarkably, the framework identifies these without explicitly encoding, see RQ 4.

4.2   Is Coevolution Necessary?

The complexity of a coevolutionary approach motivates exploring whether something simpler can achieve the same outputs. Outside the framework, it is possible to simply prompt the LLM for novel iBOB strategies given a static set of audit patterns. We compared this baseline LLM approach to two non-coevolutionary approaches (using the framework) that use "standalone" evolution of strategies or audit patterns respectively, given a static set of counterparts. We used $C_1 = 2, C_2 = C_3 = 1, C_4 = 100$ while each option used 10 transaction sequences and 4 audit patterns.

When an LLM is used without being combined with the framework (s **One-Shot**), i.e. evolution (or coevolution), we could not discover iBOB strategies. Details are in Table 3 When we only evolve strategies without coevolving audit patterns (s-**Evo**), iBOB strategies quickly emerge but we observe multiple oscillations between abusive and compliant strategies. When we "ablate" coevolution to only evolve audit patterns (d-**Evo**), strategy fitness does not improve, while pattern fitness steadily climbs. The coevolutionary framework outperforms the three ablations, producing the best performing strategy and audit pattern fitness, the highest number of abusive discoveries, and more moderate oscillations. Overall, coevolution pushes both populations toward their most adaptive strategies, generating richer dynamics than any isolated evolutionary path.

4.3   Run Length and Audit Pattern Population Size

We next explore increasing the number of generations in a run and the size of the audit pattern population. We vary the latter because it directly controls the detection pressure. In contrast, varying the strategy population size primarily affects sampling density and was less influential in preliminary experiments. Our expectation is that more audit patterns improve discrimination between abusive and compliant strategies, delaying iBOB strategy discovery and population dominance. We use $C_1 = 2, C_2 = C_3 = 1, C_4 = 100$ and extend two runs each for 20 generations. We compare using a population of 4 audit patterns to one of 10 audit patterns.

Table 4 shows that increasing the number of audit patterns reduces abusive strategy emergence (88.8 vs. 24.5) and raises the number of oscillations (1.17 vs. 2.83). Strategy fitness rises when the runs use 10 audit patterns, while audit fitness declines due to lower pattern frequency.

Table 3. Ablation of coevolution. **Red cells** mark row-wise **maximums, blue cells** mark **minimums**

Metric	s **One Shot**	s-**Evo**	d-**Evo**	**Coev**
Abusive strategies	0.0 (0.0)	5.67 (4.13)	0.0 (0.0)	38.17 (24.27)
Generation abuse Discovered	–	2.6 (0.89)	–	4.17 (2.32)
Number of Oscillations	0.0 (0.0)	2.33 (3.14)	0.0 (0.0)	0.33 (0.82)

136   J. S. Bhattacharya et al.

Table 4. Length of Run and Audit Population Size Experiment. Red cells mark rowwise maximums.

Metric	4 Pattern Pop Size	10 Pattern Pop Size
Abusive strategies	88.83 (53.6)	24.5 (30.02)
Avg Abusive Stategies Per Gen	6.06 (2.68)	1.94 (2.3)
Generation 1st abuse found	5.67 (4.37)	4.25 (2.06)
Number of Oscillations	1.17 (2.86)	2.83 (3.13)

We next extend another run of each setting to 60 generations and again contrast the audit population sizes while considering the average number of abusive strategies emerging and normalized, average fitness of each population per generation. This is visualized in Figs. 4 and 5. With **4 patterns**, abusive strategies emerge quickly, stabilizing by generation 20 with only a brief early oscillation. With **10 patterns**, oscillations persist longer and occur more frequently, delaying strategy population convergence until around generation 40. Average fitness continues fluctuating until that point. These findings suggest that expanding the pattern matching capacity disrupts early convergence and drives richer evolutionary dynamics, pointing to the importance of auditing in the real world.

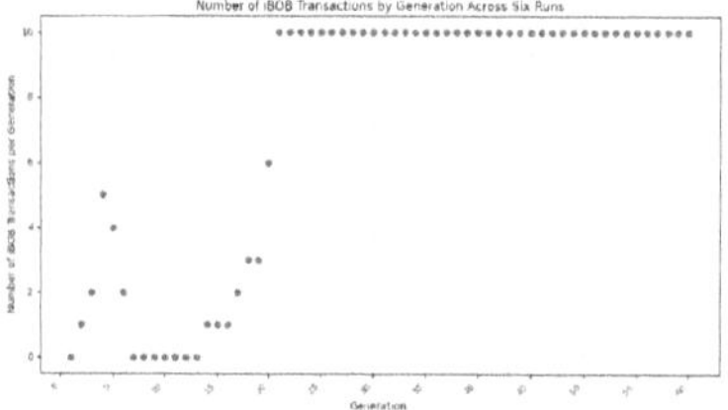

(a) Avg number of abusive strategies

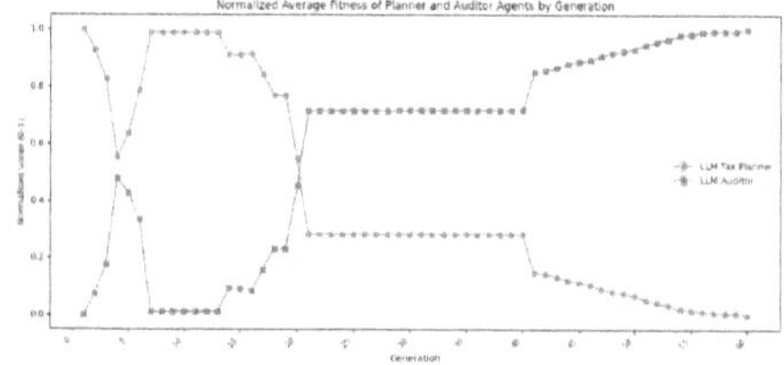

(b) Avg fitness across generations. Red line is strategies, blue line is audit patterns

Fig. 4. Four Patterns: (a) Avg number of abusive strategies and (b) Avg fitness

4.4   Exploration Discussion

Across these explorations, three findings stand out. First, the fitness functions' attributes can steer outcomes. Second, coevolution is essential: with only evolution of strategies, strategies are limited (the population rediscovers similar abusive strategies but lacks adaptability; the pattern matching tracks frequent patterns but is unable to isolate an abusive strategy). Third, increasing the auditing capacity by increasing the audit pattern population size drives rich, co-adaptive dynamics.

Currently transaction sequences require specific syntactic structure and consistent variant naming to ensure that they can be parsed and executed by the tax

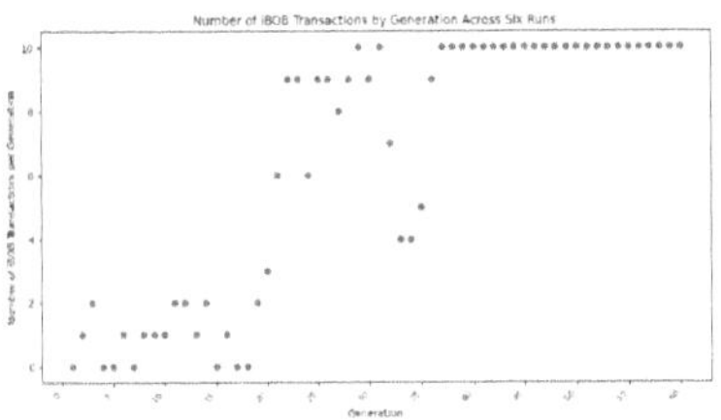

(a) Avg number of abusive strategies

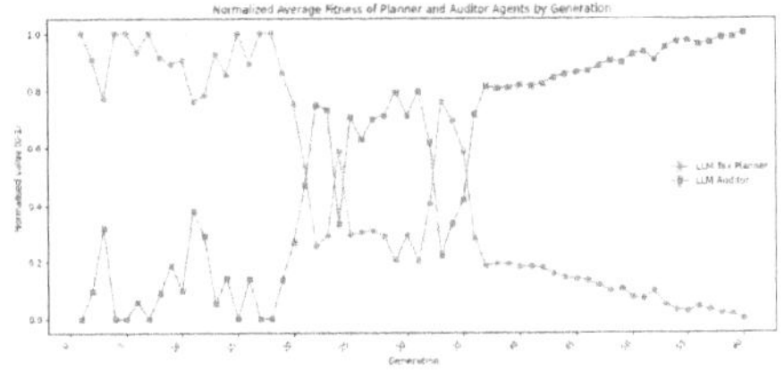

(b) Avg fitness across generations. Red line is strategies, blue line is audit patterns

Fig. 5. Ten Patterns: (a) Avg number of abusive strategies and (b) Avg fitness

calculator. Audit patterns, in contrast, avoid these requirements because they are represented in natural language. This highlights the tradeoff between executability and expressiveness: syntactic structure is necessary for executability, while the unbounded natural language representation serves pattern comparison and strategy matching with the LLM.

5   Conclusion

In future work, systematic sampling of additional linguistic cues around suspicion, detection, auditing, evasion, abuse and non-compliance may improve the nature of the LLM responses. Preventing the LLM from generating commonly prohibited strategies (with task design) would improve the realism of the model. Experiments across different LLMs would reveal the extent of the framework's robustness. An LLM-based system that reasons about taxation directly from statutory text would make the analysis more general and less effort to design. It may however incur heavy verification costs.

The framework affirmatively answers the paper's high-level question. An LLM integrated with a coevolutionary algorithm offers an alternative approach to collecting abusive tax strategies. By combining a rule-based tax calculator with iterative adversarial training, the framework provides a new approach for studying avoidance behavior, developing more flexible audit strategies, and ultimately informing tax policy. It offers a practical setting for the GP community to examine the tradeoffs between different representation bounds and executability. It also intriguingly shows that even as the LLM does not understand the meaning and operational semantics of abusive strategies, it nonetheless can help discover abusive tax strategies.

References

1. Arthur Vanderbilt: Tax Quote (2025), https://www.inc.com/geoffrey-james/130-inspirational-quotes-about-taxes.html
2. Blair-Stanek, A., Holzenberger, N., Durme, B.V.: Can gpt-3 perform statutory reasoning? (2023), https://arxiv.org/abs/2302.06100

3. Blair-Stanek, A., Holzenberger, N., Durme, B.V.: Openai cribbed our tax example, but can gpt-4 really do tax? (2024), https://arxiv.org/abs/2309.09992
4. Bloomquist, K.: Tax compliance as an evolutionary coordination game: an agent-based approach. Public Finan. Rev. **39**(1), 25–49 (2011). https://doi.org/10.1177/1091142110381640, https://doi.org/10.1177/1091142110381640
5. Dahl, M., Magesh, V., Suzgun, M., Ho, D.E.: Hallucinating law: legal mistakes with large language models are pervasive. HAI, Stanford University, January 2024, https://hai.stanford.edu/news/hallucinating-law-legal-mistakes-large-language-models-are-pervasive
6. Field, H.M.: A taxonomy for tax loopholes. Hous. L. Rev. **55**, 545 (2017)
7. Fratrič, P., Holzenberger, N., Amariles, D.R.: Can ai expose tax loopholes? towards a new generation of legal policy assistants (2025), https://arxiv.org/abs/2503.17339
8. GAO: Gao-14-453 (2013), http://www.youtube.com/watch?v=O8VDUStvxMY
9. Hemberg, E., Rosen, J., Warner, G., Wijesinghe, S., O'Reilly, U.M.: Detecting tax evasion: a co-evolutionary approach. Artif. Intell. Law **24**(2), 149–182 (2016). https://doi.org/10.1007/s10506-016-9181-y, version: Author's final manuscript
10. Holzenberger, N., Blair-Stanek, A., Durme, B.V.: A dataset for statutory reasoning in tax law entailment and question answering (2020), https://arxiv.org/abs/2005.05257
11. Jung, S., Jung, J.: Courtroom-LLM: a legal-inspired multi-LLM framework for resolving ambiguous text classifications. In: Rambow, O., Wanner, L., Apidianaki, M., Al-Khalifa, H., Eugenio, B.D., Schockaert, S. (eds.) Proceedings of the 31st International Conference on Computational Linguistics, pp. 7367–7385. ACL, Abu Dhabi, UAE, January 2025, https://aclanthology.org/2025.coling-main.493/
12. Lee, K.H., et al.: Evolving deeper llm thinking (2025), https://arxiv.org/abs/2501.09891
13. Lehman, J., Gordon, J., Jain, S., Ndousse, K., Yeh, C., Stanley, K.O.: Evolution through large models (2022), https://api.semanticscholar.org/CorpusID:249848020
14. Liventsev, V., Grishina, A., Härmä, A., Moonen, L.: Fully autonomous programming with large language models. In: Proceedings of the Genetic and Evolutionary Computation Conference, GECCO 2023, pp. 1146–1155. ACM, July 2023. https://doi.org/10.1145/3583131.3590481, http://dx.doi.org/10.1145/3583131.3590481
15. Meyerson, E., Nelson, M., Bradley, H., Moradi, A., Hoover, A.K., Lehman, J.: Language model crossover: Variation through few-shot prompting. ACM Trans. Evol. Learn. **4**, 1–40 (2023), https://api.semanticscholar.org/CorpusID:257102873
16. Novikov, A., et al.: Alphaevolve: a coding agent for scientific and algorithmic discovery. ArXiv **abs/2506.13131** (2025), https://api.semanticscholar.org/CorpusID:278658695
17. O'Neill, M., Ryan, C.: Grammatical evolution. IEEE Trans. Evol. Comput. **5**(4), 349–358 (2002)
18. Pertierra, M.A., Lawsky, S., Hemberg, E., O'Reilly, U.M.: Towards formalizing statute law as default logic through automatic semantic parsing. In: ASAIL@ICAIL (2017), https://api.semanticscholar.org/CorpusID:33245110
19. Redacted: Redacted (Redacted)
20. Rosen, J.: Computer aided tax avoidance policy analysis, January 2015
21. Scott, E.O., Srivastava, A., Pusateri, A., Taylor, A., Jones, K., Ortiz, H.: Realistic tax planning with evolutionary algorithms. In: 2024 IEEE Congress on Evolutionary Computation (CEC), pp. 1–8 (2024). https://doi.org/10.1109/CEC60901.2024.10612201

22. U.S. Congress: Definition of economic substance doctrine, 26 u.s.c. §7701(o)(5)(a). https://www.law.cornell.edu/definitions/uscode.php?width=840&height=800& iframe=true&def_id=26-USC-1846184229-1495237516&term_occur=2&term_ src=title:26:subtitle:F:chapter:79:section:7701 (1954), Accessed 15 May 2025
23. Wu, X., Wu, S.H., Wu, J., Feng, L., Tan, K.C.: Evolutionary computation in the era of large language model: Survey and roadmap. IEEE Trans. Evol. Comput. (2024)
24. Zhong, H., Xiao, C., Tu, C., Zhang, T., Liu, Z., Sun, M.: How does NLP benefit legal system: a summary of legal artificial intelligence. In: Jurafsky, D., Chai, J., Schluter, N., Tetreault, J. (eds.) Proceedings of the 58th Annual Meeting of the Association for Computational Linguistics, pp. 5218–5230. ACL, July 2020. https://doi.org/10.18653/v1/2020.acl-main.466, https://aclanthology. org/2020.acl-main.466

Sinking the Bloat in Genetic Programming Using Equality Saturation

Lucas Miranda$^{(\boxtimes)}$, Matheus Fernandes[iD], Emilio Francesquini[iD], and Fabricio Olivetti de Franca[iD]

Federal University of ABC, Santo André, SP, Brazil
lucas.miranda@aluno.ufabc.edu.br,
{fernandes.matheus,e.francesquini,folivetti}@ufabc.edu.br

Abstract. Program Synthesis (PS) aims to automatically generate computer programs from high-level specifications. Genetic Programming (GP) is a prominent metaheuristic for PS, treating synthesis as a search problem over the vast space of all possible programs. A recent approach, Higher-Order Typed Genetic Programming (HOTGP), uses a purely functional and typed paradigm to constrain this search space and improve performance. However, like many search techniques, the evolved programs often suffer from "bloat", an unnecessary growth of code that makes them complex and difficult to read. This paper investigates the integration of Equality Saturation, a term rewriting optimization technique, into the HOTGP framework. The objective of this integration is to simplify generated programs and reduce bloat, thereby enhancing both computational efficiency and program readability. Experimental results on a set of benchmarks indicate that equality saturation has a varied impact on the final synthesis success rate. Conversely, qualitative analysis confirmed a reduction in code growth (bloat), which manifested as simpler solutions in some benchmarks and a more frequent convergence toward minimal program forms.

Keywords: Genetic Programming · Program Synthesis · Equality Saturation · Bloat Reduction

1 Introduction

Program Synthesis (PS) [1,5,13,18] is the process of automatically generating a computer program that satisfies a given set of specifications. These specifications can range from formal descriptions to a set of input-output examples, a technique known as Programming-by-Examples (PBE). PS can be modeled as a search problem over a vast space where each point represents a valid program under a specific grammar.

Genetic Programming (GP) [9], introduced by John Koza toward the end of the 1980s [8,10], is a *program-by-example* technique, inspired by biological evolution, that has proven effective for PS. It evolves a population of candidate

© The Author(s), under exclusive license to Springer Nature Switzerland AG 2026
L. Manzoni et al. (Eds.): EuroGP 2026, LNCS 16521, pp. 140–155, 2026.
https://doi.org/10.1007/978-3-032-23005-8_9

```
if (if ((length x1) == (length x0))
    then ((length x1) < (length x2))
    else ((length x0) > (length x1)))
then False
else ((length x1) < (length x2))
```

(a) Bloated solution example from *compare-string-lengths*.

```
((length x1) > (length x0)) &&
((length x2) > (length x1))
```

(b) The same program after simplification.

Fig. 1. An example of code bloat (a) that was found in GP solutions for the *compare-string-lengths* benchmark and its simplified and equivalent form (b).

programs through operators like crossover and mutation, guided by a fitness function that measures how well a program meets the specifications.

The main variations of GP for PS are stack-based [6] and grammar-guided [16]. In stack-based approaches, such as PushGP, programs are linear lists of instructions that explicitly manage data on typed stacks. This allows for a very flexible and robust system where code can easily be manipulated. However, this flexibility comes at the cost of readability, as the resulting programs can be convoluted and difficult for humans to analyze. On the other hand, grammar-guided GP uses a formal grammar to constrain the search space, ensuring that only syntactically correct programs are generated. While effective, this approach often requires a problem-specific grammar to be engineered, which can bias the search and limit the generality of the solutions.

Recent approaches, such as HOTGP [2] (Higher-Order Typed Genetic Programming), propose the use of statically typed, purely functional programs with support for higher-order functions to restrict the search space and enhance synthesis performance. Despite these advances, programs generated by GP often contain unnecessarily large and complex code, a phenomenon known as "bloat" [12]. This excess code not only hinders readability but can also impair the ability to generalize.

The problem of bloat in the generated code (see Fig. 1) is that it reduces the readability and, sometimes, the generalization capability of evolved programs. This phenomenon can often be even more severe, as illustrated by the nearly indecipherable program in Fig. 2, which underscores the need for simplification techniques to reduce such bloat. A noteworthy technique is Equality Saturation (EqSat) [21], an approach originally created to optimize programs during the compilation process. It uses a data structure called e-graph to systematically apply equivalence rules while keeping all the equivalent programs compactly stored in the structure. The optimal program can be extracted from the structure by applying a heuristic cost function.

```
(((((length x0) - (product (filter (\\y -> 0 < (min y 0))
↪ x0))) - (product (range (min (head (reverse (filter
↪ (\\y -> not (y > 0)) x0))) (length (filter (\\y -> if
↪ (y == 0) then (not (y < 0)) else False) (filter (\\y
↪ -> 0 < (min y 0)) x0))))) 0 0))) - (product (range
↪ (min (product (filter (\\y -> (min y 0) == y) (filter
↪ (\\y -> y < 0) (filter (\\y -> 0 < (min y 0)) x0))))
↪ (min (product (filter (\\y -> y < 0) (filter (\\y ->
↪ y > 0) x0))) (((product (filter (\\y -> (sum (range 0
↪ y y)) > 1) (filter (\\y -> (not (0 == y)) && (if (0 <
↪ y) then (y > 0) else (0 == y))) x0))) + (0 - (product
↪ (take 0 x0)))) * (min (head (reverse (filter (\\y ->
↪ y == (if (y > 0) then y else 0)) x0))) (length
↪ x0))))) 0 0))) - (product (range (min ((min (head
↪ (reverse (filter (\\y -> y == (if (y > 0) then y else
↪ 0)) x0))) (length x0)) * (head (reverse (filter (\\y
↪ -> not (y > 0)) x0)))) 0) 0 0))) - (product (range
↪ (min (product (filter (\\y -> False) (reverse x0)))
↪ ((sum x0) * (min (product (range (min (if ((length
↪ x0) > 0) then (head x0) else (sum x0)) ((length x0) *
↪ (head (reverse x0)))) 0 0)) (min (product (filter
↪ (\\y -> y < 0) (filter (\\y -> y > 0) (singleton (snd
↪ (head (zip x0 x0))))))))) (((length (filter (\\y -> if
↪ (0 > y) then (if (0 < y) then (0 > y) else True) else
↪ ((y > 0) && (0 > y))) (0 : x0))) - (product (take 0
↪ (map (\\y -> 0) x0)))) * (head (reverse (filter (\\y
↪ -> y == (if (y > 0) then y else 0)) x0))))))))) 0 0))
```

(a) Bloated solution example from last-index-of-zero.

```
(length x0) - 5
```

(b) The same program after simplification.

Fig. 2. Example of code bloat (a) that could be found in GP solutions for the *last-index-of-zero* benchmark and its simplified, functionally equivalent form (b).

This paper investigates the effects of integrating EqSat within HOTGP's evolutionary process[1], evaluating its impact on both synthesis success rate and the quality (simplicity and readability) of the generated programs as an approach to reduce code bloat in GP.

The remainder of this paper is organized as follows. Section 2 presents background and related work. Section 3 describes the integration of EqSat into HOTGP. Section 4 describes how the experiments were conducted. Section 5 presents and discusses the results obtained, and we conclude in Sect. 6.

---

[1] The full source code for HOTGP with integrated EqSat can be downloaded from: https://github.com/Lucasgm22/hotgp/tree/equality-saturation.

2 Background and Related Work

This section reviews the fundamental concepts essential to our work. We begin by tracing the evolution of GP for program synthesis, culminating in a description of the HOTGP framework, which serves as the baseline for this work. We then introduce Equality Saturation, a technique used to optimize the order in which simplification rules are applied to computer programs during the compilation process. This sets the stage for the integration with HOTGP detailed in the following section.

2.1 Genetic Programming and HOTGP

The use of evolutionary techniques to search for functional programs has a long history. One of the first examples is ADATE [15], which synthesized programs in the ML language using incremental transformation.

A fundamental milestone was Strongly Typed Genetic Programming (STGP), proposed in [14]. STGP introduced the use of data types to constrain the search space, ensuring that only well-typed programs were generated. STGP was shown to outperform untyped GP in several problems, generating solutions more quickly and with a higher success rate. Other approaches like PolyGP [22,23] extended STGP with support for higher-order functions and polymorphism.

Building on these foundations, HOTGP [2] was proposed as a GP algorithm that synthesizes purely functional, typed, and higher-order programs. The motivation for this approach is to directly address key challenges in program synthesis: by enforcing a purely functional paradigm, HOTGP avoids undesired behaviors caused by side effects. Furthermore, the use of a strong typing system drastically constrains the vast search space, guiding the evolutionary process exclusively toward syntactically correct and type-sound programs. This combination of features aims to improve the stability and efficiency of the synthesis process. Notably, HOTGP is implemented in Haskell and leverages a general-purpose grammar extracted from Haskell's base library, allowing the synthesized code to be directly compiled by standard Haskell compilers. Among its core features, we highlight:

- **Purely Functional Paradigm:** All generated programs are pure functions without side effects, ensuring referential transparency and facilitating composition of functions.
- **Static Typing System:** The algorithm leverages static typing to reduce the search space, filtering out programs with invalid syntax or types. Type checks are performed during program construction (initialization, crossover, and mutation), ensuring only nodes matching the type expected by their parent are considered, thus guaranteeing type soundness from the start.
- **Higher-Order Functions and Lambdas:** HOTGP supports higher-order functions like `map` and `filter`, which accept other functions (lambdas) as arguments. This allows for the creation of complex constructs, such as loops and recursion, in an abstract manner.

HOTGP operates on a population of candidate programs, represented as expression trees. It employs a steady-state evolutionary model, where in each generation, a small number of individuals are replaced. The process begins with a randomly generated population of well-typed programs and iteratively applies genetic operators to evolve solutions, guided by a fitness function.

Parent selection is performed using exponential fitness normalization (also called rank selection). This method uses a Parent-Scalar hyperparameter, set to 0.9993 in the original HOTGP publication, which defines the probability of picking the nth best individual as 0.9993 times the probability of picking the $(n-1)^{th}$ best. The probability value assigned to the top-ranked individual (rank 0) serves as the base case, equivalent to 1. This specific value creates a very strong selection pressure, resulting from the fact that the multiplier between adjacent ranks (0.9993) is very close to 1. This causes the selection probability to decrease slowly across ranks, meaning top-ranked individuals retain a markedly higher likelihood of being chosen from reproduction compared to those ranked lower. The core operators are:

- **Initialization:** To create the initial population, programs are generated by recursively building expression trees. At each step, only functions or terminals that match the data type expected by the parent node are considered valid candidates, ensuring every program is well-typed from the start.
- **Crossover:** To recombine two parent solutions, a random subtree is selected from the first parent. Then, in the second parent, only subtrees that have the exact same output type are considered valid points for the swap. If no such point exists, the crossover is aborted.
- **Mutation:** The mutation operator replaces a randomly selected subtree with a new, randomly generated one. This new subtree is generated using the same type-aware initialization procedure, ensuring it matches the type required by its position in the program and maintaining its validity.

To manage program complexity and prevent excessively large solutions, HOTGP enforces a maximum depth limit on the generated expression trees during initialization and mutation. In the original experiments, this limit was set to 15 for the main program tree and 3 for lambda functions.

In its original publication [2], HOTGP demonstrated competitive performance on the General Program Synthesis Benchmark Suite. It achieved the highest success rate on 9 of the 29 benchmark problems when compared to six other algorithms and obtained at least a 50% success rate in more problems (10 out of 29) than any of the algorithms in the comparison.

However, despite its effectiveness, HOTGP suffers from a significant drawback common to many GP systems: code bloat [12,17,20]. The evolutionary process often produces programs that, while functionally correct, are unnecessarily large, complex, and difficult for humans to read. These solutions frequently contain redundant sub-expressions and convoluted logic that obscure the program's core functionality. Furthermore, this excessive complexity has been linked to overfitting [20], as the search process may synthesize intricate structures to capture specific noise in the training data rather than the underlying general rule. While the original algorithm includes a post-search refinement step to simplify the best-found solution, this process is applied only at the end and may not be sufficient to adequately reduce the resulting code.

2.2   Equality Saturation

Equality Saturation [19], recently popularized by the *egg* [21] library, has established itself as a technique for compiler optimization and theorem proving that systematically explores equivalences between sub-expressions within a program (*e.g.*, recognizing that $x*2$ and $x+x$ are semantically equivalent). It employs a data structure called an *e-graph* [19] to compactly represent a set of equivalent expressions.

An e-graph represents a set of equivalent expressions in a compact form. It is composed of e-classes, where each e-class contains a set of equivalent e-nodes. An e-node represents a single operator or value (*e.g.*, a function like $+$, a variable x, or a constant 2), and its children are pointers to other e-classes. Therefore, an entire e-class represents a set of equivalent sub-expressions.

For example, consider the expression $x+x$. Initially, this would be represented in an e-graph as shown in Fig. 3a. There are two e-classes: one for x and one for the addition operation $+$.

The core operation in an e-graph is the merge. When two e-classes are merged, the e-graph records that they are equivalent. This union operation is the fundamental mechanism for growing the set of equivalences. For instance, if we later discover that $x*2$ is equivalent to $x+x$, we would merge the e-class of $x*2$ with the e-class representing $x+x$, as seen in Fig. 3b.

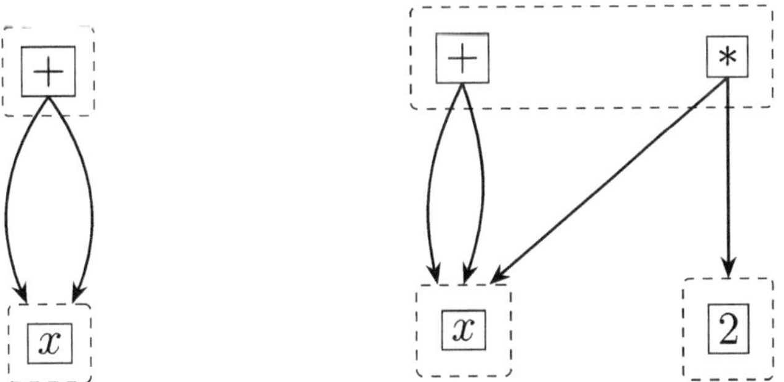

(a) Initial e-graph for the expression $x+x$.

(b) E-graph after applying the rule $x*2 \rightarrow x+x$ and merging e-classes.

Fig. 3. Illustration of the e-graph structure. Each dashed box represents an *e-class*, and each solid-bordered box is an *e-node*. (a) shows the initial state for an expression. (b) shows the state after an equivalent expression is added and merged into the same e-class.

The equivalences are provided to the EqSat algorithm through a set of rewrite rules. A rewrite rule is a directed equality, written as $LHS \rightarrow RHS$, which asserts that the left-hand side pattern can be replaced by the right-hand side pattern. These rules can encode simple algebraic identities, domain-specific optimizations, or laws from functional programming. For instance, in the context of program synthesis, we could use rules such as:

- **Algebraic Simplification:** $a + 0 \rightarrow a$.
- **List Operation Simplification:** $take\ ((len\ xs)\ xs) \rightarrow xs$.
- **Boolean Operation Simplification:** $if\ True\ then\ a\ else\ b \rightarrow a$

The EqSat algorithm works by iteratively applying the rewrite rules to the e-graph until no new expressions can be added. The process can be summarized as follows:

1. An initial e-graph is built from the input expression.
2. The algorithm enters an iterative loop. In each iteration, it searches for patterns within the e-graph that match the left-hand side of each rewrite rule.
3. For every matched pattern, the right-hand side of the rule is instantiated and inserted into the e-graph. The e-class of this new term is then marked to be merged with the e-class of the term that was originally matched.
4. This process of matching and merging is repeated until it reaches a fixed point, known as saturation. At this stage, the e-graph contains all equivalent expressions reachable through the given set of rules.
5. Finally, the saturated e-graph is analyzed to extract the optimal expression by minimizing a specific cost function (*e.g.*, node count or depth).

Since at every step the structure holds more information about the equivalent programs, a program such as `take (max(len xs, len xs), xs)`, with the rules `max(n, n)` $\rightarrow$ `n` and the one illustrated above, in a limited number of steps, the optimal program would be contained in the e-graph. On the other hand, if the simplification steps are applied in order and in a destructive manner, there would be no guarantee of reaching optimality. Despite its effectiveness, EqSat has several practical limitations. The primary challenge is its performance and memory footprint. The saturation process can lead to an exponential explosion in the size of the e-graph as new equivalences are discovered, making it computationally expensive. Furthermore, the quality of the results is highly dependent on the provided set of rewrite rules. An incomplete set may miss key optimizations while a poorly designed set can lead to infinite loops, preventing the algorithm from reaching saturation. These challenges motivate a careful and targeted application of the technique, which is a central point of our study.

The use of EqSat in GP is not new. In [3], the authors employed this approach to simplify expressions generated by symbolic regression (SR) algorithms and to assess the overparameterization bias of different approaches. They concluded that heuristic simplification methods are insufficient to correctly simplify the expressions. EqSat was also used to detect duplicates [11] in the search process of GP for SR, revealing that a standard GP on a typical run can revisit up to

60% of equivalent (*i.e.*, duplicate) expressions. Finally, in [4] the authors used EqSat to propose variation operators that enforce the creation of unique solution candidates, thus alleviating this inefficiency.

3   Integrating Equality Saturation Into HOTGP

This section details our primary contribution: the integration of EqSat directly into HOTGP's evolutionary loop. We first provide a high-level overview of our design choices and then present the specific implementation details.

HOTGP successfully generates functional programs but often suffers from code bloat, producing unnecessarily complex solutions like the *compare-string-lengths* example shown in Fig. 1. Our method addresses this by introducing a simplification step during the evolutionary process, aiming to produce more concise and readable programs, rather than relying solely on a final post-processing step.

To balance the computational cost of EqSat with the goal of impacting the next generation, we adopted a targeted approach based on HOTGP's parent selection mechanism (see Sect. 2.1), which is heavily biased towards the top-ranked individuals. This informed our optimization strategy: specifically, we chose to simplify the top two fittest individuals (Top-2) in each iteration. This number aligns with the steady-state replacement strategy of HOTGP, which also operates on two individuals per step [2]. Given the high Parent-Scalar value (0.9993), these Top-2 individuals already have the highest probability of being selected for reproduction by the steady-state mechanism, whereas lower-ranked individuals have a negligible chance of propagating their genetic material. Consequently, extending the computationally expensive EqSat process to the wider population would incur significant overhead with diminishing returns on the evolutionary trajectory. Therefore, applying EqSat simplification exclusively to these specific individuals before the parent selection phase maximizes the likelihood that the benefits of simplification are passed on to the next generation, while strictly limiting the computational overhead to only two applications of EqSat per iteration. This strategy maximizes the impact of bloat control while minimizing computational overhead.

To implement the simplification step, we employed the *hegg* library[2], a Haskell implementation of e-graphs and equality saturation. The effectiveness of EqSat is highly dependent on the quality of the provided rewrite rules. We defined a curated set of rules based on algebraic identities and common patterns in functional programming, aiming to simplify expressions without altering their semantics. The main rules used in our experiments are shown in Table 1. We acknowledge that this set of rules is not exhaustive; extending the rule set to cover more domain-specific identities could yield further optimizations, though at the cost of increased search space for the saturation engine.

A final, crucial implementation detail is the cost function used to extract the "best" program from the saturated e-graph after the rules have been applied.

---

[2] https://hackage.haskell.org/package/hegg.

Table 1. The set of rewrite rules used for the EqSat simplification process.

Rule (LHS $\rightarrow$ RHS)	Rule (LHS $\rightarrow$ RHS)
Boolean	
$True \lor a \rightarrow True$	$a \lor True \rightarrow True$
$False \lor a \rightarrow a$	$a \lor False \rightarrow a$
$True \land a \rightarrow a$	$a \land True \rightarrow a$
$False \land a \rightarrow False$	$a \land False \rightarrow False$
$a \lor a \rightarrow a$	$a \land a \rightarrow a$
$\neg(\neg a) \rightarrow a$	
Conditionals	
$if\ True\ then\ a\ else\ b \rightarrow a$	$if\ False\ then\ a\ else\ b \rightarrow b$
$if\ a\ then\ True\ else\ False \rightarrow a$	$if\ a\ then\ False\ else\ True \rightarrow \neg a$
$if\ a\ then\ b\ else\ False \rightarrow a \land b$	$if\ a\ then\ b\ else\ True \rightarrow (\neg a) \lor b$
$if\ a\ then\ False\ else\ b \rightarrow (\neg a) \land b$	$if\ a\ then\ True\ else\ b \rightarrow a \lor b$
$if\ (\neg a)\ then\ b\ else\ c \rightarrow if\ a\ then\ c\ else\ b$	$if\ a\ then\ b\ else\ b \rightarrow b$
$if\ (a > b)\ then\ b\ else\ a \rightarrow min(a, b)$	$if\ (a > b)\ then\ a\ else\ b \rightarrow max(a, b)$
$if\ (a < b)\ then\ a\ else\ b \rightarrow min(a, b)$	$if\ (a < b)\ then\ b\ else\ a \rightarrow max(a, b)$
Integer	
$min(a, a) \rightarrow a$	$max(a, a) \rightarrow a$
$a\ (mod\ 1) \rightarrow 0$	$min(a, a * a) \rightarrow a$
$a = a \rightarrow True$	$a < a \rightarrow False$
$a > a \rightarrow False$	$\neg(a = min(a, b)) \rightarrow a > b$
$\neg(a = max(a, b)) \rightarrow a < b$	$max(a, b) > a \rightarrow b > a$
$min(a, b) < a \rightarrow b < a$	$a - b > 0 \rightarrow a > b$
$a - b < 0 \rightarrow a < b$	$max(a, a * a) \rightarrow a * a$
Float and Integer	
$0 + a \rightarrow a$	$a + 0 \rightarrow a$
$a - 0 \rightarrow a$	$a - a \rightarrow 0$
$1 * a \rightarrow a$	$a * 1 \rightarrow a$
$0 * a \rightarrow 0$	$a * 0 \rightarrow 0$
$a/1 \rightarrow a$	$a/a \rightarrow 1$ (if $a \neq 0$)
List Operations	
$len(reverse(a)) \rightarrow len(a)$	$head(singleton(a)) \rightarrow a$
$reverse(singleton(a)) \rightarrow singleton(a)$	$len(singleton(a)) \rightarrow 1$
$product(singleton(a)) \rightarrow a$	$sum(singleton(a)) \rightarrow a$
$reverse(reverse(a)) \rightarrow a$	$take(len(a), a) \rightarrow a$
$take(1, singleton(a)) \rightarrow singleton(a)$	$range(a, a, a) \rightarrow singleton(a)$
List Comparison	
$min(len(a), 0) \rightarrow 0$	
Pair Operations	
$fst(toPair(a, b)) \rightarrow a$	$snd(toPair(a, b)) \rightarrow b$

As mentioned in Sect. 2.1, HOTGP enforces a maximum tree depth limit. To ensure adherence to this constraint, we defined a custom heuristic cost function as $C(tree) = 2 \times (total\_nodes) + 1$, where $total\_nodes$ is the total number of nodes in the extracted tree. This specific heuristic was chosen to strongly penalize programs with a high node count, thereby favoring smaller and shallower trees that are more likely to respect the maximum depth limit imposed by HOTGP.

4   Experimental Setup

The experimental method was designed to conduct a fair comparison between the original HOTGP algorithm and our modified version, HOTGP + EqSat. All experiments were performed using problems from the General Program Synthesis Benchmark Suite [7], a standard collection used for evaluating GP systems.

We selected a subset of five benchmarks from the suite to serve as a proof-of-concept for the proposed integration. Due to the computational overhead introduced by the e-graph saturation process combined with the base cost of the Haskell-based HOTGP implementation, we adopted a focused experimental design with 10 independent runs per benchmark. Despite this sample size, the observed reduction in code bloat was consistent and substantial across the trials (as detailed in Sect. 5), providing evidence for the method's efficacy in simplifying program structures within the scope of this study.

- ***compare-string-lengths***: A baseline problem where the original HOTGP already achieves 100% success, which consists of determining if a string's length is between the lengths of two other strings.
- ***count-odds***, *small-or-large* and *vector-average*: Problems with intermediate success rates. *count-odds* requires counting odd numbers in a list; *vector-average* involves calculating the average of a list of numbers; and *small-or-large* is a numerical classification task. These cases are ideal for observing potential performance improvements.
- ***last-index-of-zero***: A more challenging problem where HOTGP struggles, requiring the discovery of the last index of 0 in a list. It was included to test the hypothesis that EqSat's ability to explore equivalent program structures could help the search navigate complex solution spaces and find novel pathways to a correct program.

It is worth noting that while some of these problems may appear algorithmically simple from a human programmer's perspective, they pose significant challenges for GP systems, which must discover the logic without prior semantic knowledge. Moreover, we deliberately included these foundational benchmarks because they provide a controlled environment to isolate and observe the phenomenon of code bloat. In complex problems, distinguishing between necessary complexity (required to solve the task) and incidental complexity (bloat) is often ambiguous. By using these standard benchmarks, we can qualitatively and quantitatively demonstrate the bloat reduction capabilities of EqSat on structures that are well-understood.

Table 2. Results Comparison HOTGP vs. HOTGP + EqSat (% of success).

Benchmark	HOTGP + EqSat		HOTGP	
	Tr	Te	Tr	Te
compare-string-lengths	100	100	100	100
count-odds	70	70	50	50
last-index-of-zero	0	0	0	0
small-or-large	20	60	50	60
vector-average	80	80	70	70

```
if ((length x1) > (length x0)) then
  (((length x1) > (length x0)) && ((length x1) < (length
      ↪ x2)))
else
  False
```

Fig. 4. Solution by original HOTGP for *compare-string-length*.

To ensure a fair comparison, the evolutionary hyperparameters were configured to match those used in the original HOTGP publication [2]. The algorithm was configured with a steady-state of 2 individuals per step, with an initial population of 1000 individuals. The crossover and mutation rates set to 50% each. The maximum tree depth was limited to 15 for the main program and 3 for lambda functions. The search ran for a maximum of 300,000 evaluations or until a perfect solution for the training data was found.

Each algorithm (HOTGP and HOTGP + EqSat) was executed 10 times with different seeds for each benchmark. Performance was measured by the success rate on both the training (Tr) and test (Te) datasets, where success is defined as finding a program that correctly solves all cases.

5   Results and Discussion

This section presents the empirical evaluation of our proposed method, HOTGP + EqSat, comparing its performance against the baseline HOTGP algorithm. We first present the results in terms of success rates across the selected benchmarks and then we discuss the results, analyzing the impact of EqSat. Finally, we provide a qualitative analysis comparing the structure of programs generated by both approaches to illustrate the effect EqSat has in the reduction of bloat.

The main quantitative results, comparing the success rates of the original HOTGP and our modified version (HOTGP + EqSat) on both the training (Tr) and test (Te) datasets, are summarized in Table 2. Success is defined as the percentage of runs (out of 10) that found a program correctly solving all provided examples within the evaluation limit.

The analysis of the results indicates that the integration of Equality Saturation had a mixed impact. For the *count-odds* benchmark, the modified version showed a 20% improvement in the success rate, from 50% to 70%. A more modest 10% improvement was observed for *vector-average*. For benchmarks like *count-odds*, where success rates increased, HOTGP + EqSat consistently removed significant code bloat present in the original HOTGP solutions, producing more concise and readable programs.

To quantify the bloat reduction effect, we also measured the size (in node count) of the final program from each run, regardless of its correctness. Table 3 presents the mean size and standard deviation for both algorithms.

The data show a clear trend: HOTGP + EqSat produced, on average, smaller programs across all five benchmarks. This effect is most pronounced on benchmarks like *count-odds* and *vector-average*, where the mean solution size was reduced by more than 50%. This provides strong quantitative evidence that integrating EqSat actively reduces bloat during the evolutionary search, even on problems where a perfect solution was not found (*last-index-of-zero*).

Beyond the differences in success rates, we also observed a difference in the consistency with which the algorithms converged to concise solutions. For instance, consider the *compare-string-lengths* benchmark, where a known minimal solution is `((length x1) > (length x0)) && ((length x2) > (length x1))` that involves comparing the lengths of the input strings.

In our 10 successful runs for this benchmark, HOTGP + EqSat converged to this exact minimal structure (or a trivially equivalent one with the same node count) 6 out of 10 times. In contrast, the original HOTGP, while also achieving 100% success, found this minimal structure only 3 out of 10 times, often producing larger, functionally equivalent programs containing redundant checks or operations, as shown in Fig. 4.

This bloated structure explicitly checks the first condition `((length x1) > (length x0)))` twice and uses a verbose `if-then-else` False construct instead of a simple conjunction.

Table 3. Mean solution size (± standard deviation) of the best program from each of the 10 runs, per benchmark. Size is measured in node count.

Benchmark	HOTGP + EqSat	HOTGP
	(Mean Size ± Std Dev)	(Mean Size ± Std Dev)
compare-string-lengths	14.5 ± 4.8	19.8 ± 10.9
count-odds	112.5 ± 98.4	226.1 ± 212
last-index-of-zero	104 ± 136.2	146.9 ± 75.5
small-or-large	30.6 ± 33.8	48.1 ± 45
vector-average	45.3 ± 83.4	105.1 ± 164.8

```
if ((min (max -766 (length x0)) -1) == (1 + (length x0)))
then 0
else (length (filter (\y -> (mod y 2) > 0)
                (map (\y -> y) x0)))
```

(a) Solution by original HOTGP for *count-odds*.

```
length (filter (\y ->
    (product (singleton 0)) < (mod y (2 * 1))) x0)
```

(b) Simplified solution by HOTGP + EqSat.

Fig. 5. Comparison of generated solutions for the *count-odds* benchmark, illustrating the bloat reduction effect of integrating EqSat.

```
if (671 < ((div -5332 (length (showInt ((length (showInt
    ↪ ((length (showInt (5 * x0))) * x0))) * x0)))) + x0))
    ↪ then "large" else (take ((length (if (x0 == 2054)
    ↪ then "small" else "large")) - ((div -3345 (length
    ↪ (showInt x0))) + x0)) (take ((1679 - x0) - 671)
    ↪ "small"))
```

(a) HOTGP solution for training and test datasets.

```
if (((div x0 (5 - x0)) + x0) > ((max (5 - x0) -4779) +
    ↪ 1992)) then (take (4 - ((5 - x0) + 1992)) (if (((div
    ↪ x0 (max (5 - x0) -4779)) + x0) > 1997) then "large"
    ↪ else (take (((5 - x0) + 5) - 5) "large"))) else
    ↪ "small"
```

(b) HOTGP + EqSat solution for training and test datasets.

Fig. 6. Comparison of correct solutions for training and test datasets on the *small-or-large* benchmark.

Furthermore, EqSat demonstrates simplification effect on more complex solutions. Figure 5 contrasts typical correct solutions found for the *count-odds* benchmark. The original HOTGP solution (Fig. 5a) exhibits significant bloat, including a complex conditional likely evaluating to false and a redundant identity mapping using `map (\y -> y)`. While not always producing the absolute minimal code, HOTGP + EqSat consistently removed such obvious redundancies. The example solution found via HOTGP + EqSat (Fig. 5b) eliminates the unnecessary conditional and map, presenting much clearer logic, equivalent to `length (filter (\y -> 0 < (mod y 2)) x0)`. This highlights EqSat's role in simplifying code during evolution, leading to more readable, though not always perfectly minimal, programs.

For the *vector-average* benchmark, where success also improved (from 70% to 80%), our analysis showed that EqSat did not interfere with the optimal solution `sum x0/(fromIntegral (length x0))`. The original HOTGP was already capable of finding the minimal, human-readable solution. Our HOTGP + EqSat algorithm consistently converged to this same structure. This is a positive result, as it demonstrates that the simplification process correctly identifies an already-optimized program and does not negatively impact the search.

The most complex result occurred on the *small-or-large* benchmark. This benchmark is challenging since it requires the guessing of two integer values that might not be identifiable with the choice of training set. To achieve these constants, the search often resorts to complicated expressions that escape the simplification capabilities of EqSat, unless additional information is provided (e.g., the range of x) (Fig. 6).

Finally, for the *last-index-of-zero* benchmark, both algorithms failed to find a correct solution, each achieving a 0% success rate on both training and test sets. As no correct solutions were generated, a qualitative comparison of program structure is not applicable.

6   Conclusion

This work investigated the integration of the Equality Saturation into HOTGP to optimize and simplify programs during the evolutionary process. The hypothesis was that simplifying the most promising individuals in each generation could guide the search for better solutions.

The results show that the approach is promising. On the quantitative side, it had a mixed impact on success rates, with notable improvements on benchmarks like *count-odds* (50% to 70%) and *vector-average* (70% to 80%). More significantly, our size analysis in Table 3 confirmed a bloat reduction effect, with HOTGP + EqSat producing smaller programs on average across all benchmarks. This quantitative finding was supported by our qualitative analysis, which showed that EqSat not only simplified bloated code structures (as seen for *count-odds*) but also guided the search more consistently towards minimal forms (as observed for *compare-string-lengths*).

However, the degradation of performance on the *small-or-large* benchmark indicates that the indiscriminate application of rewrite rules can be detrimental. In this case, the simplification process appears to have negatively impacted the evolutionary search itself, causing the training set success rate to drop significantly (from 50% to 20%). This suggests that the logical simplifications, while correct, may have prematurely guided the search into local optima or eliminated pathways in the search space that were necessary to solve the training problem. This highlights a complex trade-off where simplification can sometimes hinder, rather than help, the discovery of a correct solution.

Acknowledgements. F.O.F. is supported by Conselho Nacional de Desenvolvimento Científico e Tecnológico (CNPq) grant 301596/2022-0.

References

1. David, C., Kroening, D.: Program synthesis: challenges and opportunities. Philos. Trans. Roy. Soc. A: Math. Phys. Eng. Sci. **375**(2104), 20150403 (2017)
2. Fernandes, M.C., De FranÇa, F.O., Francesquini, E.: HotGP - higher-order typed genetic programming. In: Proceedings of the Genetic and Evolutionary Computation Conference. GECCO '23, pp. 1091–1099. Association for Computing Machinery, New York, NY, USA (2023). https://doi.org/10.1145/3583131.3590464
3. de Franca, F.O., Kronberger, G.: Reducing overparameterization of symbolic regression models with equality saturation. In: Proceedings of the Genetic and Evolutionary Computation Conference, pp. 1064–1072 (2023)
4. de Franca, F.O., Kronberger, G.: Improving genetic programming for symbolic regression with equality graphs. In: Proceedings of the Genetic and Evolutionary Computation Conference, pp. 989–998 (2025)
5. Gulwani, S.: Dimensions in program synthesis. In: Proceedings of the 12th International ACM SIGPLAN Symposium on Principles and Practice of Declarative Programming, pp. 13–24 (2010)
6. Helmuth, T., Spector, L.: General program synthesis benchmark suite. In: Proceedings of the 2015 Annual Conference on Genetic and Evolutionary Computation, pp. 1039–1046 (2015)
7. Helmuth, T., Spector, L.: General program synthesis benchmark suite. In: Proceedings of the 2015 Annual Conference on Genetic and Evolutionary Computation. GECCO '15, pp. 1039–1046. Association for Computing Machinery, New York, NY, USA (2015). https://doi.org/10.1145/2739480.2754769
8. Koza, J.: Genetic programming: a paradigm for genetically breeding populations of computer programs to solve problems. Technical report STAN-CS-90-1314, Department of Computer Science, Stanford University (1990)
9. Koza, J.R.: Genetic programming as a means for programming computers by natural selection. Stat. Computi. **4**(2) (1994). https://doi.org/10.1007/BF00175355
10. Koza, J.R., et al.: Hierarchical genetic algorithms operating on populations of computer programs. In: IJCAI, vol. 89, pp. 768–774 (1989)
11. Kronberger, G., Olivetti de Franca, F., Desmond, H., Bartlett, D.J., Kammerer, L.: The inefficiency of genetic programming for symbolic regression. In: Affenzeller, M., et al. (eds.) PPSN 2024. LNCS, vol. 15148, pp. 273–289. Springer, Cham (2024). https://doi.org/10.1007/978-3-031-70055-2_17
12. Luke, S., Panait, L.: A comparison of bloat control methods for genetic programming. Evol. Comput. **14**, 309–44 (2006). https://doi.org/10.1162/evco.2006.14.3.309
13. Manna, Z., Waldinger, R.J.: Toward automatic program synthesis. Commun. ACM **14**(3), 151–165 (1971)
14. Montana, D.J.: Strongly typed genetic programming. Evol. Comput. **3**(2), 199–230 (1995). https://doi.org/10.1162/evco.1995.3.2.199
15. Olsson, R.: Inductive functional programming using incremental program transformation. Artif. Intell. **74**(1), 55–81 (1995). https://doi.org/10.1016/0004-3702(94)00042-Y
16. Pantridge, E., Spector, L.: Code building genetic programming. In: Proceedings of the 2020 Genetic and Evolutionary Computation Conference, pp. 994–1002 (2020)
17. Poli, R.: A simple but theoretically-motivated method to control bloat in genetic programming. In: Ryan, C., Soule, T., Keijzer, M., Tsang, E., Poli, R., Costa, E. (eds.) EuroGP 2003. LNCS, vol. 2610, pp. 204–217. Springer, Heidelberg (2003). https://doi.org/10.1007/3-540-36599-0_19

18. Srivastava, S., Gulwani, S., Foster, J.S.: From program verification to program synthesis. In: Proceedings of the 37th Annual ACM SIGPLAN-SIGACT Symposium on Principles of Programming Languages, pp. 313–326 (2010)
19. Tate, R., Stepp, M., Tatlock, Z., Lerner, S.: Equality saturation: a new approach to optimization. In: Proceedings of the 36th Annual ACM SIGPLAN-SIGACT Symposium on Principles of Programming Languages, pp. 264–276 (2009)
20. Vanneschi, L., Castelli, M., Silva, S.: Measuring bloat, overfitting and functional complexity in genetic programming. In: Proceedings of the 12th Annual Conference on Genetic and Evolutionary Computation. GECCO '10, pp. 877–884. Association for Computing Machinery, New York, NY, USA (2010). https://doi.org/10.1145/1830483.1830643
21. Willsey, M., Nandi, C., Wang, Y.R., Flatt, O., Tatlock, Z., Panchekha, P.: egg: Fast and extensible equality saturation. Proc. ACM Program. Lang. **5**(POPL) (2021). https://doi.org/10.1145/3434304
22. Yu, T.: Polymorphism and genetic programming. In: Miller, J., Tomassini, M., Lanzi, P.L., Ryan, C., Tettamanzi, A.G.B., Langdon, W.B. (eds.) EuroGP 2001. LNCS, vol. 2038, pp. 218–233. Springer, Heidelberg (2001). https://doi.org/10.1007/3-540-45355-5_17
23. Yu, T.: Polymorphism and genetic programming. In: Proceedings of the 4th European Conference on Genetic Programming. EuroGP '01, pp. 218–233. Springer, Heidelberg (2001)

Revisiting SLIM: Improved Learning Dynamics and Model Compactness in Symbolic Regression

Gorka Silva[1], Lachlan Stewart[2], Illya Bakurov[5], Mauro Castelli[4],
Davide Farinati[4,6,7(✉)], Jose Manuel Muñoz Contreras[3], Leonardo Trujillio[3],
and Leonardo Vanneschi[4]

[1] Universidad Complutense de Madrid, Madrid, Spain
[2] Australian National University, Canberra, Australia
[3] Tecnológico Nacional de México/Instituto Tecnológico de Tijuana, Tijuana, Mexico
[4] NOVA Information Management School (NOVA IMS), Universidade Nova de
Lisboa, Campus de Campolide, 1070-312 Lisbon, Portugal
[5] Michigan State University, Milan, USA
[6] Vita-Salute San Raffaele University, Milan, Italy
farinati.davide@hsr.it
[7] Comprehensive Cancer Center/Unit of Urology; URI; IRCCS Ospedale San
Raffaele, Milan, Italy

Abstract. Geometric Semantic Genetic Programming (GSGP) induces unimodal error surfaces for supervised learning problems, enabling efficient search within semantic space. However, its conventional operators cause rapid model growth, limiting interpretability. The Semantic Learning algorithm based on Inflate and deflate Mutations (SLIM) addresses this limitation through size-reducing deflate mutations, while preserving GSGP's theoretical properties. This work presents a systematic enhancement of SLIM through two sets of improvements: (i) theoretically grounded refinements, including optimal mutation step and linear scaling; and (ii) heuristic extensions, including Pareto-based tournament selection, multi-objective model identification, bounded mutation steps for implicit regularization and automatic algebraic simplification. A comprehensive evaluation on regression tasks demonstrates that all of the proposed SLIM enhancements, when considered separately, match or exceed baseline performance on test data, while achieving a significant reduction of model size. However, the best results were achieved when all enhancements are employed together. These results position the enhanced SLIM as a promising step toward more accurate, efficient, and interpretable symbolic regression for real-world data modeling.

Keywords: GP · Semantics · SLIM · Pareto Tournament · Linear Scaling · Optimal Mutation Step

1 Introduction

Many machine learning (ML) techniques have been proposed to derive data-driven regression models, including deep learning approaches [15] and state-

G. Silva and L. Stewart—Contributed equally to this work.

© The Author(s), under exclusive license to Springer Nature Switzerland AG 2026
L. Manzoni et al. (Eds.): EuroGP 2026, LNCS 16521, pp. 156–172, 2026.
https://doi.org/10.1007/978-3-032-23005-8_10

of-the-art ensembles of decision trees [25]. However, standard ML approaches present an important trade-off: they achieve high accuracy but produce black-box models [1,13]. While it is possible to derive useful explanations of how these models make inferences, achieving model interpretability is usually not feasible, given the size of the models or the way in which they are expressed [1].

When both model interpretability and accuracy are required, as is the case in Symbolic Regression (SR), Genetic Programming (GP) methods, such as Operon [3], GP-GOMEA [31] or PySR [5], among others, can be suitable approaches [12]. In fact, GP methods can potentially achieve this interpretability-accuracy balance by simultaneously optimizing model performance, structure and size. However, the syntactic search performed by GP can be inefficient and difficult to guide compared with conventional approaches used for ML training [12]. This happens because GP models are transformed through stochastic structural modifications (such as subtree crossover and mutation), that often result in abrupt or unpredictable changes in performance [21]. One way to overcome search inefficiency is to use geometric semantic operators (GSOs) [16]. These operators modify model syntax in such a way that the corresponding changes in the output (semantic) space are bounded. Such is the strategy of Geometric Semantic Genetic Programming (GSGP), which has been shown to achieve strong performance, thanks to its ability to induce a unimodal error surface for supervised learning problems [17]. The main issue with this method, however, is that the original mutation operator always produces large models (number of nodes in a syntax tree representation). While modeling accuracy is competitive, and GSGP can be implemented efficiently using parallel processing [27], it produces significantly larger models than other GP methods, thereby eliminating GP's core advantage compared to other ML techniques. However, a recent variant of GSGP called Semantic Learning algorithm based on Inflate and deflate Mutation (SLIM), has shown that it is possible to maintain the semantic properties while also using genetic operators that reduce model size. This variant has achieved good accuracy while dramatically reducing model size [28].

This work builds on SLIM and presents an enhanced version which accelerates convergence, increases accuracy and achieves a significant reduction in model size. To obtain this, several improvements are proposed. First, we incorporate simple enhancements that have already been shown to help GSGP in other works [18,27]. In particular, by computing an optimal mutation step for each mutation event and performing linear scaling of the outputs of each model. Second, by incorporating other heuristics that have been used with other GP or ML systems, but not specifically with GSGP, such as Pareto tournament selection, Pareto-based selection of the best model found, and a bounded mutation step for regularization. These simple yet powerful enhancements to SLIM yield state-of-the-art performance among GSGP algorithms, moving it a step closer to becoming a leading symbolic regression framework by uniting strong predictive accuracy with the potential for improved interpretability.

The paper is organized as follows. Background is presented in Sect. 2, while our proposal is outlined in Sect. 3. Experimental work and results are discussed in Sect. 4, while conclusions and future work are outlined in Sect. 5.

2   Background on GSGP and SLIM

This section provides and overview of GSGP, focusing on Geometric Semantic Mutation (GSM). and outlines SLIM, its recently developed variant.

Geometric Semantic Genetic Programming. In supervised learning, GP individuals typically represent models that perform input-to-output mappings. For a GP individual P and an input set $X = \{x_1, x_2, \ldots, x_n\}$, we define the output vector $s_P = \{P(x_1), P(x_2), \ldots, P(x_n)\}$ as the *semantics* of individual P [16]. Classical GP employs search operators that randomly modify the syntactic structure of individuals. The semantic effects of these operators, however, are difficult to predict. Minor syntactic modifications may result in substantial semantic variations, degrading locality and hindering the identification of optimal solutions [20]. To overcome this limitation, semantic awareness has been incorporated into GP [29]. Among various approaches, GSGP is as a GP variant that substitutes conventional syntax-based operators with Geometric Semantic Operator (GSO)s [16]. GSOs, despite altering individual syntax, have predictable geometric properties within the semantic space and induce a unimodal error surface. For a parent $T : \mathbb{R}^n \to \mathbb{R}$, GSM produces the function $T_M = T + \mathrm{ms} \cdot (T_{R1} - T_{R2})$, where ms is a parameter called mutation step and T_{R1} and T_{R2} denote random real functions with outputs constrained to the interval $[0, 1]$. This constraint is typically imposed by wrapping the two random expressions within a sigmoid function. GSM produces an individual located within a hyper-sphere of radius ms centered at the parent's semantics within semantic space.

Semantic Learning Algorithm Based on Inflate and Deflate Mutation. While several variants of GSGP have been proposed [4, 19, 23], it has one important limitation. The offspring produced by GSOs are always larger than their parents, leading to larger models with poor interpretability [16]. Various solutions have been developed to mitigate this problem, including dynamic population strategies [7] and gradient descent integration [23]. Nevertheless, most current methods still rely on the GSO developed by [16] or employ new operators that generate offspring exceeding parent size [2]. Recently, a novel GSM was developed [28] that preserves the characteristics of conventional semantic mutation (now termed Inflate Geometric Semantic Mutation (IGSM)), while simultaneously generating individuals smaller than their parents. This novel operator, designated Deflate Geometric Semantic Mutation (DGSM), owes its introduction to two intuitions. First, the IGSM definition from the previous section can trivially be reformulated as:

$$IGSM\,(T) = T + \mathrm{ms} \cdot (T_{R1} - T_{R2}) = T - \mathrm{ms} \cdot (T_{R2} - T_{R1}),$$

Second, the two random programs, T_{R1} and T_{R2}, are exchangeable as they constitute independent random variables sharing identical probability distributions. Consequently,the definition of IGSM may be rewritten as:

$$IGSM\,(T) = T - \mathrm{ms} \cdot (T_{R1} - T_{R2}).$$

Consider now an individual T that has undergone IGSM application, say, three times, yielding the new individual:

$$T + \text{ms} \cdot (T_{R1} - T_{R2}) + \text{ms} \cdot (T_{R3} - T_{R4}) + \text{ms} \cdot (T_{R5} - T_{R6}).$$

Because the two definitions are equivalent, we may now apply IGSM again employing subtraction, producing the individual:

$$T + \text{ms} \cdot (T_{R1} - T_{R2}) + \text{ms} \cdot (T_{R3} - T_{R4}) + \text{ms} \cdot (T_{R5} - T_{R6}) - \text{ms} \cdot (T_{R7} - T_{R8}).$$

Individual size continues increasing with each IGSM application. However, by applying IGSM with subtraction and reusing a random program pair from a prior mutation event, such as for instance T_{R3} and T_{R4}, we can simplify the expression to:

$$T + \text{ms} \cdot (T_{R1} - T_{R2}) + \cancel{\text{ms} \cdot (T_{R3} - T_{R4})} + \text{ms} \cdot (T_{R5} - T_{R6}) - \cancel{\text{ms} \cdot (T_{R3} - T_{R4})}.$$

The resulting individual possesses a smaller genotype compared to its parent. This motivates the DGSM, which simply removes a previously added term. DGSM maintains the same semantic characteristics as IGSM, perturbing each semantic coordinate within $[-\text{ms}, \text{ms}]$, and inducing a unimodal error surface for supervised learning problems, exactly as IGSM.

The SLIM algorithm, as its name indicates, utilizes both IGSM and DGSM as two independent operators, each having its own application probability, with the sole constraint that when a program contains only one term, solely IGSM may be applied. Specifically, the application frequency of each GSM is governed by a parameter that regulates the probability of applying DGSM rather than IGSM. Setting this parameter to zero results in an exclusive IGSM application, rendering SLIM entirely equivalent to conventional GSGP.

A common criticism directed at SLIM, and in particular at the DGSM, concerns the fact that, during the evolutionary process, parts of the genotype that were previously accepted may later be removed. At first glance, this behavior appears counterintuitive: SLIM operates on a unimodal fitness landscape, and genetic material that was added and accepted earlier during evolution should have contributed to moving the solution closer to the target. Removing such code should always push the solution further away. However, this intuition is wrong.

At each generation, new pieces of code are introduced, and components that were once beneficial may become detrimental in the presence of newly added genotypic elements. This phenomenon is addressed and empirically demonstrated by Farinati et al. [8].

Variants of SLIM. In [22], the authors examined three distinct methods (SIG2, ABS, SIG1) for defining a function that produces values within the range [-ms, ms], along with two approaches for perturbing an individual to accomplish ball mutation in the semantic space. The three functions that produce values within [-ms, ms] are:

1. Utilizing two random trees T_{R1} and T_{R2}, each passed through the sigmoid function S (mapping outputs to $[0, 1]$) as follows (standard GSM):

$$\text{SIG2} = ms \cdot (S(T_{R1}) - S(T_{R2})) \,.$$

2. Employing a single random tree T_R normalized using the absolute value:

$$\text{ABS} = ms \cdot (1 - 2/(1 + |T_R|)) \,.$$

3. Utilizing a single random tree T_R passed through the sigmoid function S:

$$\text{SIG1} = ms \cdot (2 \cdot S(T_R) - 1) \,.$$

Also, a small perturbation can be accomplished by adding a positive or negative quantity close to zero, or by multiplying by a factor slightly smaller or larger than one. Correspondingly, there are two methods to mutate individuals in SLIM: (1) IGSM adds the function, while DGSM subtracts it; and (2) IGSM multiplies by $(1 + \text{function})$, while DGSM divides by it.

Combining both methods for perturbing an individual with the three functions defined for mutation, we obtain six SLIM variants. These are designated as SLIM+ or SLIM*, depending on whether they employ addition or multiplication, followed by SIG2, ABS, or SIG1, based on the function applied. This work focuses on (SLIM+ABS) due to its strong performance, clearly demonstrated in previous work [22], and its simple construction.

Linked-List Implementation of SLIM. In order to implement SLIM, it is useful to represent each individual's genotype as a linked list of subexpressions [22]. The IGSM appends one element to the list's end, whereas the DGSM removes one element from the list. Depending on the variant employed, the individual's semantics can be computed by summing or multiplying the semantics of all blocks. A visual representation of the linked list implementation and the application of the IGSM and DGSM are reported in Fig. 1.

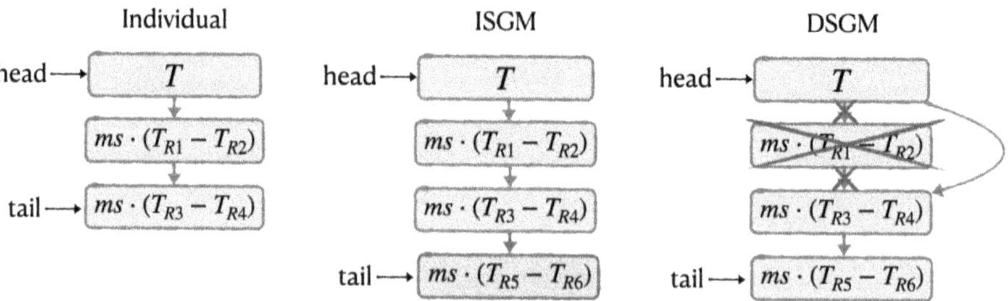

Fig. 1. An example genotype of a SLIM individual, with a visual representation of the effect of the IGSM and DGSM on its genotype. IGSM appends an item at the end of the list, while DGSM removes an item.

An open-source implementation[1] of SLIM was introduced in [24], and it was used in this study. The code for the full reproducibility of the experiments ran in this work can be found at https://github.com/lachlanjs/species-ss-slim-gsgp.

3   Proposed Extensions to SLIM

Initial results with SLIM have been promising [8,24,28,30]. However, this work hypothesizes that performance can still be improved by enhancing the search process. First, to improve the predictive accuracy of the models. Second, to reduce model size and increase the potential for interpretability.

3.1   Improving Model Accuracy

The first two algorithmic enhancements are expected to improve performance. Standard GSGP and SLIM use mutation steps ms that are randomly generated at each mutation event, as suggested in [9]. However, the ms need not be be random. Given the unimodal fitness landscape induced by the GSGP operators [16], the mutation step can be optimally computed based on the parent, the target, and the random programs involved. On the other hand, it is trivial to adjust the scale and the bias of a model's output to better fit the target variable, using a simple adjustment known as linear scaling [10,18]. Both approaches have been shown to improve the accuracy of GSGP, and should have a positive effect on SLIM performance. Moreover, these heuristics improve the convergence of the algorithm, allowing it to achieve low errors with fewer iterations. This can also have a positive effect on model size, by reducing the amount of inflate mutations.

Optimal Mutation Steps with Regularization. A key hyperparameter of IGSM is the mutation step ms, which can be statically or randomly set at each generation or mutation event. Another strategy is to compute an optimal mutation step (OMS) at each mutation event [9,14]. Let $\mathbf{s}$ be the semantic vector of program T, $\mathbf{t}$ the desired or target semantic vector, and $\mathbf{s}_R$ the semantic vector produced by the random subtree added during mutation, in particular considering, for example, the +SIG or +ABS variants. Then, the goal is to find a program T such that

$$\mathbf{t} = \mathbf{s} + \mathrm{ms} * \mathbf{s}_R .\tag{1}$$

For any $\mathbf{s}$ and $\mathbf{s}_R$, then we can try to find the best possible mutation step, by

$$\mathrm{ms} = (\mathbf{t} - \mathbf{s}) \cdot \mathbf{s}_R^{-1},\tag{2}$$

where $\mathbf{s}_R^{-1}$ is the pseudo-inverse of $\mathbf{s}_R$, which can be efficiently computed using the Moore-Penrose method [9]. One drawback is that under certain conditions, the pseudo-inverse can become unstable, producing either very large or very small values. Therefore, we include a simple regularization step to the OMS (ROMS)

---

[1] https://github.com/DALabNOVA/slim.

by bounding the values from above and from below, such that the ms is clipped to ms_l ($-ms_l$) when it goes beyond this limit. Moreover, it is set to 0 if it falls below a certain threshold $ms < ms_\epsilon$. Practically, this means that the mutation event is canceled when it has a negligible effect on semantics.

Linear Scaling (LS). It modifies the fitness function by calculating the slope and intercept of each GP model [10,18], such that the sum of the squared errors between the target and the model outputs is minimized. The slope and intercept are computed on the training data, and used to make predictions at inference time. By computing the slope and intercept of the data, GP can search for the model that best fits the target shape without accounting for general trends.

3.2 Reducing Model Size

The main limitation of GSGP is its tendency to produce large models. While the deflate mutation helps addresses this issue, this work proposes additional heuristics to curtail model growth. The ROMS should already have a positive effect on model size when it sets ms to 0, effectively cancelling some mutation events. Several other improvements are presented below.

Pareto Selection. Pareto criteria are used for parent selection and for choosing the best model at the end of a run. Pareto dominance, in terms of model size and accuracy, has been used to reduce code bloat in GP before [6,11,26], but not with GSGP. Pareto tournaments (PT) are for parent selection [11,26]. Like in regular tournament selection, a PT takes as input n randomly chosen individuals from the population. Afterward, the non-dominated individuals are identified, and one is randomly chosen as the tournament winner. Solutions are evaluated based on error (fitness) and size (number of nodes). Moreover, to select the final model to return as the best solution found, three steps are performed. First, the set of non-dominated solutions in the final population is identified. Second, we eliminate outlier individuals using the standard $1.5 \times IQR$ rule, considering either fitness or model size. Finally, the solution in the non-dominated set that is closest to the ideal solution (the origin of objective space) is computed, referred to as the optimal compromise model. The method also returns the model with the best fitness (lowest error) and the best (smallest) size.

Automatic Model Simplification (AS). Given the way that GP evolves models, relying on random syntactic modifications, it is desirable to simplify the models found, which is feasible because they are expressed symbolically. Therefore, before choosing the best solution from the final population, all non-dominated models are simplified. Each SLIM individual is first converted into a tree structure in PyTorch, and three categories of simplification rules are applied. First, identity simplifications (e.g., $x + 0 \rightarrow x$, $x \times 1 \rightarrow x$, $x - x \rightarrow 0$), constant folding (e.g., $5.0 + 3.0 \rightarrow 8.0$), and nested constant optimization (e.g., $c_1 \times (c_2 \times$

$x) \to (c_1 \times c_2) \times x)$. The process is performed recursively via bottom-up traversal. However, given the type of mutations used with SLIM (ABS or SIG), there is a limit to the amount of viable simplifications that can be made given their use of absolute value or exponential functions [2]. Nonetheless, given the use of a regularized OMS, where some values are set to zero, at least in those cases, model simplification is expected to have a non-negligable impact on model size.

4 Experimental Study

The experimental study evaluates the combined effect of the proposed enhancements to SLIM and the impact of each one individually. For this, 14 regression problems that were used to evaluate SLIM in previous work are used [28,30]. The domain, number of samples and number of features of the employed datasets are reported in Table 1. For evaluation, the algorithm is executed 30 times on each dataset, using 60% for training, 20% for validation and 20% for testing. From each run, three models are chosen, based on training performance, from the final dominated set in the final population (for all variants, including the baseline): (i) the model with the best fitness (BF); (ii) the one with the optimal compromise between size and fitness (OC); and (iii) the one with the best (smallest) size (BS).

To quantify the impact of each proposed improvement to SLIM, we analyzed nine new variants that integrate the proposed extensions, namely: optimal mutation step (OMS), Linear Scaling (LS), Pareto Tournament (PT), and automatic simplification (AS), using the following naming convention:

- First, we consider the baseline SLIM using +ABS inflate mutation, denoted as "BASE + .
- Second, we study the "ALL" variant, that uses all of the proposed extensions, joined together in a unique algorithm.
- Third, we evaluate the impact of adding single improvements to the baseline method, denoted as "BASE + [PT | OMS | LS | AS] ", to indicate which component was used; these are four variants.
- Finally, we consider four further variants, removing one of the extensions from the "ALL" method, denoted as "ALL - [PT | OMS | LS | AS] " to indicate which component was not used.

The experiments evaluate the effect produced by each enhancement, but also perform an ablation study to assess what happens when each one is omitted. Results are presented in four ways, considering the median over all 30 runs of the test Root Mean Squared Error (RMSE) and the size of the model (number of nodes): (1) plots that compare the proposed variants with the baseline SLIM implementation, to report the reduction in test RMSE and size as a percentage of the SLIM baseline; (2) median performance of the BF, BS and OC model found by each variant, shown in numerical tables; (3) critical difference diagrams showing the statistical significance of the observed results, separated by the final choice of model; (4) critical difference diagrams, which compare each final model

Table 1. Summary of datasets used for experimental evaluation. S: Number of samples. F: Number of features

ID	Dataset	S	F	ID	Dataset	S	F
1	Airfoil	1502	5	8	Diabetes	442	10
2	Bike_sharing	730	14	9	Efficiency_cooling	768	8
3	Bioavailability	359	241	10	Efficiency_heating	768	8
4	Boston	506	13	11	Forest_fires	517	12
5	Breast_cancer	569	30	12	Parkinson_updrs	5875	19
6	Concrete_slump	103	7	13	PPB	131	626
7	Concrete_strength	1030	8	14	Resid_build_sale_price	1460	79

Table 2. Parameter settings.

Param.	Value	Param.	Value	Param.	Value	Param.	Value
Gens.	100	$p(\text{inflate})$	0.5	$p(\text{cross})$	0.0	Train %	60
Reps.	30	ms_l	100.0	Binary Tourn. size	2	Val %	20
|pop|	100	ms_ϵ	0.1	Pareto Tourn. size	5	Test %	20

choice for the ALL variant and the BF model found by SLIM. Table 2 reports the used hyperparameters, based on those found in previous works with SLIM [28]. The function set uses arithmetic operators with protected division.

4.1   Results and Discussion

Figure 2 summarizes the comparative results, relative to the baseline SLIM (BASE). Figure 2(a) shows a curve for each variant, considering each of the three models chosen from the non-dominated set of solutions in the final population (BF, BS and OC). The curves show the change in median performance (RMSE on the left and size on the right) achieved on each problem, while the performance of BASE is a horizontal line at zero. Negative values indicate a reduction in error or size, which are the desired outcomes. Figure 2(b) summarizes the average (over all problems) of the median (over all runs) change in performance, compared to the BF model found with BASE; this is the model normally returned by GSGP and SLIM. Again, lower values (left-button corner) are preferable in both axis. A line connect the three models (BF,BS and OC) found by each variant.

A statistical comparison is presented in Fig. 3 based on the average rank (considering median performance) of each variant, shown as critical difference diagrams. These are based on the Friedman multigroup test and the Wilcoxon signed-rank test for pairwise comparisons with an $\alpha = 0.05$ and p-value correction using the Holm method. Horizontal lines indicate that there is no statistically significant difference between a subset of connected variants. Table 3 presents the

Revisiting SLIM   165

Table 3. Median fitness results. Rows for each dataset are ordered: Best Fitness, Best Size, Optimal Compromise. Values in bold, underline, and green represent the best in row, dataset and improvement over SLIM respectively.

Prob.	BASE	B. + PT	B. + OMS	B. + LS	B. + AS	A. - PT	A. - OMS	A. - LS	A. - AS	ALL
	(57.01)	58.15	5.03	5.55	57.01	4.94	5.59	5.39	4.87	<u>4.87</u>
1	59.18	64.76	6.69	5.62	59.18	**5.49**	5.71	66.14	5.71	5.71
	57.53	60.57	5.27	5.56	57.53	**5.00**	5.62	16.14	5.03	5.03
	(1035.19)	1036.01	592.57	583.86	1035.19	**<u>375.87</u>**	584.29	603.32	385.04	389.37
2	1037.27	1041.48	664.97	584.57	1037.25	**444.81**	585.31	1041.97	585.68	585.68
	1036.19	1039.02	618.45	584.17	1036.19	**380.70**	584.55	759.09	452.93	456.54
	(46.81)	47.42	31.54	31.22	46.81	31.38	31.18	<u>30.05</u>	31.37	31.46
3	48.26	50.35	31.74	30.89	48.11	**30.65**	30.97	54.80	30.74	30.97
	47.40	48.68	**30.61**	31.05	47.40	30.61	31.07	31.88	31.23	31.33
	(9.27)	9.67	5.18	5.33	9.27	4.81	5.51	5.60	4.80	<u>4.80</u>
4	9.70	10.79	7.59	5.66	9.70	**5.47**	5.84	10.79	5.84	5.84
	9.43	10.09	5.79	5.40	9.43	**4.86**	5.61	6.47	5.07	5.07
	(0.31)	0.34	0.27	<u>0.26</u>	0.31	0.26	0.26	0.27	0.27	0.26
5	0.47	0.54	0.32	0.28	0.47	**0.27**	0.31	0.55	0.31	0.31
	0.36	0.38	0.28	**0.26**	0.36	0.26	0.27	0.31	0.28	0.28
	(9.56)	9.57	7.77	8.20	9.56	8.28	7.97	<u>7.57</u>	7.70	8.16
6	9.97	11.70	8.78	8.06	9.97	8.33	7.95	11.70	7.95	**7.95**
	9.72	10.26	**7.70**	7.99	9.72	8.15	7.92	8.35	7.95	8.11
	(22.81)	23.56	<u>8.66</u>	8.99	22.81	8.94	10.87	9.42	9.64	9.69
7	24.97	28.38	12.33	11.62	24.97	**10.43**	13.55	29.06	13.63	13.63
	23.58	25.70	9.64	9.58	23.58	**9.26**	12.29	13.90	10.60	10.76
	(154.05)	154.94	69.61	58.19	154.05	<u>56.84</u>	57.99	71.43	57.40	57.20
8	155.87	159.64	79.12	**58.93**	155.85	60.38	61.39	158.90	61.39	61.39
8	154.71	157.26	71.01	58.16	154.71	57.80	59.05	78.94	57.68	**57.66**
	(5.75)	6.59	3.53	3.46	5.75	<u>3.29</u>	3.48	3.66	3.30	3.30
9	6.50	10.23	4.29	3.58	6.50	**3.51**	3.88	10.38	3.84	3.84
	5.98	7.63	3.69	3.54	5.98	**3.33**	3.61	4.58	3.48	3.47
	(6.08)	6.33	3.32	3.12	6.08	**<u>2.98</u>**	3.56	3.42	3.09	3.08
10	6.29	8.97	4.00	3.47	6.29	**3.15**	4.16	9.79	4.11	4.11
	6.15	7.04	3.54	3.22	6.15	**3.00**	3.74	4.96	3.39	3.33
	(1.39)	1.39	1.39	1.41	1.39	**1.38**	1.40	1.40	1.39	1.39
11	1.41	1.38	1.41	1.39	1.41	1.39	1.39	<u>1.38</u>	1.39	1.39
	1.40	1.40	1.39	1.41	1.40	1.38	1.39	1.39	1.39	1.39
	(11.68)	11.82	10.07	10.03	11.68	<u>9.89</u>	10.05	10.23	9.90	9.90
12	11.92	12.77	10.43	10.06	11.92	**10.02**	10.14	14.03	10.14	10.14
	11.81	12.14	10.17	10.04	11.81	**9.91**	10.08	10.43	10.00	9.99
	(36.87)	37.17	<u>29.59</u>	30.93	36.87	30.69	29.80	30.05	31.83	32.00
13	37.09	38.88	33.30	**30.20**	37.09	30.75	30.89	39.65	31.24	31.16
	36.98	37.73	**29.94**	30.21	36.98	30.40	30.08	30.80	31.32	31.67
	(106.15)	106.24	66.76	80.05	106.15	65.74	81.09	71.22	60.20	<u>60.20</u>
14	107.61	110.88	77.43	81.02	107.61	**66.44**	86.05	110.65	83.33	82.37
	106.89	108.29	69.20	80.98	106.89	65.06	81.81	79.80	63.68	**63.68**

median performance results of all variants based on test RMSE, while Table 4 shows the same for model size, for all models (BF, BS and OC).

The results show that the proposed extensions provide significant improvements over canonical SLIM, for both predictive accuracy and model size. In particular, the test RMSE shows a marked reduction when either OMS or LS are applied. Importantly, the regularized version of OMS (ROMS) does not negatively affect performance. This suggests that omitting small, semantically

Table 4. Median size results. Rows for each dataset are ordered: Best Fitness, Best Size, Optimal Compromise. Values in bold, underline, and green represent the best in row, dataset and improvement over SLIM respectively.

Prob.	BASE	B. + PT	B. + OMS	B. + LS	B. + AS	A. - PT	A. - OMS	A. - LS	A. - AS	ALL
1	(58.00)	57.00	68.00	65.00	55.00	58.00	54.00	**50.00**	53.00	53.00
	36.00	3.00	35.00	29.00	34.00	29.00	7.00	**3.00**	6.00	6.00
	45.00	29.00	44.00	45.00	42.00	37.00	27.00	**11.00**	21.00	20.00
2	(49.00)	49.00	51.00	51.00	47.00	47.00	49.00	48.00	45.00	**43.00**
	31.00	4.00	35.00	28.00	31.00	27.00	4.00	3.00	3.00	**3.00**
	38.00	25.00	41.00	37.00	37.00	33.00	23.00	21.00	19.00	**19.00**
3	(40.00)	39.00	39.00	40.00	39.00	39.00	39.00	**39.00**	42.00	42.00
	27.00	3.00	27.00	27.00	27.00	27.00	3.00	3.00	3.00	**3.00**
	31.00	21.00	31.00	33.00	31.00	31.00	21.00	**9.00**	19.00	20.00
4	(49.00)	47.00	47.00	48.00	48.00	45.00	45.00	45.00	44.00	**43.00**
	33.00	3.00	27.00	28.00	33.00	23.00	3.00	3.00	3.00	**3.00**
	41.00	23.00	33.00	35.00	39.00	31.00	18.00	**14.00**	17.00	17.00
5	(39.00)	**26.00**	44.00	45.00	39.00	42.00	41.00	39.00	41.00	40.00
	15.00	3.00	27.00	25.00	13.00	26.00	3.00	3.00	3.00	**3.00**
	25.00	11.00	31.00	31.00	23.00	31.00	15.00	**11.00**	15.00	14.00
6	(57.00)	51.00	55.00	56.00	53.00	57.00	53.00	53.00	51.00	**48.00**
	31.00	3.00	31.00	30.00	31.00	27.00	6.00	3.00	4.00	**3.00**
	41.00	23.00	39.00	39.00	39.00	36.00	27.00	**11.00**	17.00	21.00
7	(52.00)	49.00	53.00	49.00	51.00	49.00	52.00	48.00	48.00	**48.00**
	35.00	3.00	29.00	27.00	35.00	27.00	5.00	**3.00**	5.00	5.00
	43.00	26.00	35.00	35.00	41.00	36.00	24.00	**11.00**	21.00	19.00
8	(60.00)	54.00	55.00	44.00	57.00	44.00	44.00	54.00	38.00	**36.00**
	39.00	6.00	32.00	23.00	37.00	15.00	3.00	5.00	3.00	**3.00**
	48.00	33.00	41.00	32.00	43.00	26.00	19.00	**11.00**	17.00	16.00
9	(56.00)	55.00	49.00	56.00	53.00	50.00	47.00	51.00	47.00	**47.00**
	35.00	5.00	29.00	31.00	35.00	29.00	**3.00**	5.00	5.00	5.00
	45.00	25.00	35.00	39.00	43.00	37.00	19.00	**11.00**	18.00	19.00
10	(53.00)	53.00	52.00	53.00	52.00	52.00	**47.00**	51.00	53.00	49.00
	35.00	5.00	30.00	29.00	34.00	30.00	4.00	**4.00**	5.00	5.00
	43.00	26.00	37.00	37.00	41.00	38.00	20.00	**14.00**	23.00	23.00
11	(41.00)	**37.00**	43.00	43.00	40.00	43.00	42.00	43.00	42.00	41.00
	18.00	3.00	28.00	25.00	16.00	27.00	3.00	3.00	3.00	**3.00**
	27.00	**16.00**	35.00	32.00	27.00	34.00	17.00	20.00	19.00	19.00
12	(51.00)	46.00	46.00	46.00	49.00	45.00	48.00	45.00	46.00	**45.00**
	31.00	3.00	27.00	29.00	31.00	27.00	3.00	3.00	3.00	**3.00**
	37.00	21.00	33.00	37.00	37.00	33.00	21.00	**11.00**	17.00	17.00
13	(39.00)	41.00	39.00	39.00	39.00	39.00	39.00	39.00	39.00	**39.00**
	31.00	3.00	27.00	27.00	29.00	27.00	3.00	3.00	3.00	**3.00**
	35.00	23.00	33.00	31.00	34.00	31.00	19.00	**15.00**	19.00	19.00
14	(39.00)	40.00	39.00	39.00	39.00	39.00	39.00	39.00	37.00	**36.00**
	27.00	3.00	27.00	27.00	27.00	24.00	3.00	3.00	3.00	**3.00**
	35.00	22.00	31.00	31.00	35.00	30.00	15.00	**11.00**	12.00	13.00

insignificant, moves can help produce more compact models without sacrificing accuracy. The combination of OMS and LS does not degrade performance in any of the studied test problems. On the contrary, very strong results are typically obtained when both extensions are used together. Conversely, when either OMS or LS is removed, accuracy can slightly decrease; however, models that include one or both extensions consistently achieve a lower test RMSE than the baseline.

Furthermore, these extensions tend to slightly reduce model size, indicating that these techniques can provide an additional regularizing effect.

The proposed extensions have a pronounced impact on model size, particularly when the PT selection is employed. The baseline SLIM model already generates more compact models when PT is added, but the smallest models are obtained when PT is combined with the other proposed enhancements. It is worth noting, however, that PT alone tends to negatively affect predictive accuracy. In terms of RMSE, the highest-performing variant is ALL-PT, whereas the BASE+PT configuration ranks among the lowest. The use of PT appears to be beneficial when it is combined with other enhancements, since several variants that apply such a combination tend to achieve a low RMSE, comparable to the best performing variants, while maintaining substantially smaller models. The results achieved through PT are consistent with our intuition: by introducing model size as a secondary objective, the evolutionary search is biased away from focusing exclusively on fitness, thereby evolving more compact models.

The AS extension shows a similar, but less pronounced, effect to PT, with a consistent yet modest reduction in model size. The fact that the effect of AS on model size is modest is due to the difficulty of automatically simplifying functions that are highly non-linear, like the ones added by the IGSM. However, AS leads to a more notable reduction in size when paired with OMS, since some IGSM events are removed during the simplification step when the ms is set to zero.

In general, the proposed extensions to SLIM have a clear impact on performance and model size. But results also depend on the final model selection. While performance is clearly improved when considering the BF model, the improvements in size are less notable in that case compared to the baseline.

Finally, a comparison between the baseline SLIM and the ALL variant is presented, the variant that includes all of the proposed extensions. Analysis focuses on this variant because it achieves the best performance-size tradeoff (see Fig. 2(b)), and also because it is the simpler one to use by practitioners, applying all heuristics improvements instead of toggling them on or off. Figure 4(a) compares the median test RMSE, shown as a critical difference diagram, while Fig. 4(b) shows the same for median size. In all cases, the combined extensions clearly, and substantially, outperform the baseline SLIM. Moreover, in terms of test RMSE, even the BS model found by the ALL variant achieves competitive performance. While the average rank aligns with intuition (BF achieves the best test RMSE, followed by OC and BS), the statistical test suggests that using the smallest model can be a good strategy. This is a relevant achievement, because this means that it is feasible to choose the smallest BS, and so potentially the most interpretable one, without a substantial loss in performance.

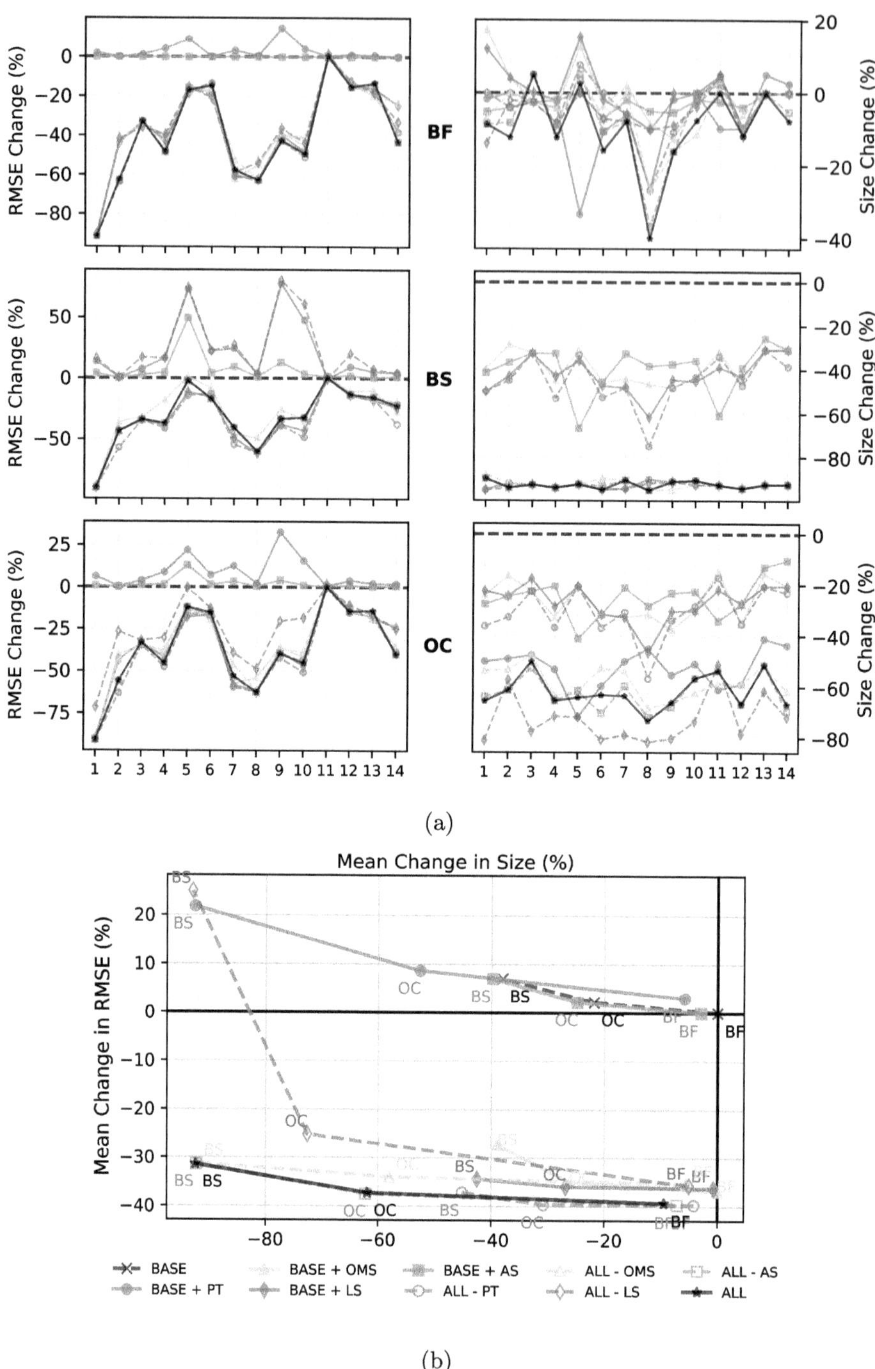

(a)

(b)

Fig. 2. (a): Percentage change of the median test RMSE and size relative to the baseline for three models: Best Fitness (BF), Best Size (BS) and Optimal Compromise (OC). (b): Average median percentage change of the same values, calculated over the fourteen datasets. In all plots lower values are better.

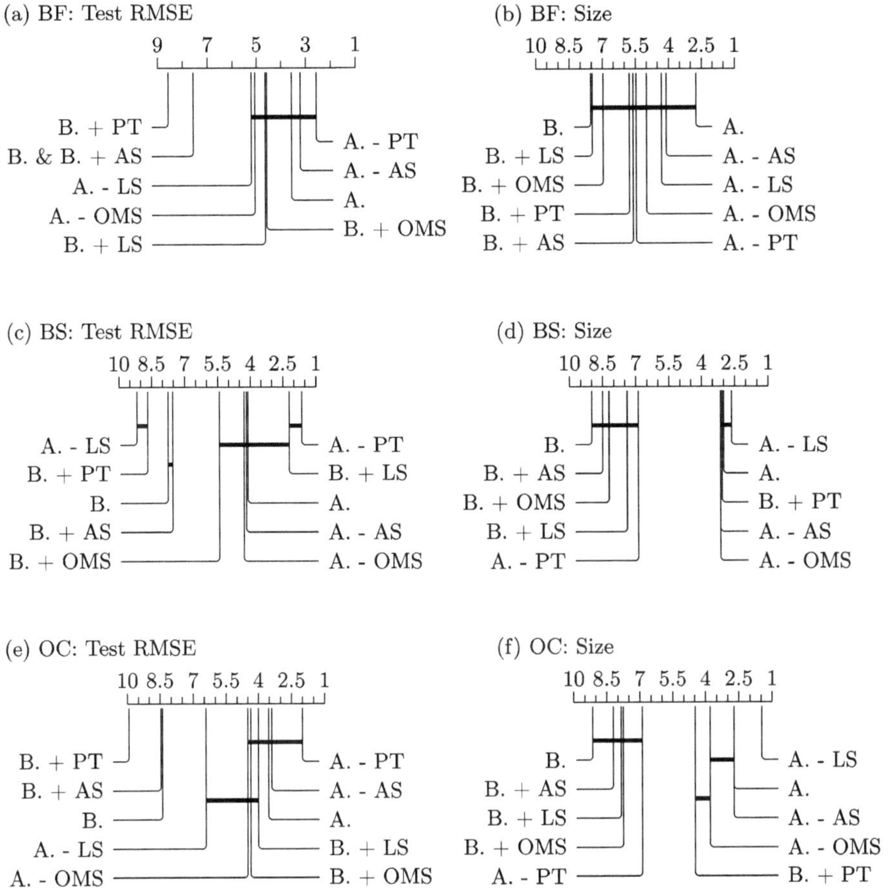

(a) BF: Test RMSE

(b) BF: Size

(c) BS: Test RMSE

(d) BS: Size

(e) OC: Test RMSE

(f) OC: Size

Fig. 3. Critical difference diagrams for the Best Fitness (BF), Optimal Compromise (OC) and Best Size (BS) models; where A.: ALL and B.: BASE.

(a) Test RMSE

(b) Size

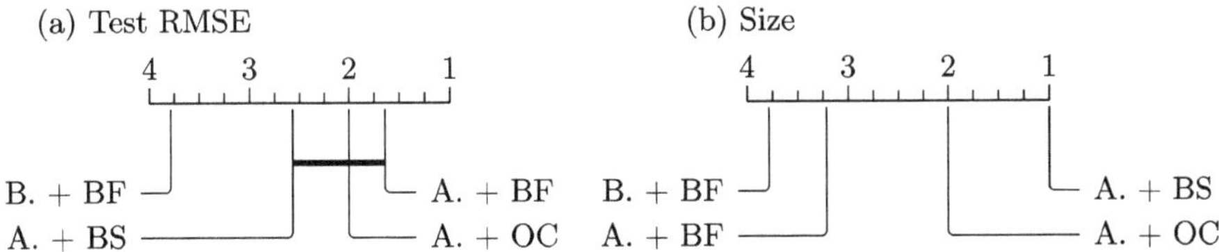

Fig. 4. Critical Difference Diagram comparing the final models (BF, BS and OC) found by the ALL variant with the SLIM (BASE) BF model.

5   Conclusions and Future Work

This study addressed a fundamental challenge for Symbolic Regression (SR): achieving strong predictive performance while maintaining compact solutions.

We presented a systematic enhancement of the Semantic Learning algorithm based on Inflate and Deflate Mutations (SLIM), a recent variant of Geometric Semantic Genetic Programming (GSGP) that combines the theoretical advantages of semantic search with mechanisms that actively control bloat.

Two complementary sets of modifications have been proposed. The first set incorporated algorithmic refinements with established theoretical foundations: (i) optimal mutation step computation with regularization bounds to ensure numerical stability, and (ii) linear scaling of model outputs. The second set introduced more direct extensions: (i) Pareto-based tournament selection, (ii) end-of-run Pareto-optimal model identification, and (iii) automatic algebraic simplification of candidate models before final selection.

Empirical validation demonstrated that the proposed extensions work synergistically to improve the predictive accuracy and compactness of SR models generated by the algorithm. The improvements are consistent across 14 test datasets, always matching or exceeding the performance of SLIM.

In future work, it would be worthwhile to study the effects of these extensions in more detail. We conjecture that our extensions (particularly those that encourage compact solutions) may cause a moderate regulatory effect, thus contributing to preventing overfitting. Additional theoretical analyses of the convergence properties and semantic landscape induced by the proposed methods may also provide further insight into the mechanisms underlying SLIM's success.

Another promising direction for future research involves the application of the improved SLIM to diverse domains, including scientific discovery and data-driven modeling in engineering and personalized medicine. Finally, it will be important to measure, in some way, if the improvements in model size are actually leading to increased interpretability in real-world scenarios.

Acknowledgments. Thanks to the SPECIES Society and the 2025 SPECIES Summer School Coordinator Anna Esparcia-Alcázar for nurturing an invaluable space where scientific ideas can flourish. This work was supported by national funds through FCT (Fundação para a Ciência e a Tecnologia), under the project - UID/04152/2025 -Centro de Investigação em Gestão de Informação (MagIC)/NOVA IMS (https://doi.org/10.54499/UID/04152/2025) - (2025-01-01/2028-12-31) and UID/PRR/04152/2025 (https://doi.org/10.54499/UID/PRR/04152/2025) (2025-01-01/ 2026-06-30). This work was supported by TecNM (Mexico) project 19400.24-P and SECIHTI (Mexico) project CF-2023-I-724.

References

1. Atzmueller, M., Fürnkranz, J., Kliegr, T., Schmid, U.: Explainable and interpretable machine learning and data mining. Data Min. Knowl. Disc. **38**(5), 2571–2595 (2024). https://doi.org/10.1007/s10618-024-01041-y
2. Bakurov, I., et al.: Geometric semantic genetic programming with normalized and standardized random programs. Genetic Program. Evol. Mach. **25**(1), 29 (2024). https://doi.org/10.1007/s10710-024-09479-1

3. Burlacu, B., Kronberger, G., Kommenda, M.: Operon C++: an efficient genetic programming framework for symbolic regression. In: Proceedings of the 2020 Genetic and Evolutionary Computation Conference Companion (2020)

4. Castelli, M., Trujillo, L., Vanneschi, L., Silva, S., Z-Flores, E., Legrand, P.: Geometric semantic genetic programming with local search. In: Proceedings of the 2015 Annual Conference on Genetic and Evolutionary Computation (Madrid, Spain) (GECCO '15), New York, NY, USA, pp. 999–1006. Association for Computing Machinery (2015). https://doi.org/10.1145/2739480.2754795

5. Cranmer, M.: Interpretable Machine Learning for Science with PySR and SymbolicRegression.jl (2023). https://arxiv.org/abs/2305.01582

6. de Jong, E.D., Pollack, J.B.: Multi-objective methods for tree size control. Genet. Program Evolvable Mach. **4**(3), 211–233 (2003). https://doi.org/10.1023/a:1025122906870

7. Farinati, D., Bakurov, I., Vanneschi, L.: A study of dynamic populations in geometric semantic genetic programming. Inf. Sci. **648**, 119513 (2023). https://doi.org/10.1016/j.ins.2023.119513

8. Farinati, D., Pietropolli, G., Vanneschi, L.: A study on the dynamics and effectiveness of the deflate geometric semantic mutation. IEEE Trans. Evol. Comput. 1–1 (2025). https://doi.org/10.1109/TEVC.2025.3611226

9. Gonçalves, I., Silva, S., Fonseca, C.M.: On the generalization ability of geometric semantic genetic programming. In: Machado, P., et al. (eds.) EuroGP 2015. LNCS, vol. 9025, pp. 41–52. Springer, Cham (2015). https://doi.org/10.1007/978-3-319-16501-1_4

10. Keijzer, M.: Improving symbolic regression with interval arithmetic and linear scaling. In: Ryan, C., Soule, T., Keijzer, M., Tsang, E., Poli, R., Costa, E. (eds.) EuroGP 2003. LNCS, vol. 2610, pp. 70–82. Springer, Heidelberg (2003). https://doi.org/10.1007/3-540-36599-0_7

11. Kotanchek, M., Smits, G., Vladislavleva, E.: Pursuing the pareto paradigm: tournaments, algorithm variations and ordinal optimization. In: Riolo, R., Soule, T., Worzel, B. (eds) Genetic Programming Theory and Practice IV, pp. 167–185. Springer, Boston (2007). https://doi.org/10.1007/978-0-387-49650-4_11

12. Cava, W.L., et al.: . Contemporary symbolic regression methods and their relative performance. In: Vanschoren, J., Yeung, S. (eds.) Proceedings of the Neural Information Processing Systems Track on Datasets and Benchmarks, Vol. 1. Curran (2021). https://datasets-benchmarks-proceedings.neurips.cc/paper/2021/hash/c0c7c76d30bd3dcaefc96f40275bdc0a-Abstract-round1.html

13. Liu, S.-Y., Cai, H.-R., Zhou, Q.-L., Ye, H.-J.: TALENT: a tabular analytics and learning toolbox. arXiv preprint arXiv:2407.04057 (2024)

14. McDermott, J., Agapitos, A., Brabazon, A., O'Neill, M.: Geometric semantic genetic programming for financial data. In: Esparcia-Alcázar, A.I., Mora, A.M. (eds.) EvoApplications 2014. LNCS, vol. 8602, pp. 215–226. Springer, Heidelberg (2014). https://doi.org/10.1007/978-3-662-45523-4_18

15. Noor, M.H.M., Ige, A.O.: A survey on state-of-the-art deep learning applications and challenges. Eng. Appl. Artif. Intell. **159**, 111225 (2025). https://doi.org/10.1016/j.engappai.2025.111225

16. Moraglio, A., Krawiec, K., Johnson, C.G.: Geometric semantic genetic programming. In: Coello, C.A.C., Cutello, V., Deb, K., Forrest, S., Nicosia, G., Pavone, M. (eds.) PPSN 2012. LNCS, vol. 7491, pp. 21–31. Springer, Heidelberg (2012). https://doi.org/10.1007/978-3-642-32937-1_3

17. Contreras, J.M.M., Trujillo, L., Hernandez, D.E., Cardenas Florido, L.A.: Benchmarking GSGP: still competitive 10 years later? Genet. Program Evolvable Mach. **26**, 1 (2024). https://doi.org/10.1007/s10710-024-09504-3
18. Nadizar, G., Garrow, F., Sakallioglu, B., Canonne, L., Silva, S., Vanneschi, L.: An investigation of geometric semantic GP with linear scaling. In: Proceedings of the Genetic and Evolutionary Computation Conference (Lisbon, Portugal) (GECCO '23), NY, USA, pp. 1165–1174. Association for Computing Machinery, New York (2023). https://doi.org/10.1145/3583131.3590418
19. Nadizar, G., Sakallioglu, B., Garrow, F., Silva, S., Vanneschi, L.: Geometric semantic GP with linear scaling: darwinian versus Lamarckian evolution. Genet. Program Evolvable Mach. **25**, 2 (2024). 17. https://doi.org/10.1007/s10710-024-09488-0
20. Nguyen, Q.U.: Examining Semantic Diversity and Semantic Locality of Operators in Genetic Programming. Ph. D. Dissertation. University College Dublin, Ireland (2011). http://ncra.ucd.ie/papers/Thesis_Uy_Corrected.pdf
21. Nordin, P., Banzhaf, W.: Complexity compression and evolution. In: International Conference on Genetic Algorithms, pp. 310–317 (1995). https://api.semanticscholar.org/CorpusID:16415863
22. Pietropolli, G., Farinati, D., Manzoni, L., Castelli, M., Silva, S., Vanneschi, L.: Introducing crossover in SLIM-GSGP. In: Xue, B., Manzoni, L., Bakurov, I. (eds.) EuroGP 2025. LNCS, vol. 15609, pp. 103–119. Springer, Cham (2025). https://doi.org/10.1007/978-3-031-89991-1_7
23. Pietropolli, G., Manzoni, L., Paoletti, A., Castelli, M.: On the hybridization of geometric semantic GP with gradient-based optimizers. Genetic Program. Evol. Mach. **24**, 2 (2023). Article number: 16. https://rdcu.be/dpWsR
24. Rosenfeld, L., et al.: Slim_gsgp: a python library for non-bloating GSGP. In: Genetic and Evolutionary Computation Conference (GECCO '25) (New York, NY, USA), Malaga, Spain, p. 9. ACM (2025). https://doi.org/10.1145/3712256.3726398
25. Shwartz-Ziv, R., Armon, A.: Tabular data: deep learning is not all you need. Inf. Fusion **81**(2022), 84–90 (2022). https://doi.org/10.1016/j.inffus.2021.11.011
26. Smits, G.F. Kotanchek, M.: Pareto-front exploitation in symbolic regression, pp. 283–299. Springer, Boston (2005). https://doi.org/10.1007/0-387-23254-0_17
27. Trujillo, L., Muñoz Contreras, J.M., Hernandez, D.E., Castelli, M., Tapia, J.J.: GSGP-CUDA — a CUDA framework for geometric semantic genetic programming. SoftwareX **18**, 101085 (2022). https://doi.org/10.1016/j.softx.2022.101085
28. Vanneschi, L.: . SLIM_GSGP: the non-bloating geometric semantic genetic programming. In: Giacobini, M., Xue, B., Manzoni, L. (eds.) Genetic Programming, pp. 125–141. Springer, Cham (2024). https://doi.org/10.1007/978-3-031-56957-9_8
29. Vanneschi, L., Castelli, M., Silva, S.: A survey of semantic methods in genetic programming. Genet. Program Evolvable Mach. **15**(2), 195–214 (2014). https://doi.org/10.1007/s10710-013-9210-0
30. Vanneschi, L., Farinati, D., Rasteiro, D., Rosenfeld, L., Pietropolli, G., Silva, S.: Exploring non-bloating geometric semantic genetic programming. Genetic Program. Theory Practice XX I **2025**, 237–258 (2025)
31. Virgolin, M., Alderliesten, T., Witteveen, C., Bosman, P.: Improving model-based genetic programming for symbolic regression of small expressions. Evol. Comput. **29**(2021), 211–237 (2021). https://doi.org/10.1162/evco_a_00278

Dynamic Vector and Matrix Memory for Tangled Program Graphs

Ali Naqvi and Stephen Kelly[✉]

McMaster University, Hamilton, Canada
{naqvia18,spkelly}@mcmaster.ca

Abstract. This paper investigates dynamic vector and matrix-memory allocation in Tangled Program Graphs (TPG) for reinforcement learning in continuous control tasks. TPG is a genetic programming framework that evolves control policies composed of multiple programs which compete to predict actions. We give each program its own memory capacity that can be mutated during evolution, and we hypothesize that adaptive capacity improves performance on higher dimensional tasks. TPG policies are evaluated on target-seeking robot control environments in the MuJoCo simulator: Reacher and Ant-Goal. Our results demonstrate that adapting the policy's memory size through interaction with the task produces significantly more proficient controllers in both tasks. Moreover, supporting diverse memory sizes in a single policy results in lower computational cost measured by the approximate number of FLOPs executed per action.

Keywords: Reinforcement Learning · Dynamic Memory · Tangled Program Graphs · Genetic Programming · MuJoCo · Continuous Control

1 Introduction

Reinforcement learning (RL) has emerged as a powerful paradigm for training autonomous agents to perform complex tasks through trial-and-error interaction with an environment [12]. Among genetic programming (GP) approaches, Tangled Program Graphs (TPG) offer a modular, evolving decision-making structure that avoids the heavy data and compute requirements associated with deep reinforcement learning (DRL) methods [6,13,15].

In continuous-control benchmarks such as MujoCo, policies must integrate high-dimensional continuous observations over time in order to maintain balance, coordinate multiple joints, and react to consequences of prior actions.

Our key contribution is an empirical study of dynamic memory allocation in TPG for continuous-control tasks in MuJoCo. We hypothesize that allowing TPG policies to adaptively adjust their memory representation based on task demands would improve both learning performance and decision-making efficiency.

© The Author(s), under exclusive license to Springer Nature Switzerland AG 2026
L. Manzoni et al. (Eds.): EuroGP 2026, LNCS 16521, pp. 173–188, 2026.
https://doi.org/10.1007/978-3-032-23005-8_11

TPG represents policies as graphs of teams of programs. Intuitively, a static memory configuration gives every program within a policy the same fixed-size "notebook", regardless of how much context is actually needed at any given time. In contrast, a dynamic memory configuration lets policies decide how to allocate and reuse their memory during an episode.

This raises two questions: (1) does dynamic memory improve performance on high-dimensional, continuous-control tasks; and (2) what is the resulting performance-cost tradeoff, and when is it favorable?

Prior work has shown that TPG policies without any persistent memory perform poorly on tasks requiring temporal integration (e.g., partially observable environments) [6], underscoring the need for effective memory strategies. Here we extend this work by introducing dynamic memory, where the dimensionality of vector and matrix registers is itself subject to evolution. We hypothesize that this allows agents to dynamically match memory complexity to task difficulty, thus maximizing performance while minimizing the computational overhead observed with static high-dimensional memory. To test this hypothesis, we run a series of experiments that examine the trade-offs between performance and computational cost.

Related work is discussed in Sect. 2. Details of our methodology and experimental setup are described in Sect. 3 and Sect. 4. Section 5 presents results and discussion, while Sect. 6 summarizes findings and future work.

2   Related Work

2.1   Vector and Matrix Operations in GP

TPG builds upon linear genetic programming (LGP), a GP variant that has traditionally been restricted to scalar memory and operations [1]. Using only scalars potentially limits expressiveness in high-dimensional problems. Because each instruction is typically reduced to a single index in the observation vector, the program can only manipulate one component at a time, which constrains the search space but also potentially makes it difficult to represent richer structures such as local correlations, patterns, or higher-order features. While this restriction can simplify the search, it also forces the LGP to approximate complex transformations through many scalar operations, which may be inefficient as the dimensionality of a task's feature space grows.

One way to strengthen GP-based policies is to introduce higher dimensions into its program space. Early work on stack-based and typed GP has explored richer data handling and heterogeneous types (for example, stack-based GP and strongly typed GP [7,9]). Building on this, [4] introduced a heterogeneous stack that can hold scalars, vectors, and matrices as first-class elements. More recently, LGPs have explicitly incorporated vector and matrix registers [3,10,11], allowing programs to read from high-dimensional observations in a single operation and to perform structured transformations over larger portions of the state space. In this work, we build on these ideas by integrating vector and matrix registers into TPG policies and studying how dynamic allocations of higher-dimensional memory affects performance on continuous-control tasks.

2.2   Tangled Program Graphs (TPG)

TPG is a GP framework used here for RL tasks. A key benefit of TPG is its inherent modularity, which allows the emergence of complex yet compact solutions while remaining competitive with deep learning approaches [5].

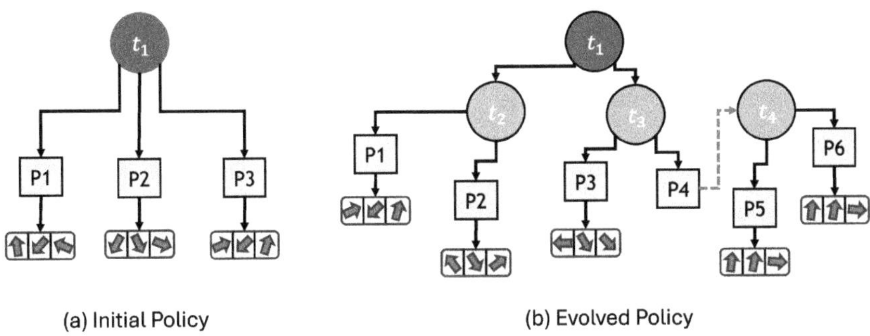

(a) Initial Policy   (b) Evolved Policy

Fig. 1. Panel (a) shows a randomly initialized TPG policy whose root teams (t_1) point to programs. Each program outputs a vector of continuous values which are used as the actions. Panel (b) illustrates how this policy can evolve over generations: programs can be mutated to point to existing teams (shown by the blue-outlined program and its outgoing edges), which then becomes a sub-team of the policy. This mutation mechanism enables modularity, allowing the root team to build complex graphs. (Color figure online)

In RL tasks, TPG policies are represented as directed graphs in which the *nodes* are teams used for decision routing and the *edges* are LGP programs that compute a dynamic bid (weight) during graph traversal [5]. At evaluation time, each action prediction begins at the root node. All programs in the root team execute relative to the current observation, and graph traversal follows the edge with the highest bid (in our implementation, taken as the program's first output register/value), recursively processing each team until an atomic/leaf program is reached, at which point its output is used as the agent's action for that timestep. Importantly, each action decision may result from a unique path through the program graph, in which a unique set of programs is executed.

Policies are initialized as a single team of programs, as illustrated in Panel (a) of Fig. 1. Modularity arises when programs are mutated to point not only to *atomic* actions interfacing directly to the task, but also to other teams, incorporating them as reusable *meta-actions* that can be shared across multiple policies, as illustrated in Panel (b) of Fig. 1. Since teams and programs are maintained as separate co-evolving populations, a policy can be composed of interconnected programs and/or teams that may be shared with other policies. These components are organized into hierarchical program graphs, enabling agents to decompose complex tasks into simpler sub-tasks [6]. Across generations, teams and programs go through evolutionary variation and selection based on episodic returns, gradually refining the structure and behavior of the program graphs.

3   Methodology

We compare two variants of TPG: a *static* memory baseline and a *dynamic* memory extension. Both use the policy representation described in Sect. 2.2 (teams, programs, and bidding), and differ only in how program memory is represented and evolved. This section describes the dynamic-memory mechanism at the program level.

3.1   Program Representation

Each TPG program is a linear sequence of instructions operating on internal memory registers and read-only variables from the observation vector $obs(t)$. Algorithm 1 provides a complete illustrative example of how individual programs operate in the proposed variant of TPG. Two critical aspects of this representation are **dynamic memory** and **observation indexing**, described in detail in this section.

Dynamic Memory. Each program contains n_m instances of scalar (s), vector (v), and matrix (m) memory, and includes an evolved parameter m_w which determines the size of its vector and matrix registers. As such, the dimensionality of these memory registers *for each individual program* can grow or shrink through mutation, rather than being fixed in advance. This lets the policy adapt how much capacity it allocates to different aspects of the observation and dynamically adjust this within an episode, instead of committing to a single static window.

Dynamic memory within program graphs consists of two primary types, vector and matrix memory:

– **Vector Memory**: One-dimensional registers that store arrays of size $[m_w, 1]$, allowing a single instruction to operate on multiple related features at once.
– **Matrix Memory**: Two-dimensional registers that store grids of size $[m_w, m_w]$, supporting higher-dimensional representations and transformations that capture relationships between features.

Figure 2 shows a policy with two programs whose internal structures differ due to the sizes of their vector and matrix registers, illustrating how evolution can produce policies with varying register dimensions.

Algorithm 1 Illustrative linear genetic program used by a TPG policy. Each program contains n_m instances of scalar (s), vector (v), and matrix (m) memory. m_w is the evolved size for vector and matrix memory. Each memory instance is paired with a data structure of the same size to store evolved constants: s_c, v_c, and m_c for scalar, vector, and matrix memory respectively. Constants are randomly initialized in the range $[-1, 1]$ when a program is created and are then left to evolve. All memory within the policy is initialized with these constants *once* at the start of each RL episode (Lines 1 to 8. See [2] for further details on evolved constants). o_i is the observation index. v_o and m_o are vector and matrix observation buffers (See Section 3.1 text for details). Note that scalar operands can access ***obs***(t) directly and thus don't require a buffer (e.g. line 25). The program returns s0 as its bid and v1 as its continuous action output for the current timestep (Line 27). If v1 is smaller than the action space (i.e. if m_w is less than the number of control variables for the task), the remaining variables are carried over from the previous timestep (This approach is assumed to provide more meaningful control signals than alternatives such as zero-padding the action vector). See [3] for the complete instruction set.

```
 1: // Initialize memory with evolved constants.
 2: if step = 0 then
 3:     for i ← 0 to n_m − 1 do
 4:         si ← s_c i
 5:         vi ← v_c i
 6:         mi ← m_c i
 7:     end for
 8: end if

 9: // Copy observation to vector memory
10: i ← o_i  mod len(obs(t))
11: for j ← 0 to m_w − 1 do
12:     v_o[j] = obs(t)[i]
13:     i = (i + 1)  mod len(obs(t))
14: end for

15: // Copy observation to matrix memory
16: i ← o_i  mod len(obs(t))
17: for j ← 0 to m_w − 1 do
18:     for k ← 0 to m_w − 1 do
19:         m_o[j, k] = obs(t)[i]
20:         i = (i + 1)  mod len(obs(t))
21:     end for
22: end for

23: // Program execution begins
24: v3 = s0*v_o                    ▷ Accessing the vector observation memory
25: v1 = obs(t)[4]*v3              ▷ Direct access to scalar observation variable
26: s0 = mean(v3)
27: return s0,v1                                            ▷ bid, action
```

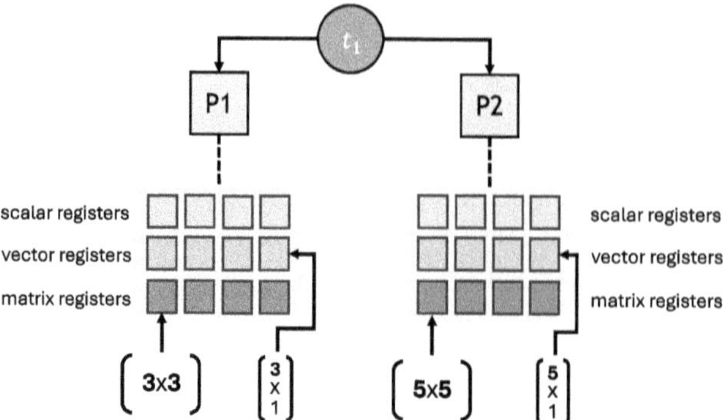

Fig. 2. Visual of a policy consisting of two programs (P1 and P2). Both belong to the same root team but use different vector and matrix register dimensions (e.g., 3×3 vs. 5×5 matrices and 3×1 vs. 5×1 vectors), illustrating how mutation can change register sizes while the overall topology remains fixed.

Observation Indexing. An additional evolved parameter o_i (observation index) controls how the current observation vector $\boldsymbol{obs}(t)$ is copied to temporary vector and matrix observation memories, v_o and m_o prior to execution. These read-only memory registers allow programs to sense $\boldsymbol{obs}(t)$ through an evolved 1D or 2D *window* of size m_w, sampled from an evolved offset index unique to each program. Figure 3 illustrates this mapping.

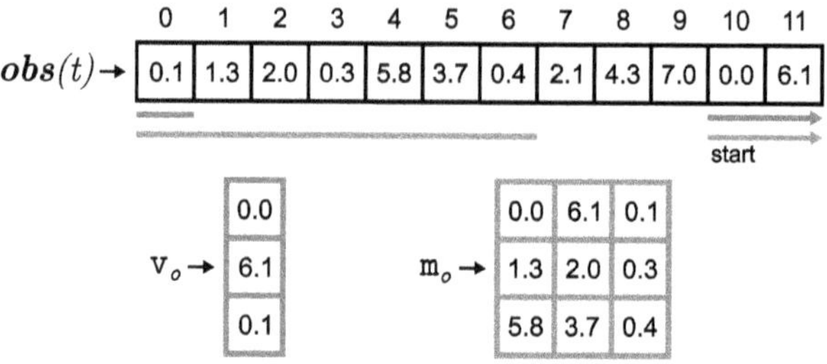

Fig. 3. Illustration of how vector observation input $\boldsymbol{obs}(t)$ is copied to vector and matrix program memories v_o and m_o, lines 9 to 22 in Algorithm 1. In this example, $\boldsymbol{obs}(t)$ is size 12, the program's evolved memory dimensionality m_w is 3, and its evolved observation index o_i is 10. This methodology allows for the observation input to be represented as any size vector or matrix. Since programs have their own mutable memory dimensionality and observation index, they are free to access any part of the observation through a *window* of this size. Colours indicate which observation variables are captured in v_o and m_o for this example.

As agents interact with the environment, they select the memory type that is most relevant to their bid and action computation. Mutation can modify both the type and dimensionality of its memory registers over the course of evolution.

4   Experimental Setup

MuJoCo (Multi-Joint dynamics with Contact) is a widely used physics engine known for its accurate and efficient simulation of complex dynamics, contact forces, and articulated bodies [14]. Its ability to simulate realistic physics and provide diverse, challenging multi-action continuous control tasks makes MuJoCo an invaluable tool for developing and evaluating RL algorithms for robotics. In this work, we investigate two tasks which require a controller to navigate part of a robot's body to a fixed target. These tasks are named Reacher and Ant-Goal, Fig. 4. Reacher is a well-known two-jointed robot arm benchmark in which the goal is to move the robot's end effector close to a target that is spawned at a random position in each episode. Ant-Goal is a custom variant of the well-known Ant quadruped robot, consisting of a torso with four legs, each with two joints. In our customized version, the goal is to apply torque to the eight hinges connecting each leg and the torso in order to navigate the ant close to a target location spawned at a random position.

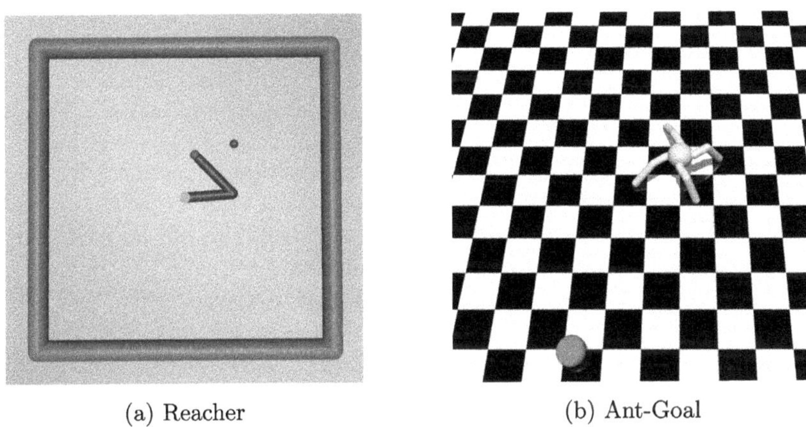

(a) Reacher   (b) Ant-Goal

Fig. 4. MuJoCo Environments.

Ant-Goal extends Ant's default observation vector to include 6 extra variables: the x,y coordinates of the torso, the x,y coordinates of the target, and the differences between the x and y positions of the torso and target (2 variables). The stepwise reward function for both tasks is:

$$-1 \times (d_w \times \sqrt{(x_a - x_t)^2 + (y_a - y_t)^2}) + (c_w \times \sum_{i=1}^{n} a_i^2) \tag{1}$$

where (x_a, y_a) and (x_t, y_t) represent the x,y positions of the agent's significant body part and target respectively, d_w and c_w are weights for the reward distance and control cost respectively, and the sum is over the agent's actions a in the current timestep. Each policy is evaluated in 20 episodes per generation, up to a maximum of 200 episodes over its lifetime. The fitness of a policy during evolution is its cumulative stepwise reward averaged over all training episodes (between 20 and 200 depending on the age). Additional task parameters are listed in Table 1a.

Table 1. MuJoCo experiment settings and environment specifications. obs_{size} and act_{size} denote the number of continuous variables in a task's observation and action space.

(a) Environment parameters

Parameter	Value
Max timestep	200
Reward dist weight (d_w)	1.0
Reward control weight (c_w)	0.05
Number of episodes (min,max)	20, 200
Number of validation episodes	100
Number of test episodes	100

(b) Observation and action spaces

Environment	obs_{size}	act_{size}
Reacher	$\mathbb{R}^{10}$	$[-1, 1]^2$
Ant-Goal	$\mathbb{R}^{33}$	$[-1, 1]^8$

In order to empirically evaluate the effects of dynamic memory, we conduct experiments for **Dynamic** and **Static** memory allocations in each task environment. In the dynamic case, vector and matrix memory size m_w for each program is initialized with uniform probability such that $m_w \in [2, obs_{size}]$ where obs_{size} is the size of the observation vector. Vector and Matrix memories for *that program* then have dimensions $m_w \times 1$ and $m_w \times m_w$, respectively. m_w is thereafter left to evolve within the limit $[2, obs_{size}]^1$. In the static baseline experiments, m_w is fixed to obs_{size} throughout evolution.

We run each experiment on a shared cluster with a 24 h time constraint. Experiments are truncated to the maximum number of generations reached by all repeats. TPG hyperparameters (Table 2) are selected based on previous research in similar continuous control environments [2,8] with no guarantee of their optimality. Prior to running formal experiments, we confirm that this parameterization leads to policies which solve $> \approx 60\%$ of test cases, indicating satisfactory task performance. Our focus in this research is to empirically evaluate the effect of dynamic memory in TPG. We are not currently interested in developing optimal controllers or comparing with state-of-the-art methods.

---

[1] Mutating integer constants such as the memory size m_w is done simply by uniformly sampling a new value within their allowed range.

Table 2. Hyperparameters for Team and Program populations

<table>
<tr><td colspan="2">(a) Team population</td><td colspan="2">(b) Program population</td></tr>
<tr><td>Parameter</td><td>Value</td><td>Parameter</td><td>Value</td></tr>
<tr><td>Root team population size</td><td>1000</td><td>Initial program size</td><td>10</td></tr>
<tr><td>New root teams per generation</td><td>100</td><td>Max program size</td><td>100</td></tr>
<tr><td>Initial team size</td><td>1</td><td>p_{delete}, p_{swap}</td><td>1.0</td></tr>
<tr><td>Max team size</td><td>30</td><td>p_{add}</td><td>0.75</td></tr>
<tr><td>Max teams per graph</td><td>30</td><td>p_{mem}</td><td>1.0</td></tr>
<tr><td></td><td></td><td>p_{atomic}</td><td>0.99</td></tr>
</table>

Note: p_{x} for $x \in \{\mathrm{add, delete, swap, mutate}\}$ are program-edit probabilities. p_{mem} is the probability of changing memory size within $[2, obs_{size}]$. p_{atomic} is the probability of a program's action pointer remaining a leaf node.

5   Results and Discussion

Dynamic and static memory experiments are compared using task performance and computational complexity as primary outcomes of interest.

5.1   Performance

Dynamic memory produces higher mean rewards throughout evolution. At intervals of 100 generations, a *validation* procedure evaluates the entire population in 100 episodes with initial conditions not seen during fitness evaluation. The individual with the best validation fitness is then evaluated in an additional 100 episodes, again with unseen initial conditions. These *champion* validation results are used for comparative analysis of evolutionary progress. Figure 5 shows the champion validation scores throughout evolution in dynamic and static memory experiments. In both Ant-Goal and Reacher, dynamic memory produces better policies early in evolution and continuously outperforms the static memory variant.

The advantage of dynamic memory is confirmed under test conditions. After evolution, validation champions are reloaded and evaluated in 100 test episodes with unseen initial conditions. Here we evaluate policies using a binary success metric which counts the number of episodes in which the agent's body reaches the target. Both TPG variants (dynamic and static) reach $> \approx 60\%$ mean success rate in both tasks, confirming the tasks are generally solvable. Comparing test episodes by mean sum of rewards (Eq. 1) confirms that dynamic memory variants provide a statistically significant advantage in both tasks (Mann-Whitney U test with a significance level of p = 0.05).

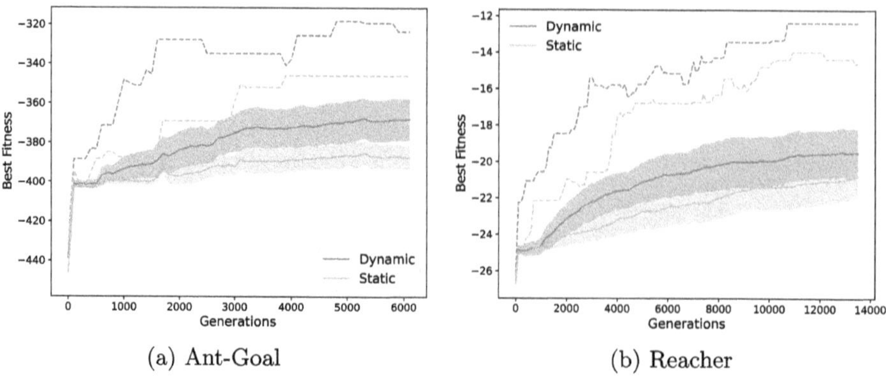

(a) Ant-Goal   (b) Reacher

Fig. 5. Dynamic memory produces significantly better validation performance throughout evolution. 30 repeats for each experiment are shown. Solid lines indicate mean validation score for champions over all repeats while shaded area represents standard deviation. Dashed lines indicate fitness for single best validation individual over all repeats.

Figure 6 shows three sample trajectories for the *dynamic* memory champion in each task. The agent is consistently able to navigate to the target in different regions of the environment, confirming task proficiency. Interestingly, the Ant-Goal agent takes slightly different trajectories for two goals in near-identical positions (Figs. 6e and 6f). This is presumably due to variations in its initial joint positions and velocities.

5.2   Policy Complexity

Dynamic memory produces significantly more efficient decision-making in Ant-Goal, which is the task with larger observation and action space. In order to measure the compute complexity of decision-making, we estimate the number of floating-point-operations (FLOPs) required for each function in the LGP operation set (See x-axis of Fig. 9 for the op set in this paper and [3] for complete details on each operation). Figure 7 reports the average flops per decision for validation champions throughout evolution. Dynamic memory size has a clear efficiency advantage because the size of vector and matrix memory (and hence their compute cost in FLOPs) can adapt through interaction with the task. In the static memory case, the dimensionality of vector and matrix memories are fixed according to the observation size. The effect is particularly apparent in Ant-Goal, Fig. 7a.

5.3   Development of Memory Size Distributions

As the number of programs in the champion policy increases, the diversity of memory sizes within the policy also increases. Over generations, the median size stabilizes to near the median of the allowed range ($[2, obs_{size}]$). Figure 8

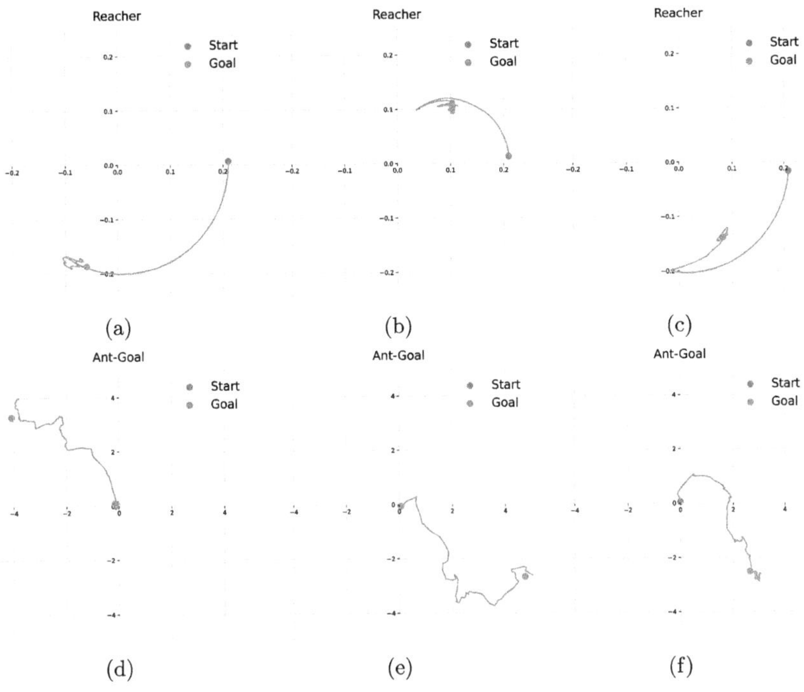

Fig. 6. In both Reacher and Ant-Goal, the agent is consistently able to navigate to target locations in different regions of the environment. The plots show example trajectories for the validation champion from dynamic memory experiments in Reacher (top) and Ant-Goal (bottom).

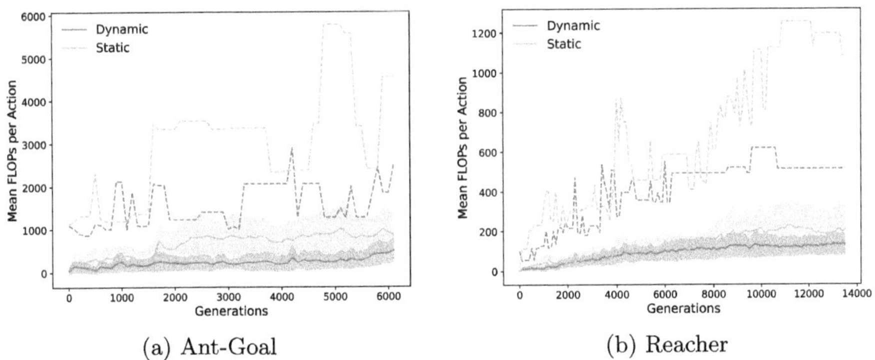

Fig. 7. Dynamic memory produces more efficient decision-making, particularly in Ant-Goal. 30 repeats for each experiment are shown. Plots shows the mean FLOPS per action over all test episodes for the single validation champion from each run. Solid lines indicate mean over repeats while shaded area represents standard deviation. Dashed line indicates FLOPS for single best individual over all repeats.

summarizes this data. Note that there is no explicit selective pressure to keep memories compact or diverse. The results in Fig. 8 confirm that diverse memory allocation within a single policy has some adaptive advantage, since all memories do not simply grow to the max size allowed.

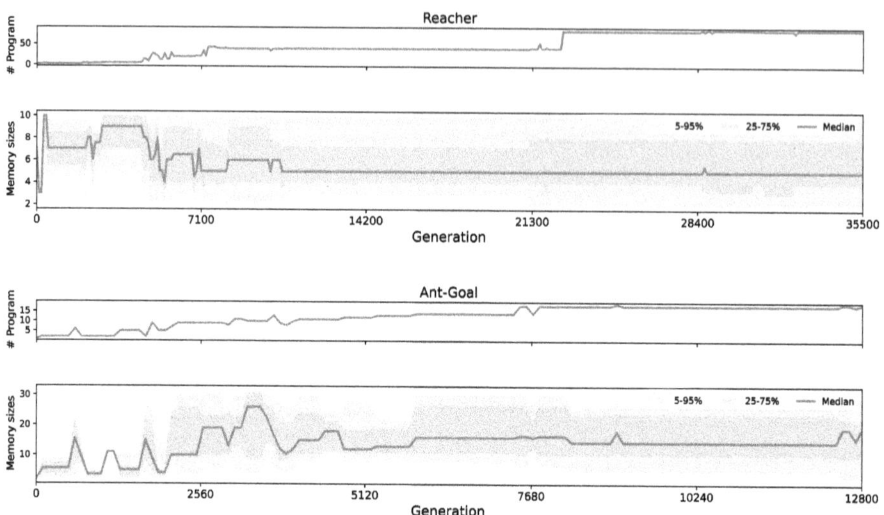

Fig. 8. Relationship between program count and diversity of memory sizes in champion policies from dynamic memory experiments. For each task, the top line plots show the number of programs in the current champion. The bottom plots show the distribution of memory sizes among these same programs.

5.4   Operation Use

The task environment strongly influences which operations are used in policies. Figure 9 shows the total number of each operation in all effective instructions for champions at the end of evolution. We note that distributions are similar between dynamic and static memory variants in each task environment, but unique sets of *useful* operations emerge for each task. This suggests that incorporating a large op set and letting task-specific subsets emerge is a realistic strategy to reduce inductive bias.

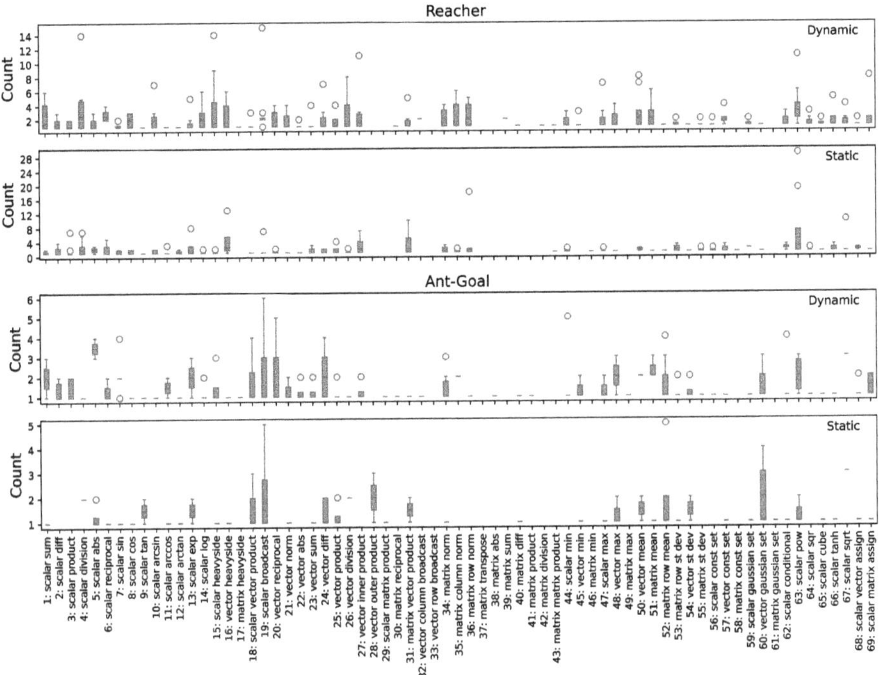

Fig. 9. The task environment impacts which operations are used in champion policies. Plot shows the total number of each operation appearing in all effective instructions for the end-of-run validation champions. Box plots show the distribution over 30 repeats.

5.5   Best Policy Structure

Figure 10 shows the structure of the best policy in the Ant-Goal task. The policy contains three teams in a shallow hierarchy: a root team t_1 that directly calls many programs, a single intermediate team t_2, and a small leaf team t_3. Because the task does not strongly reward modular sub-behaviours and the mutation probability for adding new team pointers is low ($p_{atomic} = 0.99$ in Table 2), evolution concentrates control into this compact set of teams rather than expanding the graph. Under these settings, search tends to adapt program behaviour and diverse memory usage within a few shared teams instead of building deep or highly branched policy structures.

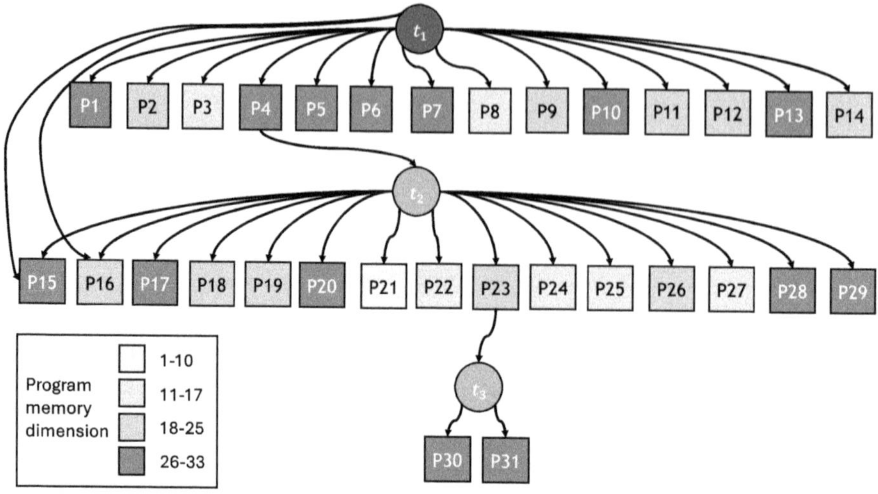

Fig. 10. Illustration of the best TPG policy in Ant-Goal. The program colours encode its total vector/matrix memory size, demonstrating how evolution produces heterogeneous memory capacities across programs within the same policy.

6   Conclusions and Future Work

This contribution establishes a baseline for the impact of dynamic vectorized memory within TPG. In challenging continuous control problems, our findings can be summarized as follows:

Performance. Dynamic memory supports better policies throughout evolution as compared to TPG with static memory structures under equivalent compute budgets. This is likely because diverse memory sizes within program graphs improves their ability to automatically decompose complex tasks.

Decision Efficiency. Dynamic memory supports more efficient decision-making as measured by the approximate number of FLOPs required per decision. Each TPG action requires execution of the programs along a *single path* through a policy's directed graph. With static memory, the size of all vectors and matrices are fixed to the number of variables in the observation vector. By contrast, evolvable memory implies that each program along the decision path potentially has a unique size, and typically 95% of these memories are smaller than the observation vector (See Fig. 8). This produces less costly decision-making.

Future work will study the potential negative impacts of dynamic memory on evolutionary search, as we suspect that further optimization of the algorithm can help compensate for the increased search space introduced by evolved memory size and observation offset (m_w and o_i parameters in Algorithm 1). We are also interested in investigating the effects of dynamic memory in multi-task RL, where enhanced problem decomposition is likely to have significant impact. In particular, prior work has demonstrated TPG's ability to overcome the well-known *distraction dilemma,* in which learning multiple dissimilar tasks simultaneously make progress more challenging [6]. However, the potential impact of dynamic memory on positive inter-task transfer in multi-task learning requires further research.

Acknowledgments. We thank Cyruss Allen Amante, Mark Angelo Cruz, Edward Gao, Richard Li, and Calvyn Siong for valuable software engineering support. This research was enabled in part by the Natural Sciences and Engineering Research Council of Canada and the Digital Research Alliance of Canada.

References

1. Brameier, M., Banzhaf, W.: Linear Genetic Programming. Springer, Cham (2007)
2. Djavaherpour, T., Naqvi, A., Norouziani, F., Vacher, Q., Kelly, S.: Genetic encoding and shared knowledge in reinforcement learning with structured memory. In: The 2025 Conference on Artificial Life (2025)
3. Djavaherpour, T., Naqvi, A., Zhuang, E., Kelly, S.: Evolving Many-Model Agents with Vector and Matrix Operations in Tangled Program Graphs, pp. 87–105. Springer, Singapore (2025). https://doi.org/10.1007/978-981-96-0077-9_5
4. Holladay, K., Robbins, K., Ronne, J.: Fifthtm: a stack based GP language for vector processing, vol. 4445, pp. 102–113 (2007). https://doi.org/10.1007/978-3-540-71605-1_10
5. Kelly, S., Heywood, M.I.: Emergent solutions to high-dimensional multitask reinforcement learning. Evol. Comput. **26**(3), 347–380 (2018). https://doi.org/10.1162/evco_a_00232
6. Kelly, S., Voegerl, T., Banzhaf, W., Gondro, C.: Evolving hierarchical memory-prediction machines in multi-task reinforcement learning (2021). https://arxiv.org/abs/2106.12659
7. Montana, D.J.: Strongly typed genetic programming. Evol. Comput. **3**(2), 199–230 (1995). https://doi.org/10.1162/evco.1995.3.2.199
8. Naqvi, A., Djavaherpour, T., Vacher, Q., Kelly, S.: Integrating neuroplasticity into genetic programming agents for adaptive decision making. In: The 2025 Conference on Artificial Life (2025)
9. Perkis, T.: Stack-Based Genetic Programming, vol. 1, pp. 148 – 153 (1994). https://doi.org/10.1109/ICEC.1994.350025
10. Praczyk, T., Szymkowiak, M.: Linear matrix genetic programming as a tool for data-driven black-box control-oriented modeling in conditions of limited access to training data. Sci. Rep. **14**(1), 12666 (2024)
11. Real, E., Liang, C., So, D.R., Le, Q.V.: Automl-zero: evolving machine learning algorithms from scratch. In: Proceedings of the 37th International Conference on Machine Learning. ICML'20, JMLR.org (2020)
12. Sutton, R.S., Barto, A.G.: Reinforcement Learning: An Introduction. A Bradford Book, Cambridge, MA, USA (2018)
13. Terven, J.: Deep reinforcement learning: a chronological overview and methods. AI **6**(3) (2025). https://doi.org/10.3390/ai6030046, https://www.mdpi.com/2673-2688/6/3/46

14. Todorov, E., Erez, T., Tassa, Y.: Mujoco: a physics engine for model-based control. In: 2012 IEEE/RSJ International Conference on Intelligent Robots and Systems, vol. 1, no. 1, pp. 5026–5033 (2012). https://api.semanticscholar.org/CorpusID: 5230692
15. Vacher, Q., et al.: Maple: multi-action programs through linear evolution for continuous multi-action reinforcement learning. In: Proceedings of the Genetic and Evolutionary Computation Conference, pp. 1062–1071 (2025)

Extending Model Selection Criteria with Extrapolation and Sensitivity Penalties for Symbolic Regression

Fitria Wulandari Ramlan[(✉)] [iD], Colm O'Riordan [iD], and James McDermott [iD]

University of Galway, Galway, Ireland
{f.wulandari1,colm.oriordan,james.mcdermott}@universityofgalway.ie

Abstract. Model selection criteria in symbolic regression have used training error and complexity measures to balance underfitting against overfitting. Standard metrics like AIC, BIC, and MDL (Minimum Description Length, which balances model complexity against fit by minimising total description length) combine these elements in different ways, based on the intuition that high complexity may allow erratic data-fitting including overfitting to noise. However, these criteria do not directly evaluate model behaviour in extrapolation regions or measure prediction stability under input perturbations. We introduce hybrid model selection criteria that extend information-theoretic measures with explicit penalties for aspects of bad model behaviour not captured by standard metrics. Our approach combines several base metrics with three additional components: (1) extrapolation divergence, which measures prediction range changes between interpolation and extrapolation regions; (2) interpolation sensitivity, which quantifies prediction variance under perturbations in dense data regions; (3) extrapolation sensitivity, which evaluates prediction stability in sparse regions. Experiments across 20 regression datasets from the PMLB repository test various weightings of these measures using both single-variable and multiple-variable perturbation methods. By testing these weightings we aim to identify model selection formulas that select models with better overall performance on unseen interpolation and extrapolation cases, balancing training fit, complexity, extrapolation stability, and prediction sensitivity.

Keywords: Symbolic Regression · Genetic Programming · Model Selection Criteria · Extrapolation · Sensitivity Analysis

1 Introduction

Symbolic regression (SR) searches for mathematical expressions that describe relationships in data [19]. In genetic programming (GP), SR evolves tree-based programs to discover transparent analytical formulas mapping inputs to outputs [8,17]. Multi-objective SR produces multiple candidate equations that trade off accuracy against complexity. These candidates form a Pareto front where

L. Manzoni et al. (Eds.): EuroGP 2026, LNCS 16521, pp. 189–204, 2026.
https://doi.org/10.1007/978-3-032-23005-8_12

simpler equations have a higher error and more complex equations fit the data better [11,18].

When selecting the optimal model from this Pareto front, researchers typically rely on measures that balance how well a model fits the training data against how complicated the model is. Training Mean Squared Error (MSE) quantifies how well a model fits observed data, while various measures of complexity attempt to capture the risk of overfitting. The intuition underlying this approach is that higher complexity may allow erratic data-fitting, including overfitting to noise rather than underlying patterns. Standard model selection metrics combine these elements in different ways. The Akaike Information Criterion (AIC) [1] penalises complexity through the number of parameters ($2k$), while the Bayesian Information Criterion (BIC) [13] applies a complexity penalty that scales with sample size ($k \log n$). The Minimum Description Length (MDL) [6,7,16] principle formalises model selection as a compression problem, and the PySR Score Metric (PSM) [3] evaluates the improvement in fit relative to added complexity, as we iterate over the Pareto front from least complex to most complex. These criteria are commonly used in SR [14] and work well for interpolation tasks where predictions remain within the training distribution.

These criteria do not directly evaluate a model's behaviour when extrapolating beyond the training domain. Symbolic expressions can show different characteristics in extrapolation regions. A polynomial model that fits the training data well may diverge rapidly when extrapolated, while simpler expressions might maintain more stable predictions across broader ranges [20]. Kronberger et al. [10] showed that SR models often have poor extrapolation behaviour and that using shape constraints can improve this by incorporating prior knowledge about expected function behaviour. The SRBench competition evaluated different SR methods on tasks including extrapolation and found differences in how methods handle predictions beyond the training distribution [5]. However, these approaches typically do not directly measure extrapolation stability or sensitivity to input perturbations, which are important for understanding model robustness in sparse regions. Fitting quality and complexity alone do not capture all aspects of model behaviour that matter for robust predictions.

To address these limitations, this work introduces hybrid model selection criteria that extend standard information-theoretic measures with explicit penalties for extrapolation divergence and prediction sensitivity. Our approach introduces three new measures to capture aspects of model behaviour not addressed by standard metrics. We evaluate models not only on their fit to training data, but also on their tendency to diverge when extrapolating and their sensitivity to small perturbations in both interpolation and extrapolation regions. The proposed hybrid criterion combines base metrics (AIC, BIC, MDL, MSE, and PSM) with three additional components: (1) extrapolation divergence, which quantifies how rapidly predictions change when moving from interpolation to extrapolation regions; (2) interpolation sensitivity, which measures prediction variance under perturbations in dense data regions; and (3) extrapolation sensitivity, which evaluates prediction variance under perturbation robustness in sparse regions where models are sparse regions where prediction stability is uncertain. These components are weighted through configurable hyperparameters to emphasise different aspects of model quality depending on application requirements.

The contributions of this work are threefold. First, we formalise a hybrid model selection framework that explicitly incorporates extrapolation and sensitivity considerations into traditional information criteria. Second, we demonstrate a practical implementation using kernel density estimation (KDE) to partition data into interpolation and extrapolation regions, along with synthetic data generation for extrapolation evaluation. Third, we provide experimental evidence across multiple datasets testing various weightings of these measures to identify configurations that select models with better overall performance, balancing interpolation accuracy, extrapolation stability, and model complexity.

2   Literature Review

Benchmarking studies have shown inconsistent relationships between interpolation and extrapolation performance in SR. Murari et al. [12] tested SR methods on physics datasets and found that models with similar training errors produced different extrapolation errors. Some researchers attribute this to the flexibility of evolved expressions, which can fit training data through mechanisms that do not transfer to new regions [20]. Others point to the search process, which optimise for training fit without considering extrapolation stability.

The SRBench competition evaluated methods across interpolation and extrapolation tasks [4]. Methods that ranked highly on interpolation tasks showed inconsistent performance on extrapolation tasks, with some maintaining their positions while others dropped substantially. This inconsistency shows that interpolation performance does not reliably predict extrapolation capability. The benchmark results also showed that no single method dominated across all task types, suggesting that different algorithmic approaches encode different inductive biases about function structure. Methods incorporating domain-specific constraints or semantic GP tended to show more stable extrapolation, though at the cost of reduced flexibility in fitting complex patterns. Several approaches have attempted to improve extrapolation through modifications to the search process rather than the selection criterion. Shape constraints restrict the space of candidate functions to those satisfying properties like monotonicity, convexity, or boundedness [10]. When domain knowledge indicates these properties should hold, enforcing them during evolution prevents the algorithm from discovering functions that violate physical or logical constraints. Empirical tests on physics datasets showed that shape-constrained SR produces models with better extrapolation behaviour compared to unconstrained search. However, this approach requires domain expertise to specify appropriate constraints and may exclude valid solutions if constraints are incorrectly specified.

Multi-objective formulations treat accuracy and complexity as separate objectives generating Pareto fronts that represent different trade-offs [11,18]. This provides multiple candidate models for selection. However, the Pareto fronts still use training accuracy and complexity measures, so they have the same limitations regarding extrapolation.

Perturbation-based sensitivity analysis measures model output response to input variations [21]. By perturbing inputs and measuring prediction variance, these methods quantify model robustness. In machine learning, sensitivity analysis has been used to identify adversarial examples and test stability. For SR, sensitivity measures can distinguish between models that capture smooth relationships and those that overfit. Models with high sensitivity produce large prediction changes from small input perturbations. Local sensitivity measures perturbation response at specific input points. Global sensitivity aggregates across the input space. Local sensitivity identifies regions where a model may be unreliable, while global sensitivity provides a summary statistic for comparing models.

Existing work has not integrated extrapolation and sensitivity considerations into model selection criteria. Shape constraints and semantic GP modify the search process but not model selection. Multi-objective approaches generate trade-offs but use training-based objectives. Density estimation and sensitivity analysis have been used as evaluation tools but not incorporated into selection metrics. There is a need for a framework for combining training fit, complexity penalties, extrapolation behaviour and, prediction sensitivity into unified criteria. This requires determining appropriate weights for each component and validating that the resulting selections improve performance on regression tasks.

3   Methodology and Experimental Setup

Our investigation addresses two main research questions: (1) Do hybrid scores with extrapolation penalties (divergence and sensitivity terms) correlate more strongly with extrapolation performance and which weight combinations produce the largest correlation differences compared to the baseline? (2) Under which conditions (weight combinations and datasets) do hybrid scores show weaker or more negative correlations than baseline metrics?

3.1   Data Preparation and Separation

We conducted experiments on 20 regression datasets from the PMLB repository (Table 1), selected to represent diverse characteristics in sample size (52 to 8192), dimensionality (2 to 124 features), and varied domains. For each dataset, we performed 30 runs to ensure statistical reliability.

Before training any models, we needed to identify which parts of the feature space contain reliable training data and which parts represent extrapolation regions. Following a similar approach to Ramlan et al. [15], we used KDE with Gaussian kernels to estimate the density of the training data in feature space. After standardising all features to have zero mean and unit variance, we applied KDE to identify high-density and low-density regions. We tested multiple bandwidth values (0.1, 0.3, 0.5, 1.0), but our main analysis focuses on bandwidth 0.3, which provided a good balance between capturing the data distribution and

Table 1. List of selected PMLB datasets used in this study (n samples, k features).

Dataset Name	n	k	Dataset Name	n	k
192_vineyard	52	2	557_analcatdata_apnea1	475	3
227_CPU_small	8192	12	561_cpu	209	7
228_elusage	52	2	584_Fri_C4_500_25	500	25
229_pwLinear	200	10	589_Fri_C2_1000_25	1000	25
519_vinnie	380	2	605_Fri_C2_250_25	250	25
522_pm10	500	7	637_Fri_C1_500_50	500	50
523_analcatdata_neavote	100	2	678_visualizing_environmental	111	3
529_pollen	3848	4	712_chscase_geyser1	222	2
542_pollution	60	15	1028_SWD	1000	10
556_analcatdata_apnea2	475	3	1096_FacultySalaries	50	4

creating meaningful separation between points inside (interpolation region) and points outside (extrapolation region). The inside region was split 80% training and 20% test, while the outside region served as the extrapolation test set.

3.2 Symbolic Regression Configuration

We used PySR [2] with populations of 200 individuals running for 300 iterations per run. Binary operators $(+, -, \times, \div)$ and unary operators (log, sin, cos, exp, square, sqrt) were allowed with maximum equation size of 200 nodes. Tournament selection used 10 candidates. Each run used a different random seed for reproducibility. An important aspect of PySR's behaviour is how it handles numerical instability during the evolutionary search. When candidate equations produce infinity or NaN values during evaluation, PySR automatically assigns them infinite loss scores[1]. This means expressions that attempt operations like taking logarithms of negative numbers or dividing by zero do not reproduce. This automatic handling applies only to expressions generated during the search.

3.3 Synthetic Data Generation for Extrapolation Training

We generated synthetic data to evaluate model behaviour in low-density regions of the training space. Using KDE, we identified the 10% of training points with lowest density scores, then generated synthetic samples by adding Gaussian noise (standard deviation 0.3) to these points. The number of synthetic samples was set to 20% of the original training set size.

3.4 Base Metrics

We evaluate five standard model selection criteria as base metrics for our hybrid framework. MSE quantifies prediction error on training data. AIC and BIC

---

[1] https://ai.damtp.cam.ac.uk/symbolicregression/dev/api.

balance model fit against complexity using different penalty terms. MDL principle incorporates parameter uncertainty through the Fisher Information Matrix (FIM). PSM evaluates improvement in fit relative to added complexity as we iterate over the Pareto front [3]. For mathematical definitions and detailed formulations of these metrics, see [9,14].

3.5   Hybrid Scoring Functions

Extrapolation Divergence is used to measure how much the model's behaviour changes when making predictions outside the range of the data it was trained on (i.e., in the extrapolation area). It works by comparing the range (max - min) of model predictions in two areas: $\hat{y}_{interp}$ is the predictions on data similar to the training data (interpolation set). $\hat{y}_{extrap}$ is the predictions on data outside the training range (extrapolation set). The Divergence is defined as:

$$\text{Divergence} = \frac{\max(\hat{y}_{extrap}) - \min(\hat{y}_{extrap})}{\max(\hat{y}_{interp}) - \min(\hat{y}_{interp})} \tag{1}$$

Here, when the *range_interp* $== 0$, it means the interpolation predictions are constant (no variance). In this case, the extrapolation divergence as defined as infinite: we discuss such cases later. Values greater than 1 indicate that the extrapolated predictions vary more than the interpolated ones, values near 1 mean similar spread. This is a stability/robustness diagnostic comparing how the prediction ranges change outside the training domain.

Sensitivity to Variable Perturbations is used to measure how sensitive an equation's predictions are to small changes in input variables. This tell us whether the model produces stable predictions or if small measurement errors in the input could lead to large changes in the output.

1. **Single Variable Perturbation (SVP)** perturbs each input variable independently while keeping all other variables fixed. For an equation with n variables, we create $2n$ perturbed versions of the data, one where each variable is increased by 3% and one where it is decreased by 3%. For instance, if an equation uses two variables x_1 and x_2, we evaluate it on four perturbed datasets: $(x_1 \times 1.03, x_2)$, $(x_1 \times 0.97, x_2)$, $(x_1, x_2 \times 1.03)$, and $(x_1, x_2 \times 0.97)$. Consider the equation $y = x_1^2 + 2x_1 + x_2$. When perturbing x_1 upward, we evaluate $y = (1.03 \cdot x_1)^2 + 2(1.03 \cdot x_1) + x_2$, where the perturbation affects all occurrences of x_1 in the equation. We then compute the variance of predictions across all these perturbations for each data point and take the mean variance across all points. This gives us a measure of how much individual variables affect the prediction.
2. **Multiple Variable Perturbation (MVP)** tests all possible combinations of perturbations. For n variables, we evaluate the equation on all 2^n combinations where each variable is either increased or decreased by 3%. For two variables, the combinations would look like: $(x_1 \times 1.03, x_2 \times 1.03)$,

$(x_1 \times 1.03, x_2 \times 0.97)$, $(x_1 \times 0.97, x_2 \times 1.03)$, and $(x_1 \times 0.97, x_2 \times 0.97)$. For the same equation $y = x_1^2 + 2x_1 + x_2$, one combination evaluates $y = (1.03 \cdot x_1)^2 + 2(1.03 \cdot x_1) + (0.97 \cdot x_2)$, showing how both variables are perturbed simultaneously while each variable's perturbation consistently affects all its occurrences. Similar to the SVP, we also compute the variance of predictions across all combinations for each data point and take the mean. This captures interaction effects between variables, and measuring how the combined effect of perturbing multiple variables affects the prediction.

Both methods return a variance value where lower values indicate more stable predictions under small input changes. We compute these sensitivity measures separately on the interpolation dataset (using training data) and the extrapolation dataset (using synthetic data). This helps us compare whether an equation's sensitivity differs between regions where it was trained and regions where it must extrapolate.

Before calculating hybrid scores, we apply preprocessing to handle outliers and enable fair comparison across metrics. This clamping step applies only to the newly introduced extrapolation measures (extrapolation divergence, and the four sensitivity metrics from SVP and MVP), as these can produce extremely large values when equations are numerically unstable. When equations fail to produce valid predictions during sensitivity testing (due to numerical errors like division by zero or overflow), we assign them a large penalty value of 1,000,000 to indicate poor stability, rather than allowing infinite or undefined predictions, variance values can be even larger. For instance, in the FacultySalaries dataset run 14, one very complex equation (with 122 operations including nested exponentials and divisions) produced a sensitivity value of about 10^{238} during extrapolation testing. Such huge values happen when small changes to input variables cause predictions to explode because of mathematical instability in the equation structure. For each run separately, we clamp these metrics using a constant upper bound of 10. This simple approach provides consistent treatment across all runs regardless of their individual value distributions. Base metrics are excluded from this clamping step because they come from well-known statistical formulas that naturally produce values within reasonable ranges.

After clamping outliers, we normalise all metrics, both base metrics and the clamped extrapolation measures using min-max normalisation to put each metric into the range [0,1]. For each metric separately, we find the minimum and maximum values across all valid observations, then transform each value using this formula:

$$\text{normalised value} = \frac{\text{value - minimum}}{\text{maximum - minimum}} \tag{2}$$

This makes sure that all metrics contribute equally when combined into hybrid scores, no matter what their original scales were. Without normalisation, metrics with naturally larger number ranges would take over the hybrid score calculation. For instance, AIC values might range from hundreds to thousands while sensitivity values might range from 0.1 to 10. Normalisation brings both onto a common 0 - 1 scale where they can be meaningfully combined.

Hybrid Score is a combination of the base metric with the three extrapolation measures described above. For each base metric (MSE, AIC, BIC, MDL, PSM), we computed a hybrid score:

$$\text{Hybrid Score} = \text{Base Metric} + \lambda_1 \cdot \text{Divergence}_{\text{extrap}}$$
$$+ \lambda_2 \cdot \text{Sensitivity}_{\text{interp}} + \lambda_3 \cdot \text{Sensitivity}_{\text{extrap}} \tag{3}$$

The weights λ_1, λ_2, and λ_3 control how much each penalty term contributes to the final score. We tested multiple weight combinations. Each combination was calculated using both SVP and MVP sensitivity methods.

4   Results

We tested both SVP and MVP sensitivity measures as described in the Methodology section. Due to space constraints, we focus on MVP results, which generally produce slightly stronger effects due to capturing interactions between variables.

4.1   Aggregate Performance Across Datasets

To answer RQ1, we calculated Spearman correlations between training metrics (base and hybrid) and test MSE for both interpolation and extrapolation test sets, across all 20 datasets and 30 runs per dataset. Tables 2 and 3 summarise the aggregate performance. Each row shows results for a specific weight combination $(\lambda_1, \lambda_2, \lambda_3)$ controlling divergence, interpolation sensitivity, and extrapolation sensitivity penalties. The "+/=/-" columns count datasets that improved (+), remained equal (=), or disimproved (-) compared to baseline. The med$\Delta\rho$ shows median correlation change across datasets.

For interpolation tasks (Table 2), hybrid scores consistently showed negative median correlation changes. The baseline configuration $(0.0, 0.0, 0.0)$ shows zero change with all 20 datasets equal. MSE-based hybrids shows the most balanced outcomes, with weight $(1.0, 0.0, 0.0)$ focusing on divergence only, producing 10 improved and 10 disimproved datasets. However, all other weight combinations showed majority disimproving, with $(1.0, 1.0, 1.0)$ showing 8 improved vs 12 disimproved datasets. AIC-based hybrids shows small negative trends, with most configurations showing 6-9 improved datasets against 11-14 disimproved. The interpolation sensitivity penalty $(0.0, 1.0, 0.0)$ performed best with 9 improved vs 11 disimproved. BIC-based hybrids shows the most negative effects, with only 4-6 datasets improving across most weight combinations. The full penalty $(1.0, 1.0, 1.0)$ resulted in just 5 improved vs 15 disimproved datasets. Then MDL-based hybrids shows the most severe disimprovement, with only 1-5 datasets improving under most configurations. Even small penalties like $(0.5, 0.5, 0.5)$ showed just 2 improved vs 18 disimproved datasets. PSM-based hybrids shows relatively better but still predominantly negative results, with 7-9 datasets improving under most configurations.

However, extrapolation tasks (Table 3) shows substantially different patterns, with many configurations producing positive median correlation changes and

Table 2. Performance of MVP hybrids on interpolation test data across 20 datasets. Weight combinations $(\lambda_1, \lambda_2, \lambda_3)$ control the divergence, interpolation sensitivity, and extrapolation sensitivity penalties. The +/=/- columns count datasets that improved, equal, or disimproved versus the base metric. medΔ shows the median correlation change.

Weight Combination	MSE		AIC		BIC		MDL		PSM	
	+/=/−	medΔ	+/=/−	medΔ	+/=/−	medΔ	+/=/−	medΔ	+/=/−	medΔ
(0.0, 0.0, 0.0)	0, 20, 0	0.00	0, 20, 0	0.00	0, 20, 0	0.00	0, 20, 0	0.00	0, 20, 0	0.00
(1.0, 0.0, 0.0)	10, 0, 10	0.01	7, 1, 12	−0.02	4, 0, 16	−0.03	1, 0, 19	−0.03	8, 0, 12	−0.02
(0.0, 1.0, 0.0)	8, 0, 12	−0.01	9, 0, 11	−0.01	5, 1, 14	−0.01	4, 2, 14	−0.01	9, 0, 11	−0.01
(0.0, 0.0, 1.0)	9, 0, 11	−0.01	9, 0, 11	−0.02	6, 0, 14	−0.04	4, 2, 14	−0.02	8, 0, 12	−0.02
(1.0, 1.0, 0.0)	8, 0, 12	−0.02	6, 0, 14	−0.03	4, 0, 16	−0.05	1, 0, 19	−0.04	7, 1, 12	−0.05
(1.0, 0.0, 1.0)	9, 0, 11	−0.01	8, 0, 12	−0.03	6, 0, 14	−0.07	2, 0, 18	−0.03	7, 1, 12	−0.05
(0.0, 1.0, 1.0)	7, 0, 13	−0.02	7, 0, 13	−0.03	5, 0, 15	−0.05	3, 0, 17	−0.02	7, 0, 13	−0.02
(1.0, 1.0, 1.0)	8, 0, 12	−0.03	7, 0, 13	−0.04	5, 0, 15	−0.06	2, 0, 18	−0.04	7, 0, 13	−0.07
(0.3, 0.3, 0.3)	7, 0, 13	−0.01	7, 0, 13	−0.02	5, 0, 15	−0.03	3, 0, 17	−0.02	8, 0, 12	−0.04
(0.5, 0.5, 0.5)	8, 0, 12	−0.02	7, 0, 13	−0.03	5, 0, 15	−0.05	2, 0, 18	−0.03	7, 0, 13	−0.05
(0.7, 0.7, 0.7)	8, 0, 12	−0.03	7, 0, 13	−0.03	5, 0, 15	−0.06	2, 0, 18	−0.03	7, 0, 13	−0.05

majority improvements. MSE-based hybrids showed strong improvements, with $(0.0, 1.0, 1.0)$ reaching med$\Delta\rho = +0.07$ (12 improved vs 8 disimproved datasets). AIC-based hybrid shows consistent improvements, with 10-13 datasets improving across most weight combinations. The interpolation sensitivity penalty alone $(0.0, 1.0, 0.0)$ was particularly effective, showing 13 improved vs 7 disimproved datasets. Combined penalties maintained improvements, with 11-12 datasets benefiting. BIC showed smaller effects, with median deltas near zero. This suggests BIC's stronger complexity penalty may already partially capture extrapolation concerns. MDL showed mixed results, with most configurations showing 1-6 improved datasets and predominantly negative median deltas. Only the interpolation sensitivity $(0.0, 1.0, 0.0)$ reached 3 improved deltas, with med$\Delta\rho$ near zero. PSM-based hybrids show strong positive results, with extrapolation sensitivity $(0.0, 0.0, 1.0)$ achieving 11 improved vs 9 disimproved, while moderate balanced weights like $(0.5, 0.5, 0.5)$ reached 11 improved cases.

To understand how individual datasets respond to hybrid penalties, we categorised datasets based on their response to hybrid penalties: (1) Insensitive (Best and Worst $\Delta\rho$ both < 0.70), (2) Selective Improvements (Best $\Delta\rho \geq 0.70$), and (3) Instability (Worst $\Delta\rho < -1.0$ with small Best $\Delta\rho$). Tables 4 and 5 present the categorisation for interpolation and extrapolation, showing baseline correlation ranges, [Best,Worst] $\Delta\rho$, and best configurations.

For interpolation (Table 4), the distribution shows 1 insensitive dataset (5%), 13 datasets with selective improvement (65%), and 6 unstable datasets (30%). The single insensitive dataset (522_pm10) showed balanced, moderate correlation changes (Best $\Delta\rho = +0.66$, Worst $\Delta\rho = -0.60$), suggesting its base metrics already provide reasonable interpolation predictions and hybrid penalties offer

Table 3. Performance of MVP hybrids on extrapolation test data. Each weight combination applies different penalties to the base metrics (MSE, AIC, BIC, MDL, PSM). The $+/=/-$ columns count how many datasets improved (+), were unchanged (=), or disimproved (-) compared to the base metric. The medΔ shows the median correlation change.

Weight Combination	MSE		AIC		BIC		MDL		PSM	
	$+/=/-$	medΔ	$+/=/-$	medΔ	$+/=/-$	medΔ	$+/=/-$	medΔ	$+/=/-$	medΔ
(0.0, 0.0, 0.0)	0, 20, 0	0.00	0, 20, 0	0.00	0, 20, 0	0.00	0, 20, 0	0.00	0, 20, 0	0.00
(1.0, 0.0, 0.0)	**11, 0, 9**	0.01	10, 0, 10	0.00	8, 0, 12	−0.02	1, 1, 18	−0.02	9, 0, 11	−0.02
(0.0, 1.0, 0.0)	**14, 0, 6**	0.02	**13, 0, 7**	0.02	9, 1, 10	0.00	3, 1, 16	0.00	**13, 1, 6**	0.03
(0.0, 0.0, 1.0)	**11, 0, 9**	0.05	**11, 0, 9**	0.02	8, 0, 12	−0.01	6, 1, 13	−0.01	**11, 0, 9**	0.04
(1.0, 1.0, 0.0)	**11, 0, 9**	0.02	**12, 0, 8**	0.01	9, 0, 11	0.00	3, 0, 17	-0.04	10, 0, 10	0.00
(1.0, 0.0, 1.0)	**11, 1, 8**	0.02	**11, 1, 8**	0.02	9, 0, 11	−0.02	2, 1, 17	−0.02	**11, 0, 9**	0.04
(0.0, 1.0, 1.0)	**12, 0, 8**	0.07	**12, 0, 8**	0.01	8, 0, 12	−0.01	4, 1, 15	−0.01	**12, 0, 8**	0.02
(1.0, 1.0, 1.0)	**12, 0, 8**	0.06	**11, 0, 9**	0.01	9, 0, 11	−0.02	3, 0, 17	−0.03	**11, 0, 9**	0.02
(0.3, 0.3, 0.3)	**12, 0, 8**	0.02	**12, 0, 8**	0.02	10, 0, 10	0.00	5, 0, 15	−0.02	**11, 1, 8**	0.04
(0.5, 0.5, 0.5)	**12, 0, 8**	0.02	**12, 0, 8**	0.02	9, 0, 11	−0.01	5, 0, 15	−0.02	**11, 0, 9**	0.04
(0.7, 0.7, 0.7)	**12, 0, 8**	0.03	**12, 0, 8**	0.01	9, 0, 11	−0.01	4, 0, 16	−0.02	**11, 0, 9**	0.04

limited additional value. The majority of datasets (65%) fall into the selective improvement category, where substantial improvements are possible with appropriate configurations. These datasets show positive improvements (Best $\Delta\rho$ ranging from +0.59 to +1.28) but also face disimprovement risks (Worst $\Delta\rho$ from −0.78 to −1.19). For example, 678_visualizing_environmental can achieve Best $\Delta\rho = +1.28$ but also Worst $\Delta\rho = -0.93$, indicating strong sensitivity to weight combination choice. This pattern aligns with our aggregate findings that interpolation outcomes depend on matching the right hybrid configuration to each dataset. The unstable datasets (30%) show severe disimprovements that substantially outweigh potential improvements. For instance, 227_cpu_small achieves only Best $\Delta\rho = +0.17$ but suffers Worst $\Delta\rho = -1.79$. Similarly, 523_analcatdata_neavote (Best = +0.53, Worst = −1.46) and 637_fri_c1_500_50 (Best = +1.00, Worst = −1.57) show that extrapolation-focused penalties frequently disrupt interpolation model selection for these datasets.

For extrapolation (Table 5), the distribution shows 1 insensitive dataset (5%), 15 datasets with selective improvement (75%), and 4 unstable datasets (20%). The single insensitive dataset (529_pollen) showed similar small changes in both directions (Best $\Delta\rho = +0.57$, Worst $\Delta\rho = -0.68$), consistent with its interpolation behaviour. The dominant selective improvement category (75%) shows that hybrid penalties successfully improve extrapolation for most datasets. Improvements range from Best $\Delta\rho = +0.61$ to +1.37, with particularly strong gains for datasets like 542_pollution (see Fig. 1) (Best = +1.37), 228_elusage (Best = +1.31), and 605_fri_c2_250_25 (Best = +1.11). Especially, several datasets that show instability for interpolation moved to selective improvement for extrapolation: 1028_SWD shifted from interpolation Instability (Best =

Table 4. MVP Interpolation: Dataset performance summary. Baseline ρ shows correlation range for base metrics. [Best,Worst] $\Delta\rho$ shows max improvement and disimprovement achievable. Best Config indicates optimal weight $(\lambda_1, \lambda_2, \lambda_3)$ and metric.

Dataset	Baseline ρ	[Best,Worst] $\Delta\rho$	Best Config
	Insensitive (n=1)		
522_pm10	[-0.32, 0.38]	[+0.66,-0.60]	(0.0,0.8,0.4) MDL
	Selective Improvement (n=13)		
1096_FacultySalaries	[-0.06, 0.10]	[+1.00,-1.01]	(0.2,0.0,0.0) MSE
192_vineyard	[-0.20, 0.15]	[+1.00,-0.84]	(0.9,0.6,1.0) AIC
228_elusage	[-0.05, 0.27]	[+0.94,-0.96]	(1.0,0.1,0.1) MSE
519_vinnie	[-0.26, 0.58]	[+1.09,-1.17]	(0.0,1.0,0.0) MSE
529_pollen	[0.18, 0.63]	[+0.59,-0.78]	(0.0,0.0,0.0) MDL
542_pollution	[-0.39, 0.40]	[+1.21,-1.15]	(0.8,0.9,1.0) MSE
556_analcatdata_apnea2	[0.03, 0.38]	[+0.67,-0.95]	(0.2,0.0,0.0) MDL
561_cpu	[-0.05, 0.52]	[+1.00,-1.19]	(0.9,0.1,0.1) MDL
584_fri_c4_500_25	[-0.04, 0.86]	[+1.00,-0.79]	(0.0,0.8,0.4) MDL
589_fri_c2_1000_25	[0.36, 0.96]	[+0.63,-0.84]	(0.3,0.1,1.0) MDL
605_fri_c2_250_25	[-0.53, 0.77]	[+1.01,-0.85]	(0.5,0.0,1.0) MDL
678_visualizing_env	[-0.54, 0.56]	[+1.28,-0.93]	(0.9,0.6,1.0) AIC
712_chscase_geyser1	[-0.26, 0.46]	[+1.15,-0.92]	(0.8,0.8,1.0) AIC
	Instability (n=6)		
1028_SWD	[-0.36, 0.63]	[+0.80,-1.33]	(0.0,1.0,1.0) MDL
227_cpu_small	[0.63, 0.98]	[+0.17,-1.79]	(0.0,0.4,0.0) PSM
229_pwLinear	[-0.23, 0.63]	[+0.51,-1.23]	(1.0,0.0,0.5) MDL
523_analcatdata_neavote	[0.05, 0.58]	[+0.53,-1.46]	(0.1,0.7,0.0) PSM
557_analcatdata_apnea1	[0.04, 0.40]	[+0.72,-1.08]	(0.1,0.0,0.0) MDL
637_fri_c1_500_50	[-0.16, 0.83]	[+1.00,-1.57]	(0.0,0.9,0.0) MDL

+0.80, Worst $= -1.33$) to extrapolation Selective Improvement (Best $= +0.66$, Worst $= -0.97$), and 523_analcatdata_neavote moved from severe interpolation disimprovement (Worst $= -1.46$) to more balanced extrapolation performance (Best $= +0.94$, Worst $= -1.00$). The instability category (20%) includes datasets where large disimprovements still happen but less severely than for interpolation. For example, 227_cpu_small improved from interpolation Worst $\Delta\rho = -1.79$ to extrapolation Worst $\Delta\rho = -1.19$, though it remains unstable (see Fig. 2). The 637_fri_c1_500_50 dataset shows the most severe extrapolation disimprovement (Worst $= -1.62$), suggesting this high-dimensional dataset $(d = 50)$ presents particular challenges for hybrid metric selection.

Regarding RQ1, the answer is **yes, for specific base metrics and weight configurations, hybrid scores with extrapolation penalties correlate with test extrapolation performance.** MSE and PSM-based hybrids with

Table 5. MVP Extrapolation: Dataset performance summary. Baseline ρ shows correlation range for base metrics. [Best,Worst] $\Delta\rho$ shows max improvement and disimprovement achievable. Best Config indicates optimal weight $(\lambda_1, \lambda_2, \lambda_3)$ and base metric.

Dataset	Baseline ρ	[Best,Worst]$\Delta\rho$	Best Config
Insensitive (n=1)			
529_pollen	[0.18, 0.63]	[+0.57,-0.68]	(0.1,0.2,0.0) AIC
Selective Improvement (n=15)			
1028_SWD	[-0.04, 0.47]	[+0.66,-0.97]	(0.6,0.7,0.8) MDL
1096_FacultySalaries	[0.01, 0.13]	[+0.84,-0.90]	(0.2,1.0,0.0) MSE
192_vineyard	[-0.15, 0.27]	[+0.83,-0.84]	(1.0,0.8,1.0) MDL
228_elusage	[-0.12, 0.26]	[+1.31,-0.81]	(0.5,0.0,0.6) AIC
519_vinnie	[-0.23, 0.46]	[+1.10,-0.84]	(0.1,0.9,0.0) MSE
522_pm10	[-0.17, 0.30]	[+0.84,-0.87]	(0.0,0.9,0.0) MSE
523_analcatdata_neavote	[0.02, 0.47]	[+0.94,-1.00]	(0.1,0.9,0.0) PSM
542_pollution	[-0.23, 0.26]	[+1.37,-0.96]	(0.8,0.9,1.0) MSE
557_analcatdata_apnea1	[0.10, 0.33]	[+0.61,-0.91]	(0.1,0.0,0.0) MDL
561_cpu	[0.00, 0.38]	[+1.02,-0.84]	(0.9,0.1,0.1) MDL
584_fri_c4_500_25	[-0.08, 0.72]	[+0.88,-0.89]	(0.0,0.8,0.4) MDL
589_fri_c2_1000_25	[0.29, 0.84]	[+0.61,-0.78]	(0.3,0.1,1.0) MDL
605_fri_c2_250_25	[-0.52, 0.66]	[+1.11,-1.24]	(0.5,0.0,1.0) MDL
678_visualizing_env	[-0.46, 0.44]	[+1.01,-1.02]	(0.9,0.6,1.0) AIC
712_chscase_geyser1	[-0.15, 0.43]	[+1.00,-0.83]	(0.8,0.8,1.0) AIC
Instability (n=4)			
227_cpu_small	[0.26, 0.46]	[+0.53,-1.19]	(0.1,0.6,0.1) MSE
229_pwLinear	[-0.18, 0.43]	[+0.82,-1.09]	(1.0,0.0,0.0) AIC
556_analcatdata_apnea2	[0.01, 0.25]	[+0.70,-1.18]	(0.2,0.0,0.0) MDL
637_fri_c1_500_50	[-0.17, 0.72]	[+1.06,-1.62]	(0.0,0.9,0.0) MDL

sensitivity penalties (particularly λ_2 and $\lambda_3 > 0$) consistently showed positive improvements for extrapolation, with 75% of datasets achieving potential gains reaching Best $\Delta\rho$ from +0.61 to +1.37. The largest positive extrapolation improvements came from $(0.0, 1.0, 1.0)$ for MSE (med$\Delta\rho = +0.07$, 12 improved datasets) and $(0.0, 1.0, 0.0)$ for AIC/PSM (13 improved datasets). BIC and MDL-based hybrids generally underperformed. These results reveal a fundamental trade-off: weight combinations that improve extrapolation correlation tend to degrade interpolation correlation.

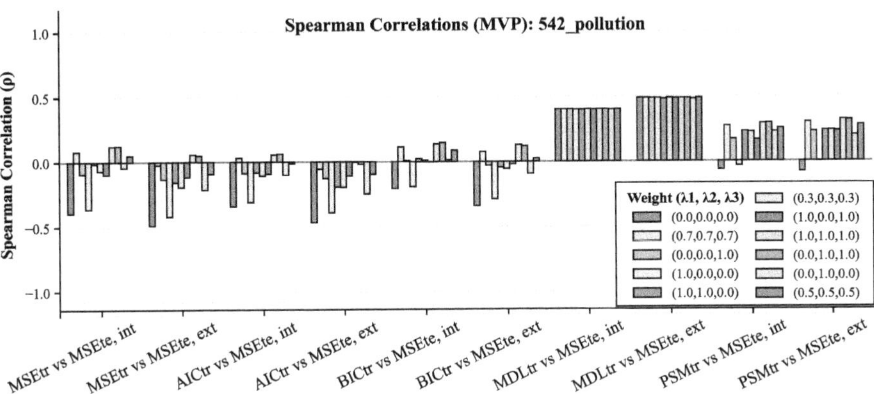

Fig. 1. MVP Spearman correlations for 542_pollution dataset. Each bar pair shows correlations between training metrics and test MSE for interpolation (int) and extrapolation (ext) test sets across different weight combinations. This dataset represents successful hybrid metric application, achieving Best $\Delta\rho = +1.37$ for extrapolation with configuration (0.8,0.9,1.0) MSE, showing substantial improvements over baseline (0.0,0.0,0.0) for both test types.

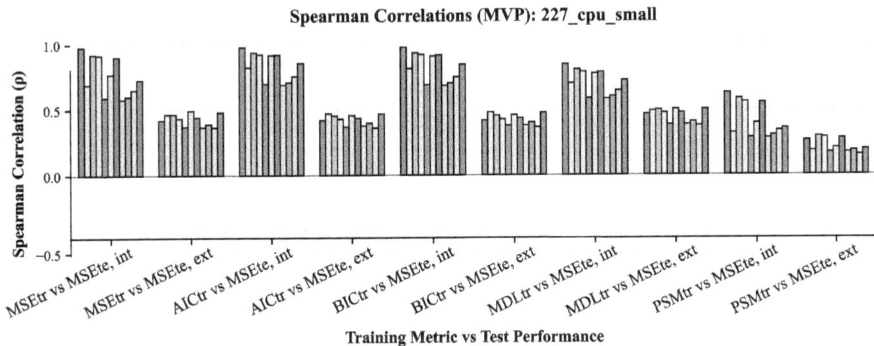

Fig. 2. MVP Spearman correlations for 227_cpu_small dataset. This dataset represents instability under hybrid penalties, with most weight combinations severely disimproving correlations. Despite achieving Best $\Delta\rho = +0.17$ for interpolation with configuration (0.0,0.4,0.0) PSM, many configurations produce Worst $\Delta\rho = -1.79$, showing the substantial risks of misapplied hybrid metrics. The baseline (0.0,0.0,0.0) shows relatively strong correlations ($\rho = 0.63$ to 0.98 for interpolation), which most hybrid configurations disrupt rather than improve.

4.2   Recommended Configurations for Practical Use

The aggregate and dataset-specific analyses above directly address RQ2 regarding which hybrid metric configurations perform best overall and under specific test conditions. As shown in Tables 2 and 3, the optimal configuration depends on priorities. When extrapolation improvement is the most important goal, MSE-based hybrids with $(0.0, 1.0, 1.0)$ achieved the strongest results (12 of 20 datasets improved, med$\Delta\rho = +0.07$), while AIC-based hybrids with $(0.0, 1.0, 0.0)$ offered consistent improvements (13 datasets) with least interpolation degradation. For balanced performance, PSM-based hybrids with $(0.5, 0.5, 0.5)$ maintained reasonable extrapolation gains (11 datasets, med$\Delta\rho = +0.04$) while limiting interpolation losses. However, BIC and MDL-based hybrids should be avoided. Individual dataset responses show that 65-75% of cases achieve selective improvement patterns where appropriate weight selection produces substantial improvements, but 20-30% show instability where hybrid configurations disrupt model selection.

5   Conclusion

This work introduced hybrid model selection criteria for symbolic regression that extend standard information-theoretic measures with explicit penalties for extrapolation divergence and prediction sensitivity. Our framework addresses a fundamental limitation of base metrics: they evaluate model quality based on training fit and complexity alone, without directly measuring how models behave when extrapolating beyond the training distribution or how sensitive their predictions are to input perturbations. Experiments across 20 PMLB regression datasets showed that hybrid metrics can substantially improve model selection for extrapolation tasks, though their effectiveness depends on the base metric and weight configuration chosen. MSE and PSM-based hybrids with sensitivity penalties (particularly λ_2 and $\lambda_3 > 0$) consistently showed positive improvements, with 75% of datasets achieving potential correlation gains reaching $\Delta\rho$ from $+0.61$ to $+1.37$. The most effective configurations were $(0.0, 1.0, 1.0)$ for MSE-based hybrids and $(0.0, 1.0, 0.0)$ for AIC and PSM-based hybrids. In contrast, BIC and MDL-based hybrids generally underperformed, suggesting their inherent complexity penalties may conflict with additional sensitivity-based penalties.

However, our results show a fundamental trade-off: weight combinations that improve extrapolation correlation tend to disimprove interpolation correlation. The optimal configuration depends on application priorities. When extrapolation performance is the most important goal, MSE-based hybrids with combined sensitivity penalties $(0.0, 1.0, 1.0)$ provide the strongest improvements. For balanced performance across both test conditions, PSM-based hybrids with moderate weights like $(0.5, 0.5, 0.5)$ offer a reasonable compromise. Importantly, while 65-75% of datasets showed selective improvement patterns where appropriate configurations produced substantial gains, 20-30% showed instability where hybrid penalties frequently disrupted rather than improved model selection.

These findings show that explicitly including extrapolation and sensitivity considerations into model selection criteria can improve the identification of models with better generalisation properties beyond the training distribution. However, the interpolation-extrapolation trade-off and the presence of unstable datasets show that hybrid metrics are not universally beneficial. Their effectiveness must be validated for specific applications rather than applied blindly. For practitioners, our results provide concrete guidance: use MSE or AIC-based hybrids with sensitivity penalties when extrapolation matters, but verify that the chosen configuration improves rather than disrupts model selection for your particular dataset.

Acknowledgement. This publication has emanated from research conducted with the financial support of Taighde Éireann – Research Ireland under Grant No. 18/CRT/6223.

References

1. Akaike, H.: A new look at the statistical model identification. IEEE Trans. Autom. Control **19**(6), 716–723 (2003). https://doi.org/10.1109/TAC.1974.1100705
2. Cranmer, M.: PySR: fast & parallelized symbolic regression in Python/Julia (2020). https://doi.org/10.5281/zenodo.4041459
3. Cranmer, M.: Interpretable machine learning for science with PySR and SymbolicRegression.jl. arXiv preprint arXiv:2305.01582 (2023). https://doi.org/10.48550/arXiv.2305.01582
4. de Franca, F.O., et al.: SRBench++âĂŕ: principled benchmarking of symbolic regression with domain-expert interpretation. IEEE Trans. Evol. Comput. (2024). https://doi.org/10.1109/TEVC.2024.3423681
5. de França, F.O., et al.: Interpretable symbolic regression for data science: analysis of the 2022 competition. arXiv preprint arXiv:2304.01117 (2023). https://doi.org/10.48550/arXiv.2304.01117
6. Grünwald, P.D.: The Minimum Description Length Principle. The MIT Press (2007). https://doi.org/10.7551/mitpress/4643.001.0001
7. Hansen, M.H., Yu, B.: Model selection and the principle of minimum description length. J. Am. Stat. Assoc. **96**(454), 746–774 (2001). https://doi.org/10.1198/016214501753168398
8. Koza, J.R.: Genetic programming as a means for programming computers by natural selection. Stat. Comput. **4**(2), 87–112 (1994). https://doi.org/10.1007/BF00175355
9. Kronberger, G., Burlacu, B., Kommenda, M., Winkler, S.M., Affenzeller, M.: Symbolic Regression. CRC Press/Taylor Francis (2024). https://doi.org/10.1201/9781315166407
10. Kronberger, G., de França, F.O., Burlacu, B., Haider, C., Kommenda, M.: Shape-constrained symbolic regression—improving extrapolation with prior knowledge. Evol. Comput. **30**(1), 75–98 (2022). https://doi.org/10.1162/evco_a_00294
11. La Cava, W., et al.: Contemporary symbolic regression methods and their relative performance. Adv. Neural. Inf. Process. Syst. **2021**(DB1), 1 (2021)
12. Murari, A., Rossi, R., Spolladore, L., Lungaroni, M., Gaudio, P., Gelfusa, M.: A practical utility-based but objective approach to model selection for regression in scientific applications. Artif. Intell. Rev. **56**(Suppl 2), 2825–2859 (2023). https://doi.org/10.1007/s10462-023-10591-4
13. Neath, A.A., Cavanaugh, J.E.: The bayesian information criterion: background, derivation, and applications. Wiley Interdisc. Rev. Comput. Stat. **4**(2), 199–203 (2012). https://doi.org/10.1002/wics.199
14. Ramlan, F.W., Kronberger, G., O'Riordan, C., McDermott, J.: Comparative analysis of model selection criteria for symbolic regression using genetic programming. In: Proceedings of the International Conference on Evolutionary Computation Theory and Applications (2025), accepted for publication

15. Ramlan, F.W., O'Riordan, C., Kronberger, G., McDermott, J.: Can synthetic data improve symbolic regression extrapolation performance? In: Proceedings of the Genetic and Evolutionary Computation Conference Companion, pp. 2548–2555 (2025). https://doi.org/10.1145/3712255.3734356

16. Rissanen, J.: Modeling by shortest data description. Automatica **14**(5), 465–471 (1978). https://doi.org/10.1016/0005-1098(78)90005-5

17. Schmidt, M., Lipson, H.: Distilling free-form natural laws from experimental data. Science **324**(5923), 81–85 (2009). https://doi.org/10.1126/science.1165893

18. Smits, G.F., Kotanchek, M.: Pareto-front exploitation in symbolic regression. In: Genetic Programming Theory and Practice II, pp. 283–299. Springer (2005). https://doi.org/10.1007/0-387-23254-0_17

19. Sun, S., Ouyang, R., Zhang, B., Zhang, T.Y.: Data-driven discovery of formulas by symbolic regression. MRS Bull. **44**(7), 559–564 (2019). https://doi.org/10.1557/MRS.2019.156

20. Uy, N.Q., Hoai, N.X., O'Neill, M., McKay, R.I., Galván-López, E.: Semantically-based crossover in genetic programming: application to real-valued symbolic regression. Genet. Program Evolvable Mach. **12**(2), 91–119 (2011). https://doi.org/10.1007/s10710-010-9121-2

21. Prabhakaran, V., Ben Hutchinson, M.M.: Perturbation sensitivity analysis to detect unintended model biases. In: Conference on Empirical Methods in Natural Language Processing (2019). https://doi.org/10.18653/v1/D19-1578

Short Presentation

Optimal Mixing in Graph-Based GP for Control: Genotypical Dependencies are Hardly Captured

Giorgia Nadizar[1,2]($\boxtimes$) (iD), Gloria Pietropolli[3] (iD), and Eric Medvet[1] (iD)

[1] Department of Engineering and Architecture, University of Trieste, Trieste, Italy
[2] IRIT - CNRS UMR5505, University Toulouse Capitole, Toulouse, France
`giorgia.nadizar@phd.units.it`
[3] Department of Mathematics, Informatics, and Geosciences, University of Trieste, Trieste, Italy

Abstract. Graph-based genetic programming (GGP) encompasses representations that evolve modular graphs or computer programs with multiple inputs and outputs, making it well suited for addressing complex real-world problems. To fully exploit this potential, variation operators need to capture and preserve structural dependencies within program graphs. The gene-pool optimal mixing evolutionary algorithm (GOMEA) is a model-based evolutionary algorithm whose strength lies in learning and exploiting such dependencies, making it a natural candidate for GGP. In this work, we investigate the integration of GOMEA with GGP. We first validate the approach on symbolic regression (SR) benchmarks, with both single and multiple outputs, where GOMEA consistently matches or outperforms a standard genetic algorithm (GA). Then, we apply GOMEA to continuous control tasks—an important application domain for GGP—and find it often struggles compared to the GA. We hypothesize that this limitation arises from the difficulty of identifying and exploiting meaningful dependencies in the inherently chaotic and high-dimensional landscapes of control problems. Thus, our findings call for further studies to improve dependency-learning mechanisms for complex, dynamic domains.

Keywords: Model-based Evolutionary Algorithms · Continuous Control · Graph-based Genetic Programming · Cartesian Genetic Programming · Linear Genetic Programming

1 Introduction and Related Work

Graph-based genetic programming (GGP) encompasses approaches which represent solutions as graphs, e.g., cartesian genetic programming (CGP) [32], or computer programs, e.g., linear genetic programming (LGP) [5], mapping multiple inputs to multiple outputs [13]. These representations allow to make the flow of information through the graph/program explicit, allowing for greater

L. Manzoni et al. (Eds.): EuroGP 2026, LNCS 16521, pp. 207–224, 2026.
https://doi.org/10.1007/978-3-032-23005-8_13

transparency, interpretability, and modularity [38]. The modular structure of graph-based representations also facilitates the decomposition of complex problems into reusable sub-components [51].

These properties make GGP particularly appealing for control tasks, where interpretability and structure are highly desirable [26]. For instance, evolved controllers can expose meaningful internal computations and provide insights into how control signals are generated—an advantage rarely available in deep learning-based methods. Recent work has demonstrated that GGP can successfully address both continuous control tasks [33,34,47] and visual control domains such as Atari games [9,18–20,52], highlighting its versatility and potential for real-world applications. Beyond control, GGP has also been applied to several domains, e.g., program synthesis [8,40], symbolic regression (SR) [16,41,53], and classification problems [4,10], where its transparent program structure aids in understanding the learned decision logic [30].

Despite these advantages, the evolutionary process in GGP often relies heavily on mutation-based search, where performance improvements depend on fortuitous random changes. Traditional crossover operators are often less effective, as they can easily disrupt the delicate graph structures that encode functional programs [21,22,31]. This raises the question of whether learning and exploiting dependencies among components of the graph, rather than relying on chance recombination, could lead to more efficient evolution [23,39]. In this context, model-based evolutionary algorithms (MBEAs) can offer a promising alternative, as they explicitly model and preserve meaningful structural relationships during variation [6].

Gene-pool optimal mixing evolutionary algorithm (GOMEA), originally introduced in [42], is a MBEA for which several variants have been developed across different domains, ranging from discrete and real-value optimization to genetic programming and grammatical evolution [1,3,15,27,29,37,48,49]. These variants have consistently shown superior performance and scalability in all of these tasks. A key feature of GOMEA is its ability to exploit the structure of the problem in the form of *linkage* (i.e., interdependent problem variables), making it particularly suited for solving complex optimization problems with excellent scalability [11]. Notably, the genetic programming (GP) variant (GP-GOMEA) achieves state-of-the-art performance in SR while evolving compact expressions, a desirable property for interpretability.

Motivated by these successes, we investigate the integration of GOMEA with two GGP representations, CGP and LGP. The integration of GOMEA principles in CGP has been explored preliminarily in [16], where authors compared the performance of CGP-GOMEA against GP-GOMEA on three SR benchmarks. Here, we first validate the proposed approach on SR benchmarks, and then assess its effectiveness in continuous control tasks, a more complex and demanding domain where deriving effective yet interpretable control laws has a direct impact on system reliability and trust. Through extensive experiments, we find that the advantage of GOMEA over a standard genetic algorithm (GA), which is evident in both single- and multi-output SR, diminishes in control tasks. We attribute this finding to the complex fitness landscapes associated with control tasks and graph-based representations, which hinder the ability of GOMEA to effectively learn and exploit dependencies within the genotype. As a consequence, its advantage over more random variation operators is largely lost.

2 Gene-Pool Optimal Mixing Evolutionary Algorithm for Graph-Based Genetic Programming

We consider the optimization of graphs with GGP, employing two distinct representations: CGP and LGP—we provide the details of these representations in Sect. 2.1. Both allow optimization of graphs with an arbitrary number of inputs and outputs, where both the functional components, i.e., the nodes, and their connectivity are evolved using an evolutionary algorithm (EA). These GGP representations share a common characteristic: a graph can be encoded as a fixed-size linear integer genotype, that is, a sequence of integers representing its structure and functions.

Traditionally, optimization in GGP has been performed using either a $(1+\lambda)$-EA—particularly in the case of CGP [32]—or a GA, which is often preferred for control tasks where evaluating larger populations in parallel can yield significant computational speedups, especially on GPUs [31,34].

In this work, we explore the use of GOMEA for graph optimization. GOMEA is a model-based EA designed to learn dependencies among genotype components and exploit them to guide variation (see Sect. 2.2). It is, in principle, representation-agnostic, provided that the genotype has a fixed-length linear encoding. However, depending on the representation, specific adjustments may be required to reduce bias and uncover the true underlying structures [50]. In our context, the goal is for GOMEA to identify dependencies within graph representations and propose informed variations among individuals, thereby avoiding disruptive crossover operations that rely solely on random chance.

We combine these two elements—graph-based representations and model-based optimization—giving rise to GGP-GOMEA. In fact, they can be naturally combined, as GOMEA is suited for fixed-size genotypes such as those used to encode graphs. However, we introduce a few minor modifications to the standard GOMEA algorithm to better adapt it to our case study, as detailed in Sect. 2.3.

2.1 Graph-Based Genetic programming (GGP)

Over time, research in GP has expanded beyond its original tree-based formulation for evolving computer programs [24], with several alternative representations being proposed, including graph-based approaches that have gained attention due to their ability to represent expressive multi-input to multi-output mappings, and their inherent modularity [13]. Within this family, we concentrate on two representations: CGP [32] and LGP [5]. Both make use of a fixed-size linear integer genotype, with a phenotype that can be naturally interpreted as a directed acyclic graph (DAG). We describe them more in detail in the following.

Cartesian genetic programming. In CGP, graphs are represented as grids of nodes in a Cartesian plane [31,32], for which evolution optimizes functionality and connectivity. Each node represents a function, which can use either the outputs of the nodes of the previous layers or the program inputs as arguments. The final outputs of the program are collected from the outputs of n_{out} selected nodes, which can also be subjected to optimization.

Here, we consider a uni-dimensional grid, i.e., a sequence of nodes of length n_{nodes}. This alternative representation does not entail any loss of generality, but it simplifies the representation and the use of GOMEA, which naturally operates on fixed-size sequences.

To fully determine a CGP graph, we need (a) n_{out} indexes to specify the nodes from which to extract the outputs and (b) n_{nodes} tuples $(h, i_1, \ldots, i_{m_{\mathrm{ar}}})$, specifying the function h and its arguments $i_1, \ldots, i_{m_{\mathrm{ar}}}$ for each node, where m_{ar} is the maximum arity of the functions available in the function set H (2 in our case). For each of the tuples, h is an index bounded by the cardinality of the function set H, i.e., by the number of available functions. Conversely, each input index can range from 0 to $n_{\mathrm{in}} + j - 1$, where j indicates the current node position, meaning that only backward connections are allowed in the graph, i.e., the information can only flow in a feed-forward manner. Thus, the genotype of a CGP graph is a sequence of bounded integers of size $l = n_{\mathrm{out}} + (1 + m_{\mathrm{ar}})n_{\mathrm{nodes}}$.

Linear genetic programming. In LGP, we consider programs, rather than graphs. Programs are lists of instructions from a programming language, where the result of each instruction is assigned to a register from a predefined set [5]. Before execution starts, the inputs of the program are copied into the first n_{in} registers; at the end of the program execution the outputs are taken from the last n_{out} registers. If we consider a language where loops are not allowed, we can interpret the information flow in a LGP program as a DAG. Hence LGP falls under the category of GGP [13].

To describe a LGP program, each instruction is determined specifying (1) the index of the register to be assigned, (2) the function to execute from those available in the function set H, and (3) the sequence of arguments to be passed to the function $(i_1, \ldots, i_{m_{\mathrm{ar}}})$, expressed in terms of register indexes ($m_{\mathrm{ar}} = 2$ being the maximum arity of functions in H). Hence, for a program of n_{lines} lines, the genotype is a sequence of bounded integers of size $l = (2 + m_{\mathrm{ar}})n_{\mathrm{lines}}$. In this case the bounds for the functions are constituted by number of available functions in the set H, whereas for register the indexes are only limited by the amount of available registers n_{reg}, as all registers are initialized to 0 and are available to select at all times.

Genetic algorithm for graph-based genetic programming. As previously mentioned, a GA has traditionally been used to optimize graph-based representations. In this work, we consider it as a baseline for all our evaluations.

Our GA works as follows. First, we initialize a population of n_{pop} individuals by (i) randomly sampling integer genotypes within the valid range for each representation, (ii) mapping each genotype to its corresponding graph representation, and (iii) evaluating it on the target problem to compute its fitness. Then, we run the evolutionary loop until the total fitness evaluation budget n_{eval} is exhausted. At each iteration, we select n_p parents using tournament selection of size n_{tour}. We generate the offspring by applying integer-flip mutation, replacing integers in the genotype with other values in the respective range. Mutation probabilities are representation and gene dependent. In CGP, we use distinct probabilities p_i, p_f, and p_o for node inputs, functions, and outputs, respectively. In LGP, we define p_a, p_f, and p_i as the probabilities for register assignments, functions, and function inputs. After evaluating the offspring, we form the next generation by promoting the $n_{\text{pop}} - n_p$ best parents along with all offspring.

While crossover can in principle be applied, it is believed to be ineffective in CGP [23]. We therefore omit it here for both representations for coherence.

2.2 Gene-Pool Optimal Mixing Evolutionary Algorithm (GOMEA)

GOMEA is a model-based EA introduced in [43] which showed to perform well in several discrete and continuous domains. Unlike traditional EAs, where mutation and crossover act blindly, GOMEA learns a model of interdependencies between variables in the genotype—the *linkage*—to estimate what patterns to propagate.

All algorithms in the GOMEA family follow a common scheme. After initializing a population of n_{pop} individuals, the algorithm enters a generational loop that continues until a termination criterion is met. In each generation, GOMEA learns a linkage model from the population; then, it applies the gene-pool optimal mixing (GOM) variation operator to every individual to generate an offspring. The linkage model guides this variation process, ensuring that offspring are generated by preserving and exploiting the dependencies discovered in the population.

The linkage model is a family of subsets (FOS) $\mathcal{F} = \{F_1, F_2, \dots\}$, with $F_i \subseteq \{1, \dots, l\}$ and l denoting the genotype length. Each F_i contains indices, also called *loci*, which represent positions in the genotype to be replaced *en bloc* during variation. Different variants to build a FOS have been proposed in the literature. Firstly, the univariate (U) FOS which assumes that there is no linkage between portions of the genotype: $\mathcal{F}_U = \{\{0\}, \{1\}, \dots, \{l-1\}\}$. The linkage tree (LT) FOS $\mathcal{F}_{LT}$ models the linkage structure using a hierarchical clustering (i.e., a tree), where nodes are sets of loci and a node is the set union of its children. Here the FOS is learned from the population, exploiting the *mutual information* between pairs of loci, as follows: firstly, a set of loci $\mathcal{F}_0$ is set to $\mathcal{F}_U$, with $\mathcal{F}_{LT} = \mathcal{F}_0$. Then the following steps are repeated until $|\mathcal{F}_0| \leq 1$: (i) identify the pair $(F_i, F_j) \in \mathcal{F}_0 \times \mathcal{F}_0$, with $i \neq j$, that maximizes the mutual information, (ii) remove F_i and F_j from $\mathcal{F}_0$, and (iii) insert $F_i \cup F_j$ into both $\mathcal{F}_0$ and $\mathcal{F}_{LT}$. More details can be found in [43], where the authors also proved that this variant of FOS is particularly beneficial for search effectiveness. Finally, the random tree (RT) FOS is a modification of LT where subsets to merge are chosen randomly instead of based on their mutual information. Note that only LT is actually learned from the population, while U and RT are not.

Once the FOS is learned, GOMEA applies the GOM variation operator to generate the offspring population. GOM integrates variation and selection, so each offspring directly replaces its parent, and no offspring is ever worse than the individual it originates from. Concretely, given a parent genotype g, GOMEA builds an identical offspring g' and then uses iteratively each F_i in the FOS $\mathcal{F}$ to try to improve g'. Specifically, GOMEA randomly picks another *donor* genotype g'' in the population and replaces the genes of g' at the loci specified by F_i with the genes of g'' at the same loci. If the resulting g' has changed and its fitness is not worse, GOMEA keeps it, otherwise it restores the genotype before the modification. Moreover, if g' remains unchanged during this phase, or if no new best fitness has been found in the last $1 + \log_{10}(n_{\mathrm{pop}})$ generations, the *forced improvement* phase begins. In this phase, GOMEA only uses the best individual found so far as donor and accepts changes only if they lead to a strict fitness improvement. Once this procedure succeeds in producing an improved offspring, it stops. On the other hand, if no improvement is achieved, the offspring is completely replaced by the best individual.

Virgolin et al. [50] showed that non-uniformity in the genotype distribution of GP populations can negatively bias the linkage learning. To mitigate this issue, they introduced an *improved linkage learning*, which we integrate into our implementation. For further details, we refer the reader to [50].

2.3   GGP-GOMEA

We apply GOMEA to GGP as follows. First, we initialize the population according to the chosen representation by randomly sampling integer values within the valid range for each gene. After initialization, the algorithm proceeds with the standard GOMEA loop: to begin, we obtain the FOS using one of the aforementioned procedures (i.e., U, RT, or LT).

Then, to perform GOM more efficiently, we modify the algorithm as described below. First, we compute a permutation of the FOS elements. Then, we iterate over the elements of the FOS, and for each of them, we iterate over the population. For each individual s, a donor d is sampled, and their genes are mixed according to the current FOS element. This process, repeated for the whole population, produces a new offspring population of size n_{pop}, which can be evaluated in parallel if desired. Selection then proceeds as in the standard GOM procedure, i.e., only promoting offspring that either match or improve the parents.

We remark that donors are always sampled from the original population until the iteration of the FOS is complete, while the offspring population is progressively updated through cumulative mutations. Once both nested iterations are complete, the FOS is recomputed, and the process is repeated until termination.

The same modifications are applied to the forced improvement phase, where donors are set to the best individuals obtained so far. This adaptation enables the parallel evaluation of the entire population, which is particularly advantageous for control tasks where simulations can be parallelized to grant a computational speed-up.

Another detail distinguishing our implementation from standard GOMEA concerns the equality check performed in GOM. When dealing with degenerate representations—i.e., cases where different genotypes map to the same phenotype—the equality check can be carried out either on the genotypes, the phenotypes, or on the corresponding behaviors. In our approach, even if both GGP representations display degeneracy, we perform the check directly on the genotype. This choice allows for neutral mutations, that is, mutations affecting only the inactive parts of the graph/program. This modification is motivated by neutrality studies in both CGP and LGP, which indicate that mutations in non-active graph regions can serve as promising stepping stones toward improved solutions [17,45].

3   Benchmark Problems

Our final objective is the optimization of continuous control policies using GGP, that is, the discovery of symbolic graphs mapping sensor inputs to actuator outputs to achieve a good robot behavior. Since this constitutes a relatively complex task, we begin with the simpler setting of SR using graph-based representations. This preliminary step allows us to assess whether GOMEA can effectively optimize graphs to produce meaningful input-output mappings.

Symbolic Regression Benchmarks. SR is a form of supervised learning that aims to learn a model in the form of a mathematical expression describing the relationship between input and output variables, given a bag D of examples. Typically, SR is a single-output problem: the examples are $D = \left\{ \left(\boldsymbol{x}^{(i)}, y^{(i)} \right) \right\}_i$, with $\boldsymbol{x}^{(i)} \in \mathbb{R}^{n_{\mathrm{in}}}$ and $y \in \mathbb{R}$, and the goal is to learn a model $f : \mathbb{R}^{n_{\mathrm{in}}} \to \mathbb{R}$.

We consider the Nguyen 9 ($n_{\mathrm{in}} = 2$) and Nguyen 10 ($n_{\mathrm{in}} = 2$) problems, commonly used in the GP literature [28], and two problems with larger input size taken from the Feynman dataset [46], Feynman 13 ($n_{\mathrm{in}} = 5$) and Feynman 43 ($n_{\mathrm{in}} = 6$). This dataset contains 100 equations from mechanical, electromagnetism, and quantum physics [2].

As continuous control tasks usually deal with multiple sensors and multiple actuators, we also consider a few multi-output SR problems, where the goal is to learn a model $f : \mathbb{R}^{n_{\mathrm{in}}} \to \mathbb{R}^{n_{\mathrm{out}}}$ from a $D = \left\{ \left(\boldsymbol{x}^{(i)}, \boldsymbol{y}^{(i)} \right) \right\}_i$. Specifically, we craft new benchmark problems by combining multiple expressions from the Feynman dataset to form systems with as many outputs as the number of merged equations. We construct Feynman A, composed of Feynman Eqs. 13 and 16 ($n_{\mathrm{in}} = 8$, $n_{\mathrm{out}} = 2$), Feynman B, composed of Feynman Eqs. 42, 43, and 48 ($n_{\mathrm{in}} = 10$, $n_{\mathrm{out}} = 3$), and Feynman C, composed of Feynman Eqs. 13, 61, 62, 70, 77, and 100 ($n_{\mathrm{in}} = 15$, $n_{\mathrm{out}} = 6$). Furthermore, we construct Composite Feynman, where we first apply Feynman Eq. 13 (having five input variables) and then Nguyen 5, 6, and 7 (each one having one input variable) in parallel to obtain the three outputs ($n_{\mathrm{in}} = 5$, $n_{\mathrm{out}} = 3$).

To reflect typical control problems, where the same input influences multiple output channels, we select equations within each system that partially share the same inputs (e.g., for Feynman a, both equations have inputs q and E_f). Moreover, for both single- and multi-output SR, we select problems where constants do not play a significant role, as we want to focus more on variable interdependencies than on optimal constant search.

Concerning the number of examples, we have $|D| = 200$ for the two Nguyen problems and $|D| = 500$ for all the other problems.

Continuous Control Benchmarks. We consider six benchmark continuous control tasks of increasing complexity in terms of input and output space from the Mujoco suite [44]. First, we examine the *inverted double pendulum* ($n_{in} = 8$, $n_{out} = 1$) a balancing task where the controller is rewarded for preventing the pendulum from falling, while being penalized for excessive movements. Additionally, we test a target-aiming task, the *reacher* ($n_{in} = 11$, $n_{out} = 2$), where a robotic arm must reach an object and is rewarded the more it gets closer, penalizing excessive movements.

The remaining four tasks—*swimmer* ($n_{in} = 8$, $n_{out} = 2$), *hopper* ($n_{in} = 11$, $n_{out} = 3$), *walker2d* ($n_{in} = 17$, $n_{out} = 6$), and *half cheetah* ($n_{in} = 18$, $n_{out} = 6$)—are locomotion problems in which the goal is to move as far as possible from the starting position. Among these, the *hopper* and *walker2d* are unstable and therefore include an additional reward component for maintaining balance; in addition, the episode terminates if the agent falls.

These control tasks have previously been addressed using both CGP and LGP, as reported in [34]. In four out of six cases, the evolved graph-based controllers achieved performance comparable to that of artificial neural networks trained with reinforcement learning (RL), while exhibiting significantly more compact structures. The two exceptions were the *hopper* and *walker2d*, both of which involve explicit stability rewards, where performance remained neatly below RL-based approaches. Further studies suggested that increasing population diversity could partially improve results on these tasks, though not to the level achieved by RL [33].

4   Experimental Evaluation

We aim at answering the following research questions:

RQ1 Is GOMEA a suitable EA for the optimization of graphs with GGP?
RQ2 Are there any benefits related to applying GOMEA for the optimization of continuous control policies with GGP?
RQ3 What are the impacts of parameters such as the employed GGP representation and of the population size?

To address these questions, we conduct a comprehensive experimental analysis comparing GOMEA with a standard GA across various problems and configurations. We consider the three GOMEA variants introduced in Sect. 2.2, which differ in the method used to learn their FOS. In the following, we refer to these as GOMEA-U, GOMEA-RT, and GOMEA-LT.

Regarding the representation parameters, for CGP we set the number of computational nodes to $n_{\text{nodes}} = 50$. For LGP, we use $n_{\text{reg}} = n_{\text{in}} + n_{\text{out}} + 10$ registers and $n_{\text{lines}} = 35$ program lines. Both GGP representations share the same function set, consisting of the following primitives: $H = \{\bullet + \bullet, \bullet - \bullet, \bullet \times \bullet, \bullet \div^* \bullet, |\bullet|, \exp\bullet, \sin\bullet, \cos\bullet, \log^*\bullet, \sqrt{\bullet}^*, \bullet < \bullet, \bullet > \bullet\}$, where $\bullet$ denotes an operand, and operators marked with $*$ are protected. The final two operators are Boolean functions that output 1 if the condition is satisfied and 0 otherwise. In addition, to allow for more expressivity, we include two constant inputs, $\{1, 0.1\}$ in every problem, which the graphs can use to build the final expression.

For all EAs, we consider two population sizes, $n_{\text{pop}} = 100$ and $n_{\text{pop}} = 1000$, referred to as small and large, respectively. We choose these values to highlight different trade-offs between the algorithms. In the case of GOMEA, a larger population can provide a more accurate estimation of the FOS, potentially improving dependency modeling. Conversely, for the GA, increasing the population size substantially reduces the number of iterations that can be performed within a fixed evaluation budget, which may limit the exploratory capability of the algorithm.

To ensure a fair comparison, we use the same computational budget n_{eval} for all algorithms, set according to the specific problem. For symbolic regression, we use $n_{\text{eval}} = 500000$. For control tasks, we use $n_{\text{eval}} = 200000$ for inverted double pendulum, reacher, and swimmer, $n_{\text{eval}} = 300000$ for hopper and walker2D; and $n_{\text{eval}} = 100000$ for half cheetah.

When using the GA as the optimization algorithm, we adopt the mutation probabilities defined in Sect. 2.1, setting $p_i = p_f = 0.1$ and $p_o = p_a = 0.3$. We use tournament selection with a tournament size of $n_{\text{tour}} = 3$, sampling a total of $n_p = n_{\text{pop}} - 10$ parents to generate the offspring population.

We execute each algorithm ten times on each problem. For analyzing the statistical significance of algorithm comparisons, we use the Mann-Whitney U test with $\alpha = 0.05$.

We release our code and experimental results at https://github.com/giorgia-nadizar/GGP-GOMEA.

4.1 Symbolic Regression

We begin by considering the SR problems, to assess whether GOMEA is suitable for optimizing graphs represented with either CGP or LGP, thereby addressing RQ1. A partial indication for CGP was previously reported in [16], where GOMEA showed no significant advantages over other EAs. However, that study employed a grid-based CGP representation, which, while conceptually similar, may affect how the FOS is learned. For this reason, we revisit the question here, extending the analysis to include a second GGP representation, namely LGP.

As the fitness metric, we use the coefficient of determination (R^2) [36], which we aim to maximize—for multi-output problems, we consider the mean $\bar{R}^2$ across output variables. A value of $R^2 = 1$ indicates a perfect fit, while $R^2 = 0$ corresponds to a performance equivalent to predicting the mean of the target variable. Since no linear scaling is applied, negative values are also possible, indicating solutions that perform worse than the mean prediction. We divide the dataset into 80% for training and 20% for testing.

We begin with the single-output problems, which represent the most classical form of SR. These problems are typically addressed using tree-based GP, and in particular, GP-GOMEA has demonstrated substantial advantages over other variants of GP across a wide range of SR benchmarks [25]. Nonetheless, GGP offers additional potential for modularity, which may further benefit performance and/or interpretability [35].

We report the distribution of the R^2 scores obtained over 10 independent runs in Fig. 1a, where higher values indicate better performance. Overall, we observe that in all cases at least one variant of GOMEA outperforms the GA. In particular, GOMEA-LT most frequently achieves the best results, although the differences are not always statistically significant. These findings confirm that GOMEA is well-suited for optimizing both GGP representations, thereby providing a positive answer to RQ1.

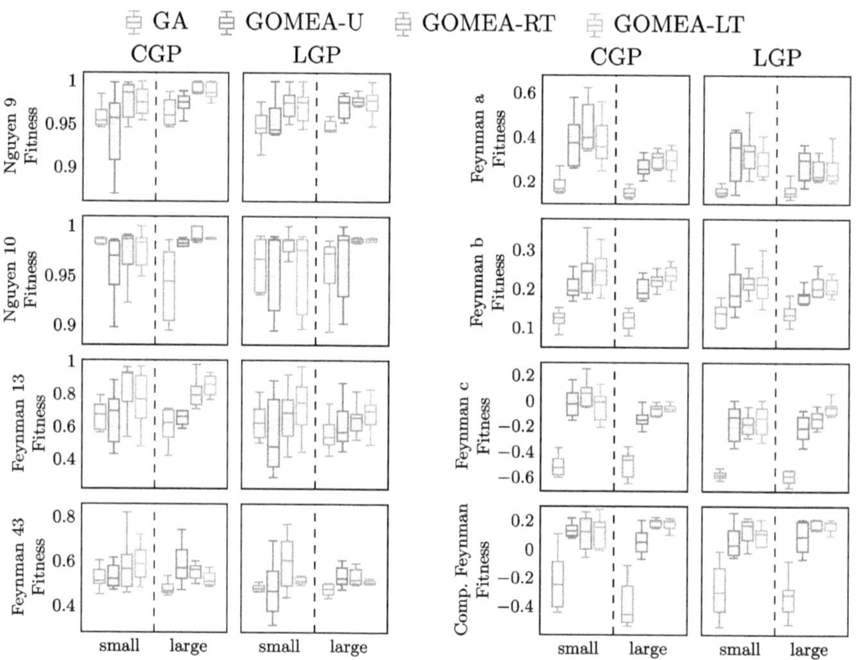

(a) R^2 score on single-output SR problems on the training set.

(b) $\bar{R}^2$ score on multi-output SR problems on the training set.

Fig. 1. Fitness (R^2 or $\bar{R}^2$ score) at the end of the optimization for each EA.

These results also allow us to begin addressing RQ3. At this stage, both GGP representations appear equally suitable, as GOMEA consistently outperforms the GA for both CGP and LGP. Drawing conclusions about the effect of population size, however, is less straightforward. Overall, the results generally support our initial intuition: for the GA, a smaller population performs at least as well as, and sometimes better than, a larger one; whereas for GOMEA, the opposite tends to hold. In most cases, a larger population appears to facilitate the learning of a more accurate FOS, which is population-dependent in GOMEA-LT.

Building on these results, we next consider the multi-output regression tasks. We show the distribution of $\bar{R}^2$ over 10 independent runs in Fig. 1b. The results clearly show that the GA performs substantially worse than all GOMEA variants across all problems. These findings further strengthen the conclusion that GOMEA is well-suited for graph optimization and effectively extend to scenarios with multiple outputs—paving the way for control tasks, where multiple outputs must be optimized simultaneously. This provides additional support for a positive answer to RQ1.

In this set of experiments, the comparison between representations tends to favor CGP over LGP, particularly when using GOMEA. This outcome is not entirely surprising, as LGP was originally developed for fundamentally different types of tasks, which may explain its comparatively lower performance in SR settings.

When comparing different population sizes, the results appear to be problem-dependent and rather mixed, preventing any definitive conclusion. However, a general trend can be observed: larger populations tend to produce results with lower variance. This suggests that the performance of the algorithm becomes more consistent and robust to stochastic factors as the population size increases.

The comparison among GOMEA variants does not always yield the expected outcome. Although GOMEA-LT performs best on several problems, it is not consistently the top performer. This raises the question of whether the algorithm can truly capture meaningful dependencies within the genotypes. One possible explanation is that the fitness landscape of the multi-output regression problems is more complex and rugged than that of the single-output tasks. If this is the case, the control problems considered next are likely to pose an even greater challenge.

A final remark concerns the observed fitness values, which are not always high; in some cases, solutions remain far from fully solving the problem. This may be partly due to the fitness metric itself. Using the mean $\bar{R}^2$ across outputs can penalize individuals that predict one output perfectly while performing poorly on others, potentially discarding partially promising solutions. Addressing this issue might require speciation mechanisms or more sophisticated multi-objective approaches, which are beyond the scope of this work. Additionally, the multi-output regression problems considered here are artificially constructed and may be inherently ill-posed: in practice, it is often more reasonable to model outputs individually rather than regressing all simultaneously if possible.

To conclude, from this preliminary analysis, we observe that GOMEA is indeed suitable for optimizing graph-based representations, providing a positive answer to RQ1. The effect of population size remains inconclusive, while

CGP appears to outperform LGP in multi-output problems, contributing toward answering RQ3. These results are based solely on training performance, as we are here evaluating the effectiveness of GOMEA as a "pure optimizer." In Sect. 4.3, we assess statistical significance and test-set performance to examine generalization ability and draw conclusions within the broader context.

4.2 Continuous Control

Following the SR experiments, we evaluate GGP-GOMEA on continuous control tasks. In this context, the fitness of a policy is defined as the average cumulative reward obtained over 5 independent simulations, each consisting of $T = 1000$ time steps. Averaging across multiple runs helps mitigate the influence of stochastic elements in the simulation and ensures more consistent results. This is particularly important given the strong selection pressure inherent in GOMEA, as it reduces the risk of propagating solutions that succeed merely due to chance.

We report the results in Fig. 2, showing the distribution of cumulative rewards over 10 independent runs, where higher values indicate better performance. These results are generally comparable to those achieved by state-of-the-art RL algorithms, although we observe some loss of performance in more complex environments such as hopper and walker2d, consistent with prior findings [34]. Compared to the SR experiments, the outcomes here are noticeably more variable, and the ranking of algorithms is less consistent across different control tasks. Overall, GOMEA no longer demonstrates a clear advantage, suggesting that it does not provide general benefits for continuous control problems. This provides a negative answer to RQ2.

In fact, while GOMEA performs comparably to the GA on some tasks, it underperforms on others. The only clear instance of improvement is observed for GOMEA-LT over the GA on the reacher task with CGP. Overall, no consistent differences are apparent between representations—sometimes GOMEA performs better with CGP, other times with LGP—nor between population sizes.

Regarding the GOMEA variants, GOMEA-U appears to perform the worst overall. This is consistent with expectations, as it is effectively a GA with high selection pressure and low mutation probability. Given that diversity is likely beneficial for these tasks [33], its poorer performance is not surprising. In contrast, there are no substantial differences between GOMEA-RT and GOMEA-LT, suggesting that the FOS learned by GOMEA-LT does not capture the correct dependencies in these control problems.

4.3 Global Picture

To draw our final conclusions, we perform a statistical analysis of the results. For each representation (i.e., CGP or LGP), algorithm configuration (i.e., population size), and problem class—namely, single-output regression (R), multi-output regression (MR), and continuous control (C)—we first identify the winning algorithm for each individual problem as the one with the highest mean performance across runs. We then conduct pairwise Mann-Whitney U tests comparing the

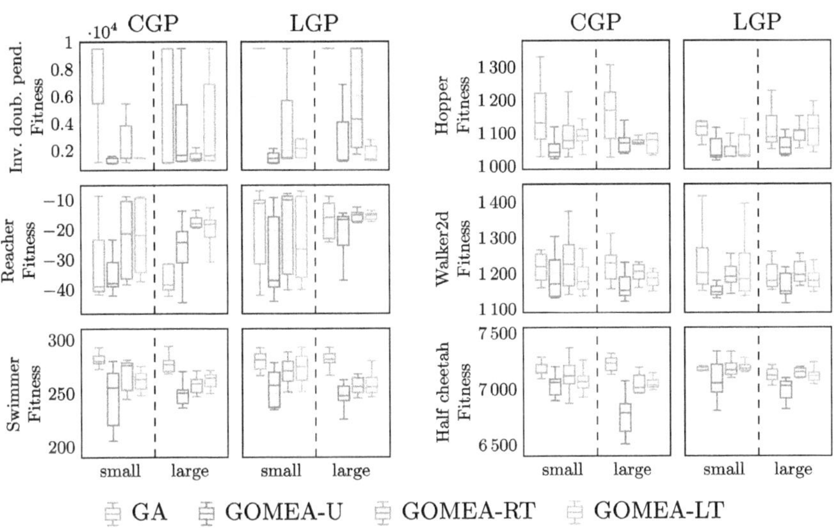

Fig. 2. Fitness (cumulative reward) at the end of optimization for each EA.

Table 1. How many times (in fraction w.r.t. the total) each algorithm is not worse than the best one on the same problem.

		CGP small			CGP large			LGP small			LGP large		
		R	MR	C	R	MR	C	R	MR	C	R	MR	C
Train	GA	0.50	0.00	1.00	0.00	0.00	0.83	0.50	0.00	1.00	0.00	0.00	1.00
	GOMEA-U	0.50	0.75	0.17	0.25	0.25	0.17	0.75	1.00	0.33	1.00	0.75	0.50
	GOMEA-RT	1.00	1.00	0.83	1.00	1.00	0.50	1.00	1.00	0.67	1.00	0.75	0.67
	GOMEA-LT	1.00	1.00	0.50	1.00	1.00	0.67	1.00	1.00	0.67	1.00	1.00	0.67
Test	GA	0.75	0.00		0.75	0.00		0.75	0.00		1.00	0.25	
	GOMEA-U	0.50	0.75		1.00	1.00		0.75	1.00		1.00	1.00	
	GOMEA-RT	1.00	1.00		0.75	0.25		1.00	1.00		0.75	0.50	
	GOMEA-LT	1.00	1.00		0.75	0.75		0.75	1.00		1.00	1.00	

winner to all other algorithms. We consider an algorithm statistically equivalent to the winner if the resulting p-value exceeds $\alpha = 0.05$. For each algorithm and problem class, we count the relative number of problems for which the algorithm is not worse than the winner. We report the resulting figures in the top part of Table 1.

The results of Table 1 clearly corroborate our previous observations from the distributions of individual runs: GOMEA performs very well on regression tasks, both single- and multi-output, but its advantage diminishes in continuous control problems. These trends are consistent across both representations and population sizes. Interestingly, for CGP with a smaller population, GOMEA-RT outperforms GOMEA-LT. This is likely due to the strong inherent bias of the CGP representation; a larger population may be required to accurately learn the true linkage structure rather than simply reflecting intrinsic bias of the representation [7,14].

From this analysis, we conclude that, for continuous control tasks, GOMEA generally underperforms relative to the GA, while it is in general better than GA in SR. We speculate that this is due to the difficulty of learning meaningful dependencies within the genotype for such complex control problems. Specifically, the fitness landscape in control tasks is highly complex, and the mapping from genotype to phenotype to behavior to fitness is poorly local. Indeed, prior work has shown that less sophisticated algorithms can achieve competitive performance [12], and methods that explicitly promote exploration and diversity may be even more effective in this domain [33].

As a final point of reflection, we recompute the same values using test-set results for regression problems (bottom part of Table 1). Here, the patterns become less clear. In particular, with larger populations, GOMEA exhibits substantially worse performance for both GGP representations. This indicates that, while the algorithm may have successfully optimized the training objective, it has overfitted the training data. Since the primary focus of this study is on the optimization capability of GOMEA, this outcome is acceptable; however, if generalization were the main concern, one could mitigate overfitting by incorporating a validation set during training.

5   Conclusion

In this paper, we investigated the use of GOMEA, a model-based EA designed to learn dependencies among genotype loci, for optimizing graphs under two GGP representations: CGP and LGP. We evaluated three variants of GOMEA, differing in how they infer linkage structures, and compared them against a standard GA. Our experiments on symbolic regression tasks—both single- and multi-output—showed that GOMEA generally outperforms the GA, confirming its effectiveness as an optimizer in these settings. However, when applied to continuous control problems, GOMEA exhibited weaker performance and was often outperformed by the GA. We attribute this to the highly complex fitness landscape of control tasks, which makes learning meaningful genotype dependencies difficult. Overall, our results suggest that sophisticated model-based EAs such as GOMEA are more suitable for structured, static optimization problems than for dynamic control tasks.

As future work, we plan to combine GOMEA with mechanisms for preserving diversity during evolution. This could involve incorporating behavior- or graph-based diversity preservation, for example through niching or speciation methods, or by adopting ideas from quality-diversity (QD). Then, a promising direction would be to restrict the dependency learning in GOMEA to individuals that are more similar, thereby allowing the algorithm to model meaningful relationships without being misled by overly heterogeneous populations.

References

1. Aalvanger, G., Luong, N.H., Bosman, P.A., Thierens, D.: Heuristics in permutation GOMEA for solving the permutation flowshop scheduling problem. In: International Conference on Parallel Problem Solving from Nature, pp. 146–157, Springer (2018)
2. Aldeia, G.S.I., de Franca, F.O.: Interpretability in symbolic regression: a benchmark of explanatory methods using the feynman data set. Genet. Program Evolvable Mach. **23**(3), 309–349 (2022)
3. Bouter, A., Alderliesten, T., Witteveen, C., Bosman, P.A.: Exploiting linkage information in real-valued optimization with the real-valued gene-pool optimal mixing evolutionary algorithm. In: Proceedings of the Genetic and Evolutionary Computation Conference, pp. 705–712 (2017)
4. Brameier, M., Banzhaf, W.: A comparison of linear genetic programming and neural networks in medical data mining. IEEE Trans. Evol. Comput. **5**(1), 17–26 (2002)
5. Brameier, M., Banzhaf, W., Banzhaf, W.: Linear Genetic Programming, vol. 1. Springer (2007)
6. Cheng, R., He, C., Jin, Y., Yao, X.: Model-based evolutionary algorithms: a short survey. Complex Intell. Syst. **4**(4), 283–292 (2018)
7. Cui, H., Heider, M., Hähner, J.: Positional bias does not influence cartesian genetic programming with crossover. In: International Conference on Parallel Problem Solving from Nature, pp. 151–167. Springer (2024)
8. De La Torre, C., Lavinas, Y., Cortacero, K., Luga, H., Wilson, D.G., Cussat- Blanc, S.: Multimodal adaptive graph evolution for program synthesis. In: International Conference on Parallel Problem Solving from Nature, pp. 306– 321. Springer (2024)
9. De La Torre, C., Nadizar, G., Lavinas, Y., Luga, H., Wilson, D., Cussat- Blanc, S.: Evolution of inherently interpretable visual control policies. In: Proceedings of the Genetic and Evolutionary Computation Conference, pp. 358–367 (2025)
10. De La Torre, C., et al.: Evolved and transparent pipelines for biomedical image classification. In: European Conference on Genetic Programming (Part of EvoStar), pp. 173 189. Springer (2025)
11. Dushatskiy, A., Mendrik, A.M., Alderliesten, T., Bosman, P.A.: Convolutional neural network surrogate-assisted GOMEA. In: Proceedings of the Genetic and Evolutionary Computation Conference, pp. 753–761 (2019)
12. El Saliby, M., Nadizar, G., Salvato, E., Medvet, E.: Eventually, all you need is a simple evolutionary algorithm (for neuroevolution of continuous control policies). In: Proceedings of the Genetic and Evolutionary Computation Conference Companion, pp. 1904–1913 (2024). https://doi.org/10.1145/3638530.3664112
13. Françoso Dal Piccol Sotto, L., Kaufmann, P., Atkinson, T., Kalkreuth, R., Porto Basgalupp, M.: Graph representations in genetic programming. Genetic Programm. Evol. Mach. **22**(4), 607–636 (2021)
14. Goldman, B.W., Punch, W.F.: Length bias and search limitations in cartesian genetic programming. In: Proceedings of the 15th Annual Conference on Genetic and Evolutionary Computation, pp. 933–940 (2013)
15. Guijt, A., Thierens, D., Alderliesten, T., Bosman, P.A.: Solving multistructured problems by introducing linkage kernels into GOMEA. In: Proceedings of the Genetic and Evolutionary Computation Conference, pp. 703–711 (2022)
16. Harrison, J., Alderliesten, T., Bosman, P.A.: Gene-pool optimal mixing in cartesian genetic programming. In: International Conference on Parallel Problem Solving from Nature, pp. 19–32. Springer (2022

17. Hu, T., Banzhaf, W.: Neutrality and variability: two sides of evolvability in linear genetic programming. In: Proceedings of the 11th Annual Conference on Genetic and Evolutionary Computation, pp. 963–970 (2009)
18. Jorgensen, S., et al.: Large language model-based test case generation for GP agents. In: Proceedings of the Genetic and Evolutionary Computation Conference, pp. 914–923 (2024)
19. Jorgensen, S., Nadizar, G., Pietropolli, G., Manzoni, L., Medvet, E., O'Reilly, U.M., Hemberg, E.: Policy search through genetic programming and LLM-assisted curriculum learning. ACM Trans. Evol. Learn. (2025)
20. Kelly, S., Heywood, M.I.: Emergent tangled graph representations for Atari game playing agents. In: Genetic Programming: 20th European Conference, EuroGP 2017, Amsterdam, The Netherlands, April 19-21, 2017, Proceedings 20, pp. 64–79. Springer (2017)
21. Kocherovsky, M., Banzhaf, W.: Crossover destructiveness in cartesian versus linear genetic programming. In: Artificial Life Conference Proceedings 36, vol. 2024, p. 20, MIT Press One Rogers Street, Cambridge, MA 02142-1209, USA journals-info. . . (2024)
22. Kocherovsky, M., Banzhaf, W.: Operators in cartesian genetic programming. In: Genetic Programming: 28th European Conference, EuroGP 2025, Held as Part of EvoStar 2025, Trieste, Italy, April 23–25, 2025, Proceedings, vol. 15609, p. 68, Springer Nature (2025)
23. Kocherovsky, M., Kianinejad, M., Bakurov, I., Banzhaf, W.: On the effectiveness of crossover operators in cartesian genetic programming. In: European Conference on Genetic Programming (Part of EvoStar), pp. 68–84. Springer (2025)
24. Koza, J.R.: Genetic programming as a means for programming computers by natural selection. Stat. Comput. **4**(2), 87–112 (1994)
25. La Cava, W., et al.: Contemporary symbolic regression methods and their relative performance. In: Advances in Neural Information Processing Systems 2021(DB1), 1 (2021)
26. Lipton, Z.C.: The mythos of model interpretability: in machine learning, the concept of interpretability is both important and slippery. Queue **16**(3), 31–57 (2018)
27. Luong, N.H., La Poutré, H., Bosman, P.A.: Multi-objective gene-pool optimal mixing evolutionary algorithms. In: Proceedings of the 2014 Annual Conference on Genetic and Evolutionary Computation, pp. 357–364 (2014)
28. McDermott, J., et al.: Genetic programming needs better benchmarks. In: Proceedings of the 14th Annual Conference on Genetic and Evolutionary Computation, pp. 791–798 (2012)
29. Medvet, E., Bartoli, A., De Lorenzo, A., Tarlao, F.: GOMGE: gene-pool optimal mixing on grammatical evolution. In: International Conference on Parallel Problem Solving from Nature, pp. 223–235. Springer (2018)
30. Mei, Y., Chen, Q., Lensen, A., Xue, B., Zhang, M.: Explainable artificial intelligence by genetic programming: a survey. IEEE Trans. Evol. Comput. **27**(3), 621–641 (2022)
31. Miller, J.F.: Cartesian genetic programming: its status and future. Genet. Program Evolvable Mach. **21**, 129–168 (2020)
32. Miller, J.F., Thomson, P.: Cartesian genetic programming. In: Genetic Programming, pp. 121–132, Springer, Berlin, Heidelberg (2000). ISBN 978-3-540-46239-2
33. Nadizar, G., Medvet, E., Wilson, D.: Searching for a diversity of interpretable graph control policies. In: Proceedings of the Genetic and Evolutionary Computation Conference, pp. 933–941 (2024), https://doi.org/10.1145/3638529.3653987

34. Nadizar, G., Medvet, E., Wilson, D.G.: Naturally interpretable control policies via graph-based genetic programming. In: European Conference on Genetic Programming (Part of EvoStar), pp. 73–89, Springer (2024)
35. Nadizar, G., Rovito, L., De Lorenzo, A., Medvet, E., Virgolin, M.: An analysis of the ingredients for learning interpretable symbolic regression models with human-in-the-loop and genetic programming. ACM Trans. Evol. Learn. Optim. 4(1), 1–30 (2024). https://doi.org/10.1145/3643688
36. Nagelkerke, N.J., et al.: A note on a general definition of the coefficient of determination. biometrika 78(3), 691–692 (1991)
37. Orphanou, K., Thierens, D., Bosman, P.A.: Learning bayesian network structures with GOMEA. In: Proceedings of the Genetic and Evolutionary Computation Conference, pp. 1007–1014 (2018)
38. Sakallioglu, B., Nadizar, G., Manzoni, L., Medvet, E.: Evolving typed token processing networks. In: Proceedings of the Genetic and Evolutionary Computation Conference Companion, pp. 2177–2181 (2025)
39. Shem-Tov, E., Sipper, M., Elyasaf, A.: Deep learning-based operators for evolutionary algorithms. In: Genetic Programming Theory and Practice XXI, pp. 51–66, Springer (2025)
40. Sobania, D., Schweim, D., Rothlauf, F.: A comprehensive survey on program synthesis with evolutionary algorithms. IEEE Trans. Evol. Comput. 27(1), 82–97 (2022)
41. Song, J., Lu, Q., Tian, B., Zhang, J., Luo, J., Wang, Z.: Symbol graph genetic programming for symbolic regression. In: International Conference on Parallel Problem Solving from Nature, pp. 221–237, Springer (2024)
42. Thierens, D., Bosman, P.A.: Optimal mixing evolutionary algorithms. In: Proceedings of the 13th Annual Conference on Genetic and Evolutionary Computation, pp. 617–624 (2011)
43. Thierens, D., Bosman, P.A.: Hierarchical problem solving with the linkage tree genetic algorithm. In: Proceedings of the 15th Annual Conference on Genetic and Evolutionary Computation, pp. 877–884 (2013)
44. Todorov, E., Erez, T., Tassa, Y.: Mujoco: A physics engine for model-based control. In: 2012 IEEE/RSJ International Conference on Intelligent Robots and Systems, pp. 5026–5033, IEEE (2012)
45. Turner, A.J., Miller, J.F.: Neutral genetic drift: an investigation using cartesian genetic programming. Genet. Program Evolvable Mach. 16(4), 531–558 (2015)
46. Udrescu, S.M., Tegmark, M.: Ai feynman: A physics-inspired method for symbolic regression. Sci. Adv. 6(16), eaay2631 (2020)
47. Vacher, Q., et al.: Maple: multi-action programs through linear evolution for continuous multi-action reinforcement learning. In: Proceedings of the Genetic and Evolutionary Computation Conference, pp. 1062–1071 (2025)
48. Virgolin, M., Alderliesten, T., Bel, A., Witteveen, C., Bosman, P.A.: Symbolic regression and feature construction with GP-GOMEA applied to radiotherapy dose reconstruction of childhood cancer survivors. In: Proceedings of the Genetic and Evolutionary Computation Conference, pp. 1395–1402 (2018)
49. Virgolin, M., Alderliesten, T., Witteveen, C., Bosman, P.A.: Scalable genetic programming by gene-pool optimal mixing and input-space entropy-based building-block learning. In: Proceedings of the Genetic and Evolutionary Computation Conference, pp. 1041–1048 (2017)
50. Virgolin, M., Alderliesten, T., Witteveen, C., Bosman, P.A.: Improving model-based genetic programming for symbolic regression of small expressions. Evol. Comput. 29(2), 211–237 (2021)

51. Walker, J.A., Miller, J.F.: The automatic acquisition, evolution and reuse of modules in cartesian genetic programming. IEEE Trans. Evol. Comput. **12**(4), 397–417 (2008)
52. Wilson, D.G., Cussat-Blanc, S., Luga, H., Miller, J.F.: Evolving simple programs for playing atari games. In: Proceedings of the Genetic and Evolutionary Computation Conference, pp. 229–236 (2018)
53. Yazdani, S., Shanbehzadeh, J.: Balanced cartesian genetic programming via migration and opposition-based learning: application to symbolic regression. Genet. Program Evol. Mach. **16**(2), 133–150 (2015)

Multi-tree Genetic Programming
with Semantic Complementarity
for Feature Construction in Symbolic
Regression

Jiayu Zhang, Qi Chen$^{(\boxtimes)}$, Bing Xue, and Mengjie Zhang

Centre for Data Science and Artificial Intelligence and School of Engineering and
Computer Science, Victoria University of Wellington, PO Box 600, Wellington 6140,
New Zealand
`{Jiayu.Zhang,Qi.Chen,Bing.Xue,Mengjie.Zhang}@ecs.vuw.ac.nz`

Abstract. Semantic complementarity among trees plays an important role in Multi-Tree Genetic Programming (MTGP) based feature construction. It enables trees to capture diverse yet relevant information, thereby reducing multicollinearity among constructed features and improving generalization. However, this complementarity is often constrained by syntax-driven initialization, which overlooks semantic diversity and leads to trees with highly correlated outputs in the initial population, and by conventional genetic operators that disregard semantic relationships and disrupt useful complementarities among trees. To overcome these limitations, this paper proposes a semantic clustering initialization method that promotes semantic complementarity by assembling individuals with trees originate from different semantic clusters. In addition, a component replacement crossover is introduced to preserve this complementarity by exchanging semantically related trees during evolution. Experiments on 12 regression datasets show that the proposed MTGP framework consistently improves predictive accuracy and convergence over MTGP baselines, confirming the benefits of incorporating semantic clustering and semantically informed variation operators.

Keywords: Multi-Tree Genetic Programming · Evolutionary Feature Construction · Symbolic Regression · Semantic Complementarity

1 Introduction

Feature construction aims to create new, informative features from raw inputs to better represent underlying relationships and improve learning performance [21]. *Genetic Programming* (GP) [8], as a gradient-free evolutionary algorithm with a variable-length representation, can automatically evolve human-interpretable symbolic models. This ability makes GP a natural framework for feature construction where symbolic transformation of input variables produce intermediate features that simplify learning tasks such as *symbolic regression* (SR) [4,9,11,15,20].

L. Manzoni et al. (Eds.): EuroGP 2026, LNCS 16521, pp. 225–240, 2026.
https://doi.org/10.1007/978-3-032-23005-8_14

In this context, *Multi-Tree Genetic Programming* (MTGP) is often employed, where each individual consists of multiple GP trees, and each representing a constructed feature [3,6,19]. MTGP-based feature construction can simultaneously evolve a set of simple yet complementary trees to enhance learning performance. Its effectiveness is strongly influenced by the *semantic complementarity* among the trees ensuring each contributes distinct, non-redundant information.

However, promoting semantic complementarity remains challenging due to *semantic redundancy* between trees, which can subsequently lead to multicollinearity among constructed features. Semantic redundancy occurs when two or more GP trees produce highly correlated outputs, reducing expressive diversity and making downstream learning models unstable or less interpretable. Such redundancy often originates from random initialization, such as ramped half-and-half [17,18], which generates trees independently without considering semantic diversity, and from semantically blind genetic operators like subtree crossover and mutation that may introduce redundant trees in GP individuals during evolutionary process.

To address these issues, this work proposes a new MTGP method to construct semantically complementary features for SR. Specifically, we design a novel initialization strategy to promote diversity among trees from the beginning, and introduce a new crossover operator designed to preserve and enhance semantic complementarity throughout the evolutionary process. The specific objectives are as follows:

1. To design a semantic clustering initialization that promotes semantic complementarity among trees within each GP individual, thereby producing a more diverse and informative feature set.
2. To develop a component replacement crossover that replaces the least important tree in an individual with a semantically aligned one, preserving overall semantic complementarity while improving individual quality during evolution.
3. To evaluate the effect of the proposed framework and to evaluate its effectiveness in constructing semantically complementary features and improving regression accuracy.

2   Related Work

2.1   Semantic-Aware Initialization in GP

Semantic-aware initialization in GP aims to improve population diversity and search efficiency by explicitly considering the semantics, that is, the output behavior, of individuals during initial population generation. Traditional initialization methods, such as grow, full, and ramped half-and-half [4], generate trees based solely on syntax, which often leads to semantic redundancy. This redundancy can hinder exploration and slow convergence, as the initial population may cover only a narrow region of the semantic space.

To address this, several semantic-aware initialization strategies have been proposed. Early studies [2, 7] emphasized that semantically diverse initial populations improve exploration efficiency and generalisation performance. Building on this, several semantic-aware initialization strategies were proposed to enhance the coverage of the semantic space. Pawlak et al. [14] introduced Semantic Geometric Initialization which ensures that the convex hull of the initial population encloses the target semantics, thus guaranteeing that geometric semantic crossover can potentially reach the desired solution. Dick [5] further advanced the idea of safe tree initialization using interval arithmetic to prevent invalid or semantically degenerate offspring, maintaining behavioral consistency during search. These works highlight that incorporating semantic information at initialization helps create a broader and more expressive foundation for search, facilitating more effective exploration in GP.

Although existing semantic-aware initialization methods mainly focus on enhancing population-level diversity in single-tree GP, their principles are highly relevant to feature construction in MTGP. Adapting semantic-aware initialization to consider intra-individual semantic diversity can therefore potentially promote complementary feature representations. However, to the best of our knowledge, no prior work has specifically addressed semantic-aware initialization for MTGP, where the goal is to encourage both inter- and intra-individual semantic diversity/complementarity to improve learning performance.

2.2 Crossover Operators in MTGP

In MTGP, each individual is composed of a set of trees. This representation introduces complex dependencies among trees, motivating a well-designed crossover that can selectively exchange meaningful components between individuals, promote semantic diversity, and preserve useful feature interactions.

A common crossover used in MTGP is the position-wise random crossover [10, 16, 22], which randomly selects an index position in the parents and then performs a standard subtree crossover to the trees at that position. However, this approach completely ignores the contribution of each tree to the individual's overall performance. As a result, highly informative trees may be disrupted during crossover, breaking the relationships they share with other trees and diminishing the overall semantic coherence of the individual.

To address this issue, several studies have leveraged the importance of GP trees to guide crossover. For instance, Al-Helali et al. [1] proposed a probabilistic mechanism where trees deemed more important are selected for crossover with a higher probability, aiming to accelerate the propagation of desirable features. Nguyen et al. [12] proposed the self-competitive crossover, which specifically identifies and modifies the lowest-contributing tree within an individual, with the goal of replacing it with a better tree.

While tree-importance awareness is a notable advancement, these methods remain unaware of the replacement tree's semantics. They determine which tree to participate in crossover but overlook the semantic compatibility of the replacement, often leading to the substitution of an important yet semantically

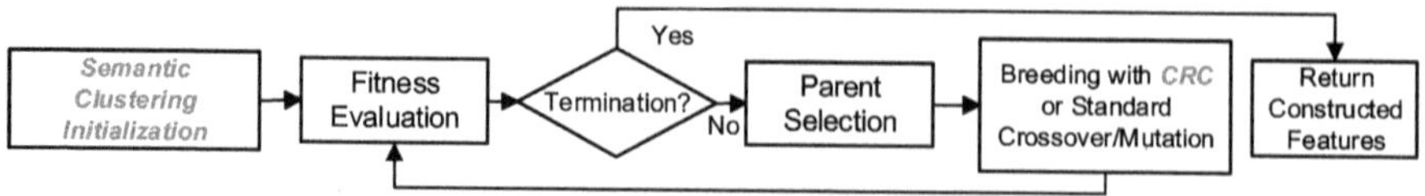

Fig. 1. The workflow of the proposed MTGP method.

redundant tree, and thus an ineffective crossover operation. To address this limitation, Zhang et al. [19] proposed a geometric semantic macro-crossover operator (GSMX) which performs semantic-guided crossover at the tree level rather than at the traditional subtree level. GSMX selects and recombines trees based on their semantic relationships. Specifically, relevance to the target semantics and complementarity among features. However, this approach requires computing and comparing the semantics of multiple trees across individuals, often using cosine similarity, which can incur substantial computational overhead on large datasets or when individuals contain many trees.

3   The Proposed Method

This paper proposes a new MTGP-based feature construction method for regression, where GP evolves multiple features that are subsequently fed into a regression model. To enhance semantic complementarity among GP trees, this method introduces two key innovations. First, a new initialization method is proposed to generate a diverse and high-quality initial population by clustering trees according to their semantic outputs and assembling individuals from clusters exhibiting semantic complementarity. Second, a new crossover operator is designed to refine the discovered cluster-level composition during evolution by replacing the least important tree with another tree from the same cluster. Together, these two mechanisms promote semantic diversity and complementarity throughout the evolutionary process, resulting in more informative features with reduced multicollinearity.

3.1   Overall Framework

The algorithm follows the standard GP procedure containing the following steps, as depicted in Fig. 1.

i) Population Initialization: The algorithm proposes the new semantic clustering initialization (see Sect. 3.2) to create the initial population. Each individual consists of m semantic complementarity GP trees, thus forming a diverse and well-structured initial population.

ii) Fitness Evaluation: For each individual with m trees $\{\varphi_1, \ldots, \varphi_m\}$, evaluate each tree on the training data X_{train} to obtain column vectors $\varphi_j(X_{\text{train}}) \in \mathbb{R}^n$. Stacking these column-wise vectors forms a feature matrix $\Phi(X_{\text{train}}) \in \mathbb{R}^{n \times m}$. A ridge regression model is then applied on $\Phi(X_{\text{train}})$ to assess the

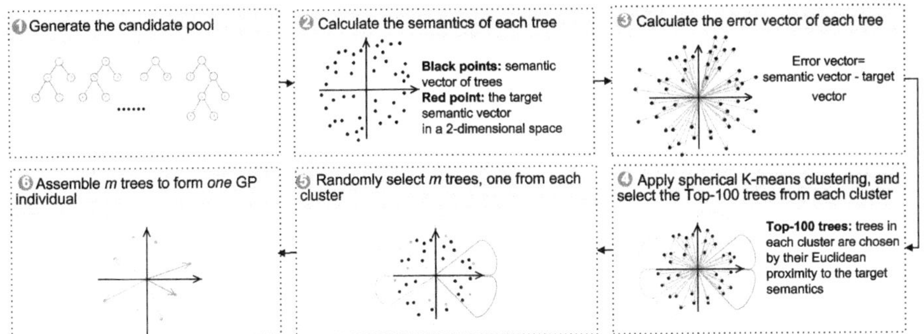

Fig. 2. The schematic diagram of SCI.

individual's predictive performance determined by the model's R^2 score which is defined as:

$$R^2 = 1 - \frac{\sum_{i=1}^{N}(y_i - \hat{y}_i)^2}{\sum_{i=1}^{N}(y_i - \bar{y})^2} \tag{1}$$

where y_i is the true value of the i-th sample, $\hat{y}_i$ is the model's prediction, $\bar{y}$ is the mean of all true values, and N is the number of instances. A higher R^2 value, closer to 1, indicates a better fit to the training data.

iii) Parent Selection: Parents are selected through tournament selection.

iv) Breeding: Parents generate offspring via a dual-crossover mechanism and a mutation operator (see Sect. 3.3). In addition to standard crossover and mutation, a new crossover operator is introduced to preserve the semantic complementary by replacing the least important trees with alternative from the same cluster.

The above steps continue until the stopping criterion is met, at which point the evolutionary run terminates and returns the constructed features. The details of the two proposed components are presented in the following sections.

3.2 Semantic Clustering Initialization Method

To promote semantic complementarity between trees in a MTGP individual, this work proposes a new initialization method named Semantic Clustering Initialization (SCI). The proposed method first generates a pool of candidate trees, then performs clustering on these trees according to their outputs/semantics similarity, then assembles individuals from distinct clusters to encourage semantic complementary. The procedure, as illustrated in Fig. 2, is detailed as follows:

1. Construction of the candidate tree pool: To enable effective semantic clustering during initialization, an oversized candidate tree pool of N_{pool} trees is generated using the ramped half-and-half, where $N_{\text{pool}} = 20,000$, an order of magnitude larger than the $N_{\text{tree}} = 2,000$ trees required to constitute the initial population ($N_{\text{pop}} = 200$).

2. Calculate the semantics and unit error vectors: For semantic clustering, the semantics and corresponding error vector of each tree are computed. For each tree i in the pool, its semantic vector $s_i(X_{train})$ is obtained on the training data X_{train}, as highlighted in Step 2 of Fig. 2. The error vector e is then calculated as the difference between the target semantics y_{train} and $s_i(X_{train}$, where $e = T(X_{train}) - y_{train}$, as highlighted in Step 3 of Fig. 2. To focus on direction rather than magnitude, the error vector is then normalized to form the unit error vector $\hat{e}_i = e_i/\|e_i\|_2$.

3. Spherical k-means clustering on unit error vectors: The candidate trees are then clustered based on the directional similarity of their error vectors. By focusing on the direction rather than the magnitude of errors, this clustering captures whether two trees make errors on the same instances in similar ways, independent of scale. This provides a principled way to distinguish between semantically redundant trees—those grouped in the same cluster—and semantically complementary ones that lie in different clusters. To this aim, Spherical k-means clustering is applied, since it better captures directional similarity among error vectors. The number of clusters k is equal to the number of trees per individual, ensuring that each tree originates from a distinct cluster. The spherical k-means clustering partitions the N_{pool} trees based on their unit error vectors $E = \{\hat{e}_1, \hat{e}_2, \ldots, \hat{e}_{N_{pool}}\}$ into k clusters $C = \{C_1, C_2, \ldots, C_k\}$. The process iteratively assigns each tree to the cluster with the nearest centroid, as measured by Euclidean distance. This iterative procedure aims to minimize the within-cluster sum of squares, defined by the objective function:

$$\underset{C}{\arg\min} \sum_{k=1}^{K} \sum_{\hat{e}_i \in C_k} \|\hat{e}_i - \mu_k\|_2^2 \tag{2}$$

Since all vectors are constrained to the unit hypersphere, minimizing the Euclidean distance between a vector $\hat{e}_i$ and a centroid μ_k is equivalent to maximizing their cosine similarity. This is achieved by updating the centroid μ_k for each cluster C_k in two steps:

(a) Mean Calculation: The preliminary centroid $\bar{\mu}_k$ is computed as the element-wise mean of all vectors within the cluster.

$$\mu'_k = \frac{1}{|C_k|} \sum_{\hat{e}_i \in C_k} \hat{e}_i \tag{3}$$

(b) Normalization: since μ'_k is generally not of unit length, it is subsequently $\mathcal{L}_2$ normalized to obtain the final centroid μ_k, projecting it back onto the unit sphere.

$$\mu_k \leftarrow \frac{\mu'_k}{\|\mu'_k\|_2} \tag{4}$$

Within each cluster, trees are ranked according to the Euclidean distance between their semantics and the target semantics, and the top 100 trees are preserved to form a global candidate pool. Step 4 in Fig. 2 illustrates how spherical k-means clustering partitions the semantic space into distinct directional clusters. Each sector represents a cluster, with trees from different clusters aligned along different semantic directions.

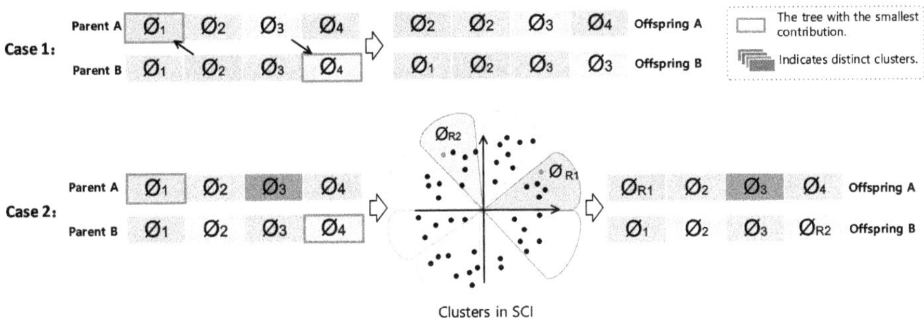

Fig. 3. Illustration of CRC.

4. Assembly of complementary individuals: To generate each GP individual in the initial population, one tree is randomly sampled from each cluster in the global candidate pool and assemble into a multi-tree individual. This ensures that each individual is composed of trees with diverse error directions as shown in Step 5 in Fig. 2.

3.3   Dual-Crossover Mechanism

To preserve semantic complementarity during evolution, we introduce a new crossover operator-Component Replacement Crossover (CRC). Unlike standard crossover, which swaps tree/subtrees without considering their semantics, CRC selectively replaces the least important tree in one parent with a semantically compatible tree. This ensures that the substituted components share similar semantic patterns while still contributing new information.

CRC first assesses the importance of each tree using the corresponding parameter values obtained from the subsequent regression model. The tree with the smallest parameter value, assumed to be the least important to the predictive model, is selected for replacement. As shown in Fig. 3, the replacement is chosen in a cluster-guided manner. The operator replaces the selected tree with another tree drawn from the same semantic cluster in the other parent. As illustrated in the *parent-based replacement* example in Fig. 3, the least important tree $\varnothing_1$ in Parent A is replaced with $\varnothing_2$ from Parent B, since both belong to the same cluster (in the same color). Similarly, the least important tree $\varnothing_4$ in Parent B is replaced with a tree from the corresponding cluster in Parent A. If the other parent does not contain a tree from that cluster, as illustrated in the *pool-based replacement* example in Fig. 3, a replacement is sampled from the corresponding cluster in the global candidate pool generated during initialization. This strategy maintains the discovered cluster-level composition while refining its internal trees, leading to more stable and effective improvements.

While CRC could be effective in preserving semantic complementarity, it operates within the constraints of cluster-guided replacement and therefore offers limited structural exploration. Standard position-wise random subtree crossover

Table 1. Datasets

No.	Dataset Name	Sample Count	Feature Count
1	628_fri_c3	1000	5
2	646_fri_c3	500	10
3	647_fri_c1	250	10
4	657_fri_c2	250	10
5	592_fri_c4	1000	25
6	591_fri_c1	100	10
7	596_fri_c2	250	5
8	637_fri_c1	500	50
9	595_fri_c0	1000	10
10	589_fri_c2	1000	25
11	622_fri_c2	1000	50
12	588_fri_c4	1000	100

is still required to maintain sufficient syntactic diversity, enabling broader structural search and preventing premature convergence. Together, these two operators form a dual-crossover mechanism in which CRC refines existing cluster-level compositions, while standard position-wise random subtree crossover explores new ones, achieving a more stable and adaptive evolutionary process.

Finally, the standard subtree mutation further supports exploration by introducing additional structural variations. In each generation, crossover and mutation are applied with probabilities of 0.7 and 0.3, respectively; During crossover, the operator is chosen with equal probability between position-wise subtree crossover and CRC, resulting in overall operation ratios of 35%/35%/30% for the two crossover types and mutation, respectively.

4   Experimental Settings

The experiments were conducted on 12 datasets from the Penn Machine Learning Benchmarks (PMLB) [13]. This subset was chosen to provide a robust evaluation, with sample counts ranging from 100 to 1,000 and feature counts from 5 to 100, as shown in Table 1. For each dataset, we randomly partition the data into a training set and a test set using an 80:20 ratio using a fixed splitting strategy. The parameter settings are listed in Table 2, following common practice in MTGP [16,21]. All experiments were conducted for 30 independent runs on each dataset.

To evaluate the effectiveness of the proposed MTGP method and to analyze the contributions of its components, we compare it with the following benchmark methods.

Table 2. Parameter settings

Parameter	Value
Number of Trees in the Pool	20000
Minimum Initial Tree Depth	2
Maximum Initial Tree Depth	4
Tree Generation Method	Ramped Half-and-Half
Population Size	200
Number of Trees in Each Individual	10
Number of Clusters (k)	10
Maximal Number of Generations	100
Maximum Tree Depth	8
Tournament Size	3
Elitism (Number of Individuals)	1
Crossover and Mutation Rates	0.7 and 0.3
CRC Rate	0.5
Functions	Add, Sub, Mul, Div, Ln, Exp, Sin, Cos, Negative

- **MTGP:** The standard MTGP employs the same algorithmic backbone as the proposed framework, with differences confined to the initialization and genetic operators. The population is initialized using the ramped half-and-half method. Each individual is composed of 10 trees generated randomly, without any semantic guidance. The evolutionary search is driven by the standard genetic operators of position-wise subtree crossover and random subtree mutation.
- **MTGP-SCI:** A variant of MTGP incorporating the SCI only. It is used in the ablation study to evaluate the effect of the initialization strategy.
- **MTGP-SCom:** The proposed MTGP with semantic complementarity algorithm extends the MTGP framework with SCI and CRC.

Notably, MTGP with the new CRC only variant is not evaluated, as its design relies on the cluster structure and global candidate pool generated by SCI. Therefore, CRC cannot be run independently of the SCI. The three MTGP methods all have 30 independent runs on the benchmark datasets, and the results are obtained for analysis.

5   Experimental Results

The performance of the three MTGP methods is evaluated and compared in terms of predictive accuracy, model size and computational cost. Tables 3 and 4 report the mean and standard deviation of R^2 scores on the training and test sets, respectively. The evolutionary convergence behaviour of different methods is illustrated in Figs. 4 and 5. Statistical significance of the performance differences is further examined using the Wilcoxon signed-rank test on the final best-of-run R^2 scores with a significance level of $\alpha = 0.05$. Finally, the comparison focuses on model size and computational cost.

5.1 Comparisons on the Training Performance

As shown in Table 3, across all datasets, all the three MTGP methods achieve high training R^2 indicating that they all fit the training data well.

Table 3. The ***Training*** R^2 (mean $\pm$ std) over the 30 independent runs (best per row in **bold**). $\uparrow$ / $\downarrow$ / $\sim$ indicate statistically significant improvement, degradation, or no difference, respectively. The symbol in the **MTGP-SCI** column compares it against **MTGP**. The two symbols in the **MTGP-SCom** column compare it against **MTGP** and **MTGP-SCI**, respectively.

	MTGP	MTGP-SCI	MTGP-SCom
628_fri_c3	0.965 ± 0.006	$0.965 \pm 0.005 \sim$	$\mathbf{0.974 \pm 0.004}$ $\uparrow$ $\uparrow$
646_fri_c3	0.958 ± 0.013	$0.958 \pm 0.012 \sim$	$\mathbf{0.974 \pm 0.007}$ $\uparrow$ $\uparrow$
647_fri_c1	0.961 ± 0.014	$0.962 \pm 0.011 \sim$	$\mathbf{0.975 \pm 0.010}$ $\uparrow$ $\uparrow$
657_fri_c2	0.951 ± 0.013	$0.948 \pm 0.016 \sim$	$\mathbf{0.969 \pm 0.006}$ $\uparrow$ $\uparrow$
592_fri_c4	0.932 ± 0.026	$0.923 \pm 0.028 \sim$	$\mathbf{0.956 \pm 0.023}$ $\uparrow$ $\uparrow$
591_fri_c1	0.971 ± 0.012	$0.968 \pm 0.011 \sim$	$\mathbf{0.984 \pm 0.007}$ $\uparrow$ $\uparrow$
596_fri_c2	0.966 ± 0.007	$0.966 \pm 0.009 \sim$	$\mathbf{0.978 \pm 0.005}$ $\uparrow$ $\uparrow$
637_fri_c1	0.917 ± 0.029	$0.903 \pm 0.035 \sim$	$\mathbf{0.939 \pm 0.023}$ $\uparrow$ $\uparrow$
595_fri_c0	0.953 ± 0.001	$0.952 \pm 0.001 \sim$	$\mathbf{0.955 \pm 0.001}$ $\uparrow$ $\uparrow$
589_fri_c2	0.929 ± 0.027	$0.934 \pm 0.021 \sim$	$\mathbf{0.961 \pm 0.017}$ $\uparrow$ $\uparrow$
622_fri_c2	0.900 ± 0.035	$0.902 \pm 0.033 \sim$	$\mathbf{0.939 \pm 0.034}$ $\uparrow$ $\uparrow$
588_fri_c4	0.890 ± 0.041	$0.881 \pm 0.037 \sim$	$\mathbf{0.937 \pm 0.025}$ $\uparrow$ $\uparrow$

MTGP-SCI and MTGP achieved comparable final performance, indicating that with the standard position-wise random crossover and mutation, the SCI did not result in a clear performance advantage.

In contrast, MTGP-SCom consistently achieves the highest training R^2 on all datasets, shown in bold. These improvements are modest but systematic, typically improve 0.01 to 0.05 (1%-5%) over MTGP, indicating that the semantic complementarity mechanism enhances the search's ability to capture complex relationships while maintaining general fitting strength. Moreover, MTGP-SCom also yields smaller standard deviations than MTGP and MTGP-SCI, indicating that the integration of semantic complementarity enhances not only accuracy but also training stability across independent runs.

The evolutionary curves in Fig. 4 show that across nearly all datasets, MTGP-SCom (blue) achieves a higher R^2 earlier in the evolution compared with MTGP (black) and MTGP-SCI (green). This indicates that semantic complementarity accelerates convergence, allowing the algorithm to find high-quality models more quickly. By the final generations, MTGP-SCom consistently reaches the highest mean R^2, in line with the numerical results in Table 3. This confirms that the combination of dual crossover and mutation guided by semantic complementarity enables the algorithm to more effectively explore new cluster-level compositions and refine existing ones.

In early generations, MTGP-SCI often starts with slightly higher R^2 than MTGP, demonstrating that SCI provides a more diverse and better-quality initial population. Yet, without the combination of dual crossover and mutation guided by semantic complementarity, its advantage diminishes in later generations, underscoring the importance of this design for maintaining sustained progress.

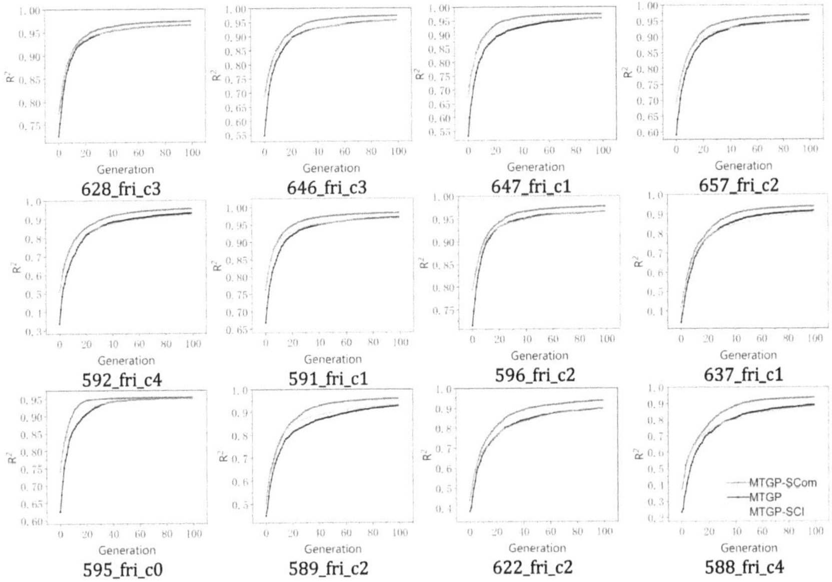

Fig. 4. The evolutionary curves on the mean R^2 of the 30 best-of-generation individuals on the ***Training Sets***.

5.2 Comparisons on the Test Performance

As shown in Table 4, all three MTGP methods maintain strong generalization, with most test R^2 scores above 0.9, indicating that models trained by the MTGP framework generalize well to unseen data.

The strong performance of the MTGP-SCom extended to the test sets, where it achieved the highest mean R^2 score on 10 of the 12 datasets. On these datasets, it also achieved the lowest standard deviation, indicating more consistent generalization and greater stability with reduced sensitivity to stochastic factors such

Table 4. The **_Test_** R^2 (mean $\pm$ std) over the 30 independent runs.

	MTGP	MTGP-SCI	MTGP-SCom
628_fri_c3	0.959 ± 0.009	$0.961 \pm 0.008 \sim$	$\mathbf{0.970 \pm 0.005}$ ↑ ↑
646_fri_c3	0.939 ± 0.018	$0.937 \pm 0.017 \sim$	$\mathbf{0.957 \pm 0.017}$ ↑ ↑
647_fri_c1	0.881 ± 0.356	$0.884 \pm 0.356 \sim$	$\mathbf{0.960 \pm 0.025}$ ↑ ↑
657_fri_c2	0.940 ± 0.025	$0.928 \pm 0.040 \sim$	$\mathbf{0.963 \pm 0.011}$ ↑ ↑
592_fri_c4	0.849 ± 0.351	$0.905 \pm 0.036 \sim$	$\mathbf{0.945 \pm 0.028}$ ↑ ↑
591_fri_c1	0.782 ± 0.153	$0.719 \pm 0.377 \sim$	$\mathbf{0.833 \pm 0.127}$ $\sim$ $\sim$
596_fri_c2	0.948 ± 0.014	$0.946 \pm 0.017 \sim$	$\mathbf{0.962 \pm 0.009}$ ↑ ↑
637_fri_c1	0.889 ± 0.047	$0.808 \pm 0.347 \sim$	$\mathbf{0.911 \pm 0.040}$ $\sim$ ↑
595_fri_c0	$\mathbf{0.950 \pm 0.005}$	$0.943 \pm 0.039 \sim$	0.949 ± 0.022 ↑ ↑
589_fri_c2	0.927 ± 0.037	$0.936 \pm 0.021 \sim$	$\mathbf{0.961 \pm 0.019}$ ↑ ↑
622_fri_c2	0.859 ± 0.224	$0.903 \pm 0.041 \sim$	$\mathbf{0.938 \pm 0.032}$ ↑ ↑
588_fri_c4	$\mathbf{0.864 \pm 0.048}$	$0.853 \pm 0.053 \sim$	0.846 ± 0.352 ↑ ↑

as initialization and crossover randomness. All of these indicate that the learning improvement of MTGP-SCom on the training sets is not due to overfitting but rather to more effective knowledge transfer and balanced explorationâĂŞexploitation.

As shown in Fig. 5, the evolutionary plots on the test sets confirms the starting advantage of MTGP-SCI and MTGP-SCom from SCI on a majority of the datasets. Both methods also exhibit noticeably smoother convergence curves, indicating more stable evolutionary dynamics and fewer oscillations in test performance on most test set. In contrast, MTGP shows frequent fluctuations in its test R^2 trajectories, suggesting a less consistent search process and potential overfitting or sensitivity to stochastic variations during evolution. Interestingly, even on datasets where the initial populations of MTGP-SCom start with comparable or slightly lower R^2 values than MTGP, such as 657_fri_c2, 591_fri_c1 and 622_fri_c2, it still consistently surpasses MTGP by the end of the evolutionary process. This observation suggests that while SCI provides a strong yet not always optimal heuristic, the subsequent adaptive exploration and correction allow the algorithm to discover and refine more effective cluster-level compositions.

5.3 Statistical Significance Test Results

When comparing MTGP-SCom against MTGP on the test sets, MTGP-SCom achieves a statistically superior or equivalent performance to MTGP on all 12 datasets, with 10 wins and 2 tie. This provides statistical evidence for the overall effectiveness of the proposed approach on enhancing generalisation performance of MTGP.

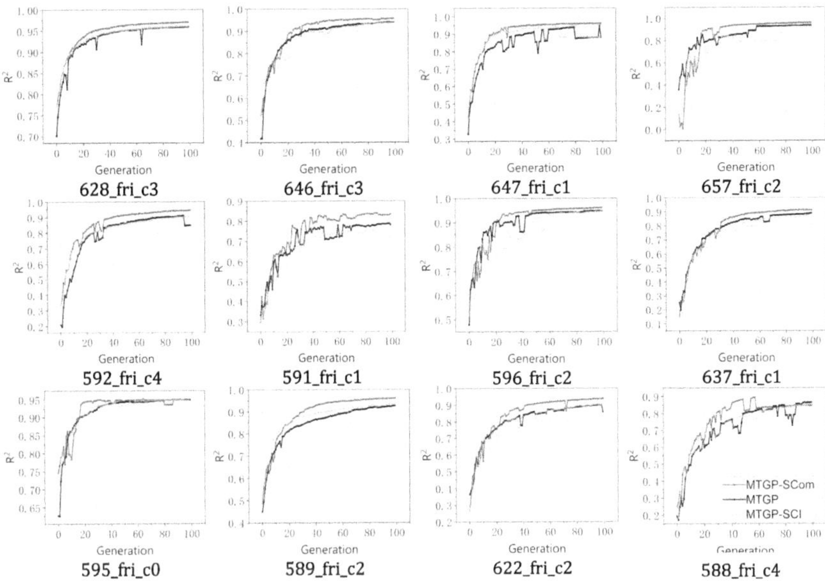

Fig. 5. The evolutionary curves on the mean R^2 of the 30 best-of-generation individuals on the **Test Sets**.

The impact of SCI is then isolated by comparing MTGP-SCI with MTGP. Despite the clear performance advantage exhibited by MTGP-SCI in the starting generations (see Fig. 5), this lead does not translate into a statistically superior final outcome. The comparison of best-of-run solutions yielded no statistically significant difference across the 12 datasets. The result indicates that although SCI provides an initial boost through a well-structured, semantically diverse population, this advantage diminishes during evolution because standard genetic operators tend to disrupt those structures.

The effect of CRC is verified by comparing MTGP-SCom against MTGP-SCI. MTGP-SCom demonstrates a statistically significant advantage over MTGP-SCI on 11 datasets, with 1 ties. This result confirms that the inclusion of the CRC is essential to capitalize on the initial performance advantage established by SCI.

On the training sets, as shown in Table 3, MTGP-SCom consistently and significantly outperforms both MTGP and MTGP-SCI across all twelve datasets, highlighting the strong effectiveness of its semantic complementarity mechanism. In contrast, MTGP-SCI shows no statistically significant improvement over MTGP, indicating that the SCI initialization alone does not lead to higher training accuracy. These results suggest that while SCI provides a useful starting point, its benefit is fully realized only when combined with the semantic refinement process in CRC, which enables more stable and effective learning of high-quality constructed features in MTGP-SCom.

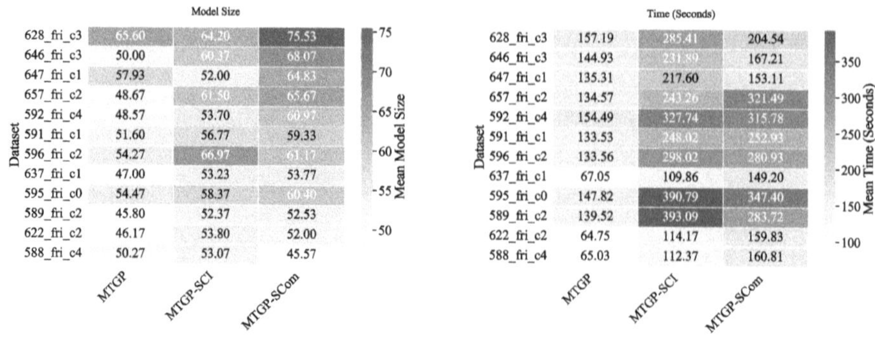

(a) Model Size Comparisons   (b) Computational Cost Comparisons

Fig. 6. Comparison of model size and computational cost across datasets.

5.4   Analysis of Model Size

We evaluated the model complexity by counting the number of nodes in the 30 best-of-run individuals obtained for each dataset. Figure 6a presents the mean number of nodes of the 30 best-of-run individuals for each dataset. MTGP generally produced the most compact models, followed by MTGP-SCI and MTGP-SCom. However, the differences were relatively small.

This modest increase in model complexity is attributed to the CRC. In MTGP, disruptive genetic operators frequently dismantle effective individuals, causing the search to converge on simpler solutions that merely survived the destructive process. In contrast, the CRC's low-disruptiveness preserves these superior yet more intricate compositions once discovered. Therefore, the marginal increase in model size is a direct consequence of a more effective search process capable of both discovering and retaining higher-quality solutions.

5.5   Computational Cost

Figure 6b reports the average computational cost. Compared with MTGP, MTGP-SCI incurs additional computational cost due to SCI. The notable increase in average cost observed for MTGP-SCI relative to MTGP can be attributed to the additional clustering and selection procedures involved in SCI. In contrast, when comparing MTGP-SCom with MTGP-SCI, we observe that the CRC substantially improves efficiency by reusing pre-computed semantics, reducing redundant computations and improving overall efficiency.

6   Conclusions

This paper presented an MTGP framework enhanced by semantic clustering initialization (SCI) and component replacement crossover (CRC) to improve feature construction for regression. SCI promotes semantic complementarity by assembling semantically diverse trees to form each individual at initialization, while CRC refines the discovered cluster-level composition, thereby preserving semantic complementarity and enabling more effective variation. Experiments on 12 regression datasets showed that MTGP-SCom, which integrates both SCI and CRC, consistently outperforms both standard MTGP and MTGP-SCI, which incorporates SCI only, achieving smoother convergence and higher predictive accuracy, with only a small increased in model size and computational cost.

Future work will explore adaptive or data-driven clustering mechanisms at initialization to further enhance semantic complementarity, as well as investigate how dynamically adjusting cluster structure during evolution may improve generalisation and scalability across more complex, high-dimensional regression tasks.

References

1. Al-Helali, B., Chen, Q., Xue, B., Zhang, M.: Multitree genetic programming with new operators for transfer learning in symbolic regression with incomplete data. IEEE Trans. Evol. Comput. **25**(6), 1049–1063 (2021)
2. Beadle, L., Johnson, C.G.: Semantic analysis of program initialisation in genetic programming. Genet. Program Evolvable Mach. **10**(3), 307–337 (2009)
3. Cárdenas Florido, L., Trujillo, L., Hernandez, D.E., Muñoz Contreras, J.M.: M5GP: parallel multidimensional genetic programming with multidimensional populations for symbolic regression. Math. Comput. Appl. **29**(2), 25 (2024)
4. Chen, Q., Xue, B., Browne, W., Zhang, M.: Evolutionary Regression and Modelling, pp. 121–149. Springer Nature Singapore (2024). https://doi.org/10.1007/978-981-99-3814-8_5
5. Dick, G.: Improving geometric semantic genetic programming with safe tree initialisation. In: European Conference on Genetic Programming, pp. 28–40. Springer (2015). https://doi.org/10.1007/978-3-319-16501-1_3
6. Gong, N., Reddy, C.K., Ying, W., Chen, H., Fu, Y.: Evolutionary large language model for automated feature transformation. In: Proceedings of the AAAI Conference on Artificial Intelligence, vol. 39, pp. 16844–16852 (2025)
7. Jackson, D.: Phenotypic diversity in initial genetic programming populations. In: European Conference on Genetic Programming, pp. 98–109. Springer (2010). https://doi.org/10.1007/978-3-642-12148-7_9
8. Koza, J.R.: Genetic programming as a means for programming computers by natural selection. Stat. Comput. **4**(2), 87–112 (1994)
9. La Cava, W., Moore, J.H.: Learning feature spaces for regression with genetic programming. Genet. Program Evolvable Mach. **21**(3), 433–467 (2020)
10. La Cava, W., Singh, T.R., Taggart, J., Suri, S., Moore, J.H.: Learning concise representations for regression by evolving networks of trees. arXiv preprint arXiv:1807.00981 (2018)
11. Muñoz, L., Trujillo, L., Silva, S., Castelli, M., Vanneschi, L.: Evolving multidimensional transformations for symbolic regression with M3GP. Memetic Comput. **11**(2), 111–126 (2019)
12. Nguyen, S., Thiruvady, D., Zhang, M., Alahakoon, D.: Automated design of multipass heuristics for resource-constrained job scheduling with self-competitive genetic programming. IEEE Transactions on Cybernetics **52**(9), 8603–8616 (2022)

13. Olson, R.S., La Cava, W., Orzechowski, P., Urbanowicz, R.J., Moore, J.H.: PMLB: a large benchmark suite for machine learning evaluation and comparison. BioData mining **10**(1), 36 (2017)
14. Pawlak, T.P., Krawiec, K.: Semantic geometric initialization. In: Genetic Programming, pp. 261–277. Springer International Publishing (2016). https://doi.org/10.1007/978-3-319-30668-1_17
15. Wang, C., Chen, Q., Xue, B., Zhang, M.: Multi-task semantic genetic programming with reinforcement learning for multi-output symbolic regression. IEEE Trans. Evol. Comput. (2025)
16. Zhang, H., Chen, Q., Xue, B., Banzhaf, W., Zhang, M.: Automatically choosing selection operator based on semantic information in evolutionary feature construction. In: Pacific Rim International Conference on Artificial Intelligence, pp. 385–397. Springer (2023). https://doi.org/10.1007/978-981-99-7022-3_36
17. Zhang, H., Chen, Q., Xue, B., Banzhaf, W., Zhang, M.: A double lexicase selection operator for bloat control in evolutionary feature construction for regression. In: Proceedings of the Genetic and Evolutionary Computation Conference, pp. 1194–1202 (2023)
18. Zhang, H., Chen, Q., Xue, B., Banzhaf, W., Zhang, M.: Bias-variance decomposition: an effective tool to improve generalization of genetic programming-based evolutionary feature construction for regression. In: Proceedings of the Genetic and Evolutionary Computation Conference, pp. 998–1006 (2024)
19. Zhang, H., Chen, Q., Xue, B., Banzhaf, W., Zhang, M.: A geometric semantic macro-crossover operator for evolutionary feature construction in regression. Genet. Program Evolvable Mach. **25**(1), 2 (2024)
20. Zhang, H., Chen, Q., Xue, B., Banzhaf, W., Zhang, M.: Improving generalization of evolutionary feature construction with minimal complexity knee points in regression. In: European Conference on Genetic Programming (Part of EvoStar), pp. 142–158. Springer (2024). https://doi.org/10.1007/978-3-031-56957-9_9
21. Zhang, H., Zhou, A., Chen, Q., Xue, B., Zhang, M.: Sr-forest: a genetic programming-based heterogeneous ensemble learning method. IEEE Trans. Evol. Comput. **28**(5), 1484–1498 (2024)
22. Zhang, H., Zhou, A., Zhang, H.: An evolutionary forest for regression. IEEE Trans. Evol. Comput. **26**(4), 735–749 (2021)

NEVO-GSPT: Population-Based Neural Network Evolution Using Inflate and Deflate Operators

Davide Farinati[1,3,4]($\boxtimes$), Frederico JBB Santos[2], Leonardo Vanneschi[1], and Mauro Castelli[1]

[1] NOVA Information Management School (NOVA IMS), Universidade Nova de Lisboa, Campus de Campolide, 1070-312 Lisbon, Portugal
`{lvanneschi,mcastelli,dfarinati}@novaims.unl.pt`
[2] Department of Engineering and Architecture, University of Trieste, Trieste, Italy
`fredericojose.jacomedebritosantos@phd.units.it`
[3] Vita-Salute San Raffaele University, Milan, Italy
`farinati.davide@hsr.it`
[4] Comprehensive Cancer Center/Unit of Urology; URI; IRCCS Ospedale San Raffaele, Milan, Italy

Abstract. Evolving neural network architectures is a computationally demanding process. Traditional methods often require an extensive search through large architectural spaces and offer limited understanding of how structural modifications influence model behavior. This paper introduces NeuroEVOlution through Geometric Semantic perturbation and Population based Training (NEVO-GSPT), a novel Neuroevolution algorithm based on two key innovations. First, we adapt geometric semantic operators (GSOs) from genetic programming to neural network evolution, ensuring that architectural changes produce predictable effects on network semantics within a unimodal error surface. Second, we introduce a novel operator (DGSM) that enables controlled reduction of network size, while maintaining the semantic properties of GSOs. Unlike traditional approaches, NEVO-GSPT's efficient evaluation mechanism, which only requires computing the semantics of newly added components, allows for efficient population-based training, resulting in a comprehensive exploration of the search space at a fraction of the computational cost. Experimental results on four regression benchmarks show that NEVO-GSPT consistently evolves compact neural networks that achieve performance comparable to or better than established methods in the literature, such as standard neural networks, SLIM-GSGP, TensorNEAT, and SLM.

Keywords: Neuroevolution · Evolutionary Machine Learning · Artificial Neural Networks · Population Based Training · Geometric Semantic Mutation

1 Introduction

The design of effective Artificial Neural Networks (ANNs) architecture represents one of the core challenges in modern machine learning. Despite ANNs' success

L. Manzoni et al. (Eds.): EuroGP 2026, LNCS 16521, pp. 241–258, 2026.
https://doi.org/10.1007/978-3-032-23005-8_15

across numerous domains, determining the optimal architecture for a given task typically requires extensive manual experimentation or computationally costly automated search [11]. Traditional optimization methods such as grid and random search must evaluate each architecture from scratch and offer no principled link between architectural modifications and their effects on network behavior, making them both inefficient and opaque. Neural Architecture Search (NAS) [9] automates this process but at a heavy computational price, often consuming thousands of GPU-hours and thus remaining accessible only to few institutions. Moreover, NAS approaches typically treat architecture optimization as a black-box problem, providing little interpretability or understanding of why certain designs outperform others. Neuroevolution (NE) [24], which uses evolutionary algorithms to optimize ANNs, offers a compelling alternative. Early NE methods, such as NeuroEvolution of Augmenting Topologies (NEAT) [29], successfully evolved both topologies and weights through principled genetic representations, but they scale poorly to modern deep architectures and remain computationally expensive because each individual must be fully trained or evaluated. More recent NE variants [7, 14] have attempted to mitigate these costs, yet the central challenge persists: efficiently exploring the architecture space while maintaining predictable relationships between structure and behavior.

To address this, the present work introduces NeuroEVOlution through Geometric Semantic perturbation and Population based Training (NEVO-GSPT), a novel NE method that enhances both computational efficiency and semantic awareness. NEVO-GSPT adapts to the evolution of ANNs the basic principles of Geometric Semantic Genetic Programming (GSGP) [17] and its recent development called Semantic Learning algorithm based on Inflate and deflate mutations (SLIM-GSGP) [32,33]. NEVO-GSPT extends a preexisting framework called Semantic Learning Machine (SLM) [12], using operators that have predictable geometric effects on the semantics of the individuals and induce a unimodal error surface for any supervised learning task. One of the key innovations of NEVO-GSPT compared to SLM is the introduction of the Deflate Geometric Semantic Mutation (DGSM). While the traditional Inflate operator (IGSM) expands networks by adding perturbation components, DGSM reduces complexity by removing previously added elements. Alternating between inflation and deflation prevents uncontrolled model growth, enabling the evolution of compact, interpretable networks while preserving the favorable properties of geometric semantic operators. NEVO-GSPT is also computationally efficient, because, following the inspiration of SLIM-GSGP, it maintains networks as linked lists of perturbation components. When mutation occurs, only the new or removed component must be evaluated, while previous evaluations are reused. This incremental mechanism eliminates the need to retrain entire networks, drastically reducing fitness computation costs. Our experimental study, conducted on four regression problems, was designed to investigate four key factors: (1) the effect of a priori backpropagation training of the initial population; (2) the benefits of a posteriori fine-tuning of evolved models; (3) the impact of different inflate–deflate probabilities; and (4) the influence of the complexity of added components.

The manuscript is organized as follows. Section 2 reviews related work in NE. Section 3 provides background information allowing the reader to frame this work. Section 4 presents our methodology, detailing the IGSM and DGSM for NE and the adopted population-based training approach. Section 5 describes our experimental study. Finally, Sect. 6 concludes the paper, also discussing directions for future work.

2 Related Work

2.1 Neuroevolution Methods

The field of NE has evolved significantly since early methods like NEAT [29], which allow evolving both network topology and weights through speciation and incremental complexity growth. Extensions such as HyperNEAT [28] and ES-HyperNEAT [23] employed indirect encodings via CPPNs (Compositional Pattern Producing Networks) to evolve large-scale networks with regularities, though primarily for shallow architectures. More recent approaches have focused on deep architectures: CoDeepNEAT [16] evolves deep neural networks by introducing a two-level evolutionary process, where one population evolves modules (groups of layers), while another population evolves how these modules are assembled into complete networks; CGP-CNN (Cartesian Genetic Programming for Convolutional Neural Networks) [30,31] uses Cartesian Genetic Programming to represent CNNs as directed acyclic graphs; methods like DENSER [2] employ dynamic structured grammatical evolution. While these approaches have demonstrated competitive results on benchmarks such as ImageNet and CIFAR, they face a common challenge: computational cost, often requiring hundreds to thousands of GPU-days [22]. This computational barrier has limited their practical application and accessibility.

2.2 Semantic-Based Approaches

A fundamentally different approach emerged from Geometric Semantic Genetic Programming (GSGP), which introduced operators with predictable effects on program semantics, inducing unimodal fitness landscapes. The Semantic Learning Machine (SLM) [12] adapted these principles to neural network construction, incrementally adding neurons through geometric semantic mutation with constrained output contributions. However, SLM has limitations: (1) it does not use gradient-based optimization (backpropagation), instead relying solely on evolution for parameter optimization; and (2) the evolved networks, while more compact than traditional Geometric Semantic Genetic Programming (GSGP) trees, can still grow substantially. SLIM - GSGP [20] addressed the growth problem by introducing deflate mutation alongside inflate mutation, allowing both expansion and reduction of program size while maintaining the unimodal fitness landscape property. This approach has shown promising results in reducing model size compared to standard GSGP, though it has primarily been applied to symbolic regression rather than neural network evolution.

All in all, the need for a NE method that can efficiently evolve compact, interpretable neural network architectures while maintaining competitive performance on modern benchmarks remains a significant gap in the literature [35]. Such a method would need to: (1) leverage gradient-based optimization for parameter learning while evolving topology; (2) maintain strict control over the size of the individuals; (3) induce favorable fitness landscape properties to facilitate efficient search; and (4) achieve practical training times measured in GPU-minutes (rather than GPU-days). The method we propose in this paper was designed taking these four guidelines into account.

3   Background

3.1   Geometric Semantic Genetic Programming

In supervised learning tasks, Genetic Programming (GP) individuals are generally represented as computer programs that translate input values into output results. For a given individual P and a dataset of n fitness cases $X = \{x_1, x_2, \ldots, x_n\}$, the resulting output vector $s_P = \{P(x_1), P(x_2), \ldots, P(x_n)\}$ is known as the individual's *semantics* [17]. Traditional GP operators like crossover and mutation modify program structure syntactically, making their semantic effects difficult to predict and potentially disrupting search locality [18].

GSGP [17] addresses this by replacing syntactic operators with Geometric Semantic Operator (GSO)s that enforce specific geometric behaviors in semantic space, inducing a unimodal error surface for any supervised learning task. For a parent function $T : \mathbb{R}^n \to \mathbb{R}$, *Geometric Semantic Mutation* (GSM) defines a new function as $T_M = T + \text{ms} \cdot (T_{R1} - T_{R2})$, where ms denotes the mutation step and T_{R1} and T_{R2} are random functions whose outputs lie within $[0, 1]$. This ensures the mutated individual lies within a hypersphere of radius ms centered at the parent's semantics.

Inspired by the success of GSGP, various extensions have been proposed in recent years [3,5,19]. However, a significant drawback of the method is that offspring tend to be larger than their parents, leading to solutions that are often difficult to interpret [17].

3.2   Semantic Learning Algorithm Based on Inflate and Deflate Mutation

To address GSGP's bloating problem, recent work [32] introduced Deflate Geometric Semantic Mutation (DGSM), which reduces individual size while preserving semantic properties. The newly proposed operator builds on two core ideas. First, the Inflate Geometric Semantic Mutation (IGSM) from the previous section can be algebraically reformulated as:

$$IGSM(T) = T + \text{ms} \cdot (T_{R1} - T_{R2}) = T - \text{ms} \cdot (T_{R2} - T_{R1}),$$

based on the identity $(T_{R1} - T_{R2}) = - (T_{R2} - T_{R1})$. Second, since T_{R1} and T_{R2} are randomly generated from the same distribution, they are interchangeable. Thus, another valid expression for IGSM is:

$$IGSM\,(T) = T - \mathrm{ms} \cdot (T_{R1} - T_{R2})\,.$$

Consider an individual T that has undergone three applications of IGSM, resulting in:

$$T + \mathrm{ms} \cdot (T_{R1} - T_{R2}) + \mathrm{ms} \cdot (T_{R3} - T_{R4}) + \mathrm{ms} \cdot (T_{R5} - T_{R6})\,.$$

Using the alternative subtraction-based formulation of IGSM, we can apply a fourth mutation:

$$T + \mathrm{ms} \cdot (T_{R1} - T_{R2}) + \mathrm{ms} \cdot (T_{R3} - T_{R4}) + \mathrm{ms} \cdot (T_{R5} - T_{R6}) - \mathrm{ms} \cdot (T_{R7} - T_{R8})\,.$$

This continues to expand the individual. However, by reusing an earlier pair of random programs (such as T_{R3} and T_{R4}) the individual can be simplified:

$$T + \mathrm{ms} \cdot (T_{R1} - T_{R2}) + \cancel{\mathrm{ms} \cdot (T_{R3} - T_{R4})} + \mathrm{ms} \cdot (T_{R5} - T_{R6}) - \cancel{\mathrm{ms} \cdot (T_{R3} - T_{R4})}.$$

The middle terms cancel, yielding a more compact individual. DGSM implements this by removing previously added terms while maintaining all semantic properties of GSM, including bounded perturbations and unimodal error surfaces. As the name implies, the Semantic Learning algorithm based on Inflate and deflate Mutation (SLIM-GSGP) algorithm incorporates both IGSM and DGSM with configurable probabilities, restricting DGSM when only one term remains.

3.3 Semantic Learning Machine

The SLM [12] applied GSGP principles to feedforward neural networks. Specifically, when mutating a parent network T, the SLM creates an offspring by adding a randomly generated network whose output is scaled by ms, ensuring that the semantic change remains bounded within a sphere of radius ms around the parent's semantics. This approach guarantees that offspring remain within a controlled region of the semantic space, similar to the ball mutation property of Geometric Semantic Mutation (GSM) in GSGP. SLM's limitations include: (1) no backpropagation for parameter optimization (the parameters of the random networks are initialized and remain fixed), relying solely on evolution for improving network performance; and (2) potential for substantial network growth despite being more compact than GSGP trees.

3.4 Tensorized NeuroEvolution of Augmenting Topologies

Tensorized NeuroEvolution of Augmenting Topologies (TensorNEAT) [34] addresses a computational bottleneck in NE: the inability to efficiently parallelize NEAT's operations across GPU architectures. TensorNEAT overcomes this limitation by tensorizing networks with diverse topologies into uniform tensors for batched GPU computation. Networks are encoded as fixed-size tensors with padding, allowing all NEAT operations (mutation, crossover, speciation, and network inference) to be vectorized across the entire population and to be executed in parallel on GPUs. TensorNEAT achieves significant speedups over traditional implementations. Experimental results demonstrate speedups exceeding $500\times$ compared to `NEAT-Python`, the most widely used open-source NEAT implementation.

4   Methodology

In this work, we present NEVO-GSPT, a novel NE approach that integrates the principles of GSGP and SLIM-GSGP to evolve a population of Neural Networks (NNs). This method simultaneously evolves the architecture, weights, and activation functions of the NNs in the population. The algorithm relies on two core operators: IGSM and DGSM, both inspired by the GSO described in Sect. 3.

The objective of this work is to reproduce the effect of the operators used in GSGP and SLIM-GSGP on NNs.

4.1   Inflate Geometric Semantic Mutation for Neuroevolution

The IGSM is a NE operator designed to introduce controlled variation into an existing neural network N, while preserving its original functionality. This is achieved by appending a new auxiliary random network R, which acts as a perturbation to the parent model. The network R is constructed under the following conditions:

- It contains exactly one neuron per layer of N, totaling n neurons for an n-layer network.
- It receives connections only from the parent network; no backward connections into N are allowed. This ensures that N's semantics remain unchanged.
- All internal weights in R are drawn uniformly from the range $[-1, 1]$, except for the final output weight.
- The output weight is drawn from $[0, ms]$, where $ms \in \mathbb{R}$ is a tunable parameter known as the mutation step.
- The output neuron of R uses a `TanH` activation function to ensure bounded outputs in $[-1, 1]$.
- Other neurons in R may use arbitrary activation functions.

Once constructed, the semantic contribution of R is scaled by the mutation step and added to the output of N, producing the final offspring network. The IGSM guarantees that the overall effect of the perturbation lies within the range $[-ms, ms]$, thereby offering a bounded, additive modification to the parent network's behavior. This operation can be concisely expressed as:

$$IGSM(N) = N + ms \cdot R \tag{1}$$

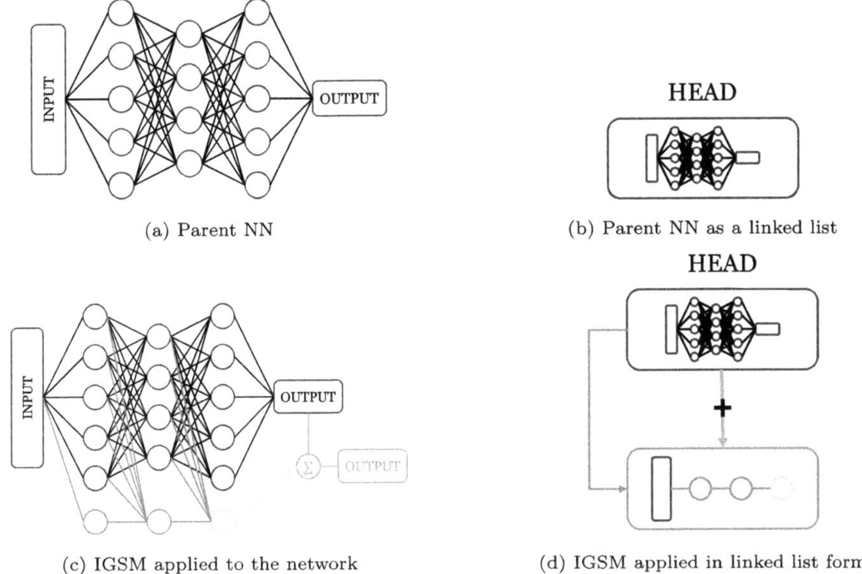

(a) Parent NN

(b) Parent NN as a linked list

(c) IGSM applied to the network

(d) IGSM applied in linked list form

Fig. 1. Illustration of IGSM using both neural network and linked list representations.

As with the SLIM-GSGP algorithm, the resulting structure of the offspring can be interpreted as a linked list. Figure 1 illustrates both the standard neural network representation and its linked list counterpart before and after applying the IGSM transformation.

4.2 Deflate Geometric Semantic Mutation for Neuroevolution

The DGSM is a complementary operator to IGSM that introduces a mechanism for semantic simplification in evolved neural networks. While IGSM incrementally adds new perturbation components to the output of a neural network, DGSM reverses this process by removing previously added perturbations. This allows for the reduction of model complexity without compromising semantic fidelity.

In the context of NE, DGSM is implemented by identifying a previously added auxiliary network R_i (introduced via IGSM) and subtracting its contribution from the current network. Given a neural network N that has undergone multiple IGSM transformations, its output can be expressed as:

$$N' = N + ms \cdot R_1 + ms \cdot R_2 + \cdots + ms \cdot R_k$$

DGSM selects a random $R_j \in \{R_1, R_2, \ldots, R_k\}$ and removes its contribution:

$$DGSM(N') = N' - ms \cdot R_j$$

This operation preserves the core semantic properties of geometric semantic mutation: it induces bounded, linear perturbations to the network output and maintains the unimodal structure of the error landscape. However, it simultaneously counters the inherent bloating problem of repeated additive perturbations by reducing the effective size of the evolved model.

To enable this behavior, a history of perturbation components must be maintained in a linked list-like structure, where each element corresponds to a specific auxiliary network previously appended via IGSM. DGSM acts by deleting an element from this list.

Figure 2 illustrates the application of DGSM in both neural network and linked list form.

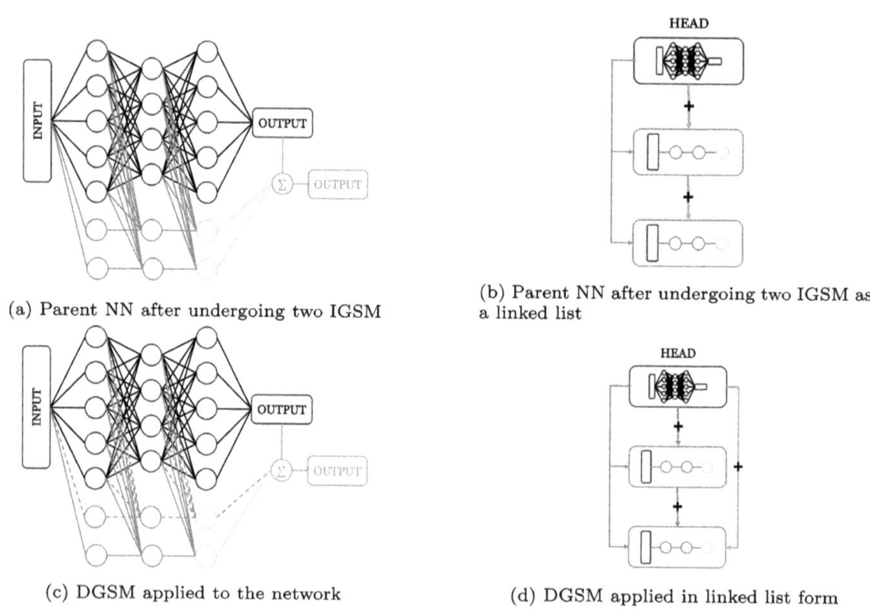

(a) Parent NN after undergoing two IGSM

(b) Parent NN after undergoing two IGSM as a linked list

(c) DGSM applied to the network

(d) DGSM applied in linked list form

Fig. 2. Illustration of DGSM using both neural network and linked list representations.

4.3 Population Based Training

The evaluation of new individuals is performed by summing the semantics of all elements in the linked list, as proposed in previous work [33]. As a result, when IGSM is applied, only a single NN with a width of one needs to be evaluated, since the semantics of all other components have already been computed. In contrast, when DGSM is used, the semantics and fitness of the new individual can be determined through a simple summation. This makes the evaluation process highly efficient, enabling the evolution of a population of individuals.

Maintaining a population allows for a more comprehensive exploration of the fitness landscape associated with a given problem.

5   Experimental Analysis

This section presents the experiments conducted to evaluate NEVO-GSPT. It is organized into three parts. Section 5.1 describes the main benchmarks used to assess the performance of the proposed algorithm. Section 5.2 provides an in-depth analysis of the various hyperparameters and configurations of NEVO-GSPT. Finally, Sect. 5.3 reports the performance of NEVO-GSPT compared to other baseline algorithms from Evolutionary Computation (EC), NE, and Machine Learning (ML).

5.1   Experimental Settings

In order to evaluate NEVO-GSPT we chose four regression datasets commonly used as benchmark to evaluate GP [1,4,6,13,15,21,27], NE [12] and NN [8]. The used datasets and their characteristics are presented in Table 1. These datasets offer different overfitting profiles [10,15], making it possible to study the generalization capabilities of NEVO-GSPT.

Table 1. Characteristics of the datasets used for model comparison.

Dataset	Number of observations	Number of features
airfoil	1502	5
concrete strength	1029	8
bioavailability	359	241
ld50	234	626

Several hyperparameters were analyzed, and their optimal values were determined in Sect. 5.2. Other hyperparameters were kept fixed throughout all experiments, with the population size set to 100 and the mutation step fixed at 2. For all comparisons, a 80-20 train–test split Monte Carlo cross-validation was employed, using consistent splits for each run across methods. The experiments presented in Sect. 5.2 were conducted over ten independent runs, whereas those reported in Sect. 5.3 were performed over thirty runs. The corresponding p-values resulting from the pairwise Wilcoxon signed-rank test for the latter experiments are provided in the Supplementary Material. The python code used for this work can be found at NEVO-GSPT[1].

---

[1] https://github.com/Fari98/NEVO-GSPT.

5.2   Ablation Studies

This Section provides an in-depth analysis of the algorithm's configurations and hyperparameters, focusing on identifying the optimal setup for each dataset and explaining the underlying reasons for its effectiveness.

A Priori Training. During population initialization, the $|P|$ NNs are randomly generated with varying structures and weights, where $|P|$ denotes the population size. Each NN can either be trained using backpropagation or directly passed to the evolutionary process. This section investigates the impact of *A priori Training* (AprT) on the evolution process. For each dataset, three configurations are evaluated: no AprT, AprT applied to half of the NNs (selected randomly), and AprT applied to all NNs. The results, presenting the RMSE on the training and test sets for each dataset, are shown in Fig. 3.

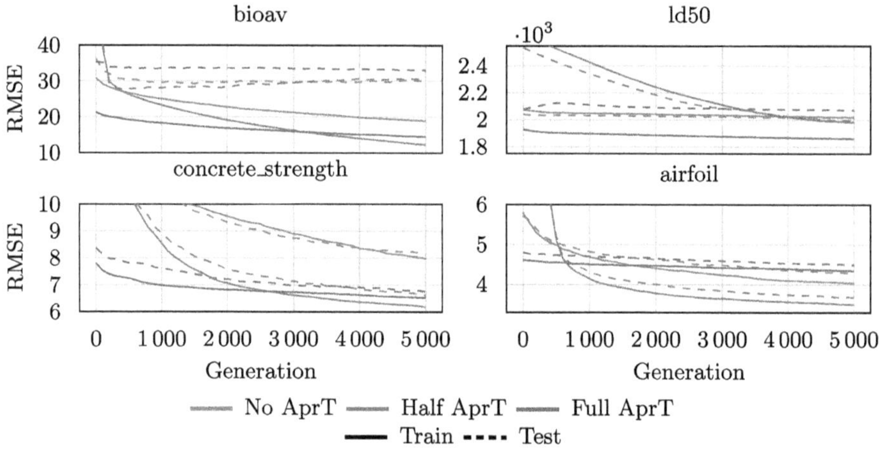

Fig. 3. Comparison of the impact of AprT on the evolutionary process, showing RMSE performance on the training and test sets across all datasets.

The results in Fig. 3 demonstrate that the Half AprT configuration consistently achieves the best balance between training and test RMSE across all datasets. While No AprT exhibits slower convergence and higher error, and Full AprT tends to overfit in some cases, Half AprT provides faster convergence and superior generalization performance. This indicates that a moderate level of AprT integration optimally guides the evolutionary process.

A Posteriori Training. The final solutions generated by NEVO-GSPT are NNs that can also be further trained to optimize their weights. This Section examines the effect of *A posteriori Training* (ApoT) on the final solutions

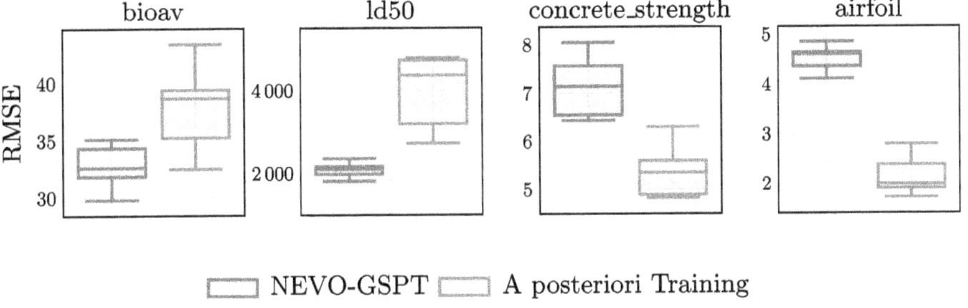

Fig. 4. Comparison of the impact of ApoT on the evolutionary process, showing RMSE performance on the test set across all datasets.

obtained at the last generation. The results, showing the test RMSE for each dataset, are presented in Fig. 4.

As shown in the results, ApoT has a positive effect on datasets that are less prone to overfitting (i.e., *concrete_strength* and *airfoil*), improving the test RMSE. However, the opposite trend is observed in datasets susceptible to overfitting (i.e., *bioav* and *ld50*), where ApoT worsens test performance and reduces the generalization capability of the algorithm.

Deflate and Inflate Probabilities. Different combinations of inflate and deflate probabilities can be used to balance the trade-off between convergence speed and program size. In this experiment, we tested four inflate probability values (0.3, 0.5, 0.7, and 1), with the corresponding deflate probability always set to one minus the inflate probability. The results of this analysis are shown in Fig. 5 and Fig. 6, presenting the RMSE performance on the training and test sets, as well as the resulting program sizes.

As expected, lower inflate probabilities produce smaller programs. However, the impact on performance is less straightforward. For datasets that are not prone to overfitting (i.e., *concrete_strength* and *airfoil*), a low inflate probability results in slower convergence. Conversely, for datasets prone to overfitting (i.e., *bioav* and *ld50*), a lower inflate probability improves the algorithm's generalization capability. These results reinforce the observation reported in [10] that, for such problems, program size tends to correlate with overfitting.

Neurons to be Added. As explained in Sect. 4.1, when the IGSM operator is applied, one neuron per layer is added to the parent NN. However, more complex or simpler networks could also be generated. In this section, we investigate the impact of adding fewer neurons per layer by testing different proportions as hyperparameters: 0.3, 0.5, 0.7, and 1. Figure 7 shows the training and test RMSE performance for each of these configurations.

As observed in Fig. 7, adding fewer than one neuron per layer consistently results in slower convergence and reduced generalization capability of the algorithm.

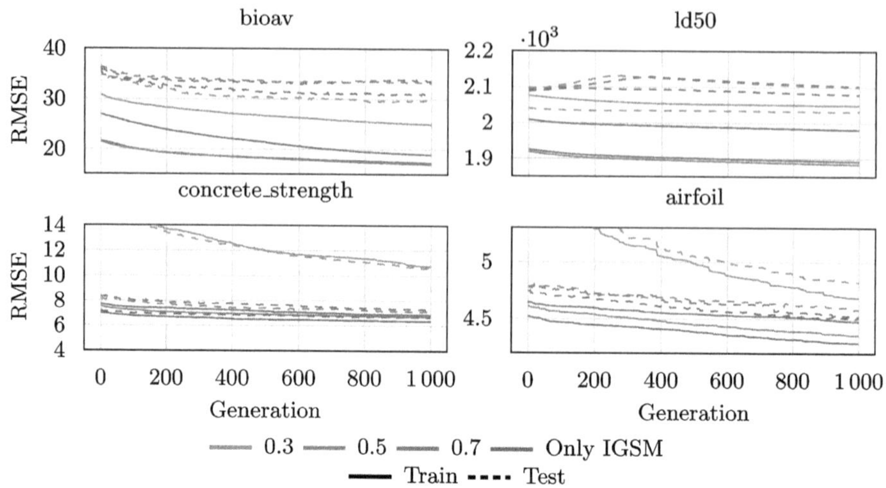

Fig. 5. Comparing the RMSE performance of difference combinations of inflate and deflate probabilities.

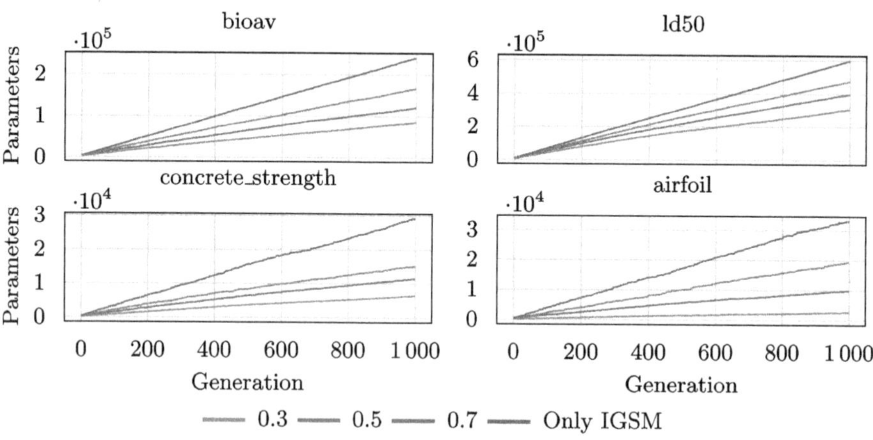

Fig. 6. Comparing the size of difference combinations of inflate and deflate probabilities.

5.3   Experimental Results

After exploring the hyperparameters of NEVO-GSPT and selecting the best configurations for each dataset, distinguishing clearly between those prone to overfitting and those that are not, we compare the algorithm against a set of baselines from EC and NE. Among the baselines, we include a NN with the same architecture as the one evolved by NEVO-GSPT, whose weights are optimized via backpropagation, as well as SLM [12], TensorNEAT [34], and SLIM-GSGP [25] (variant *SLIM*SIG1*). For fair comparison, all evolutionary algorithms are configured with consistent hyperparameters to match the number of parameters of

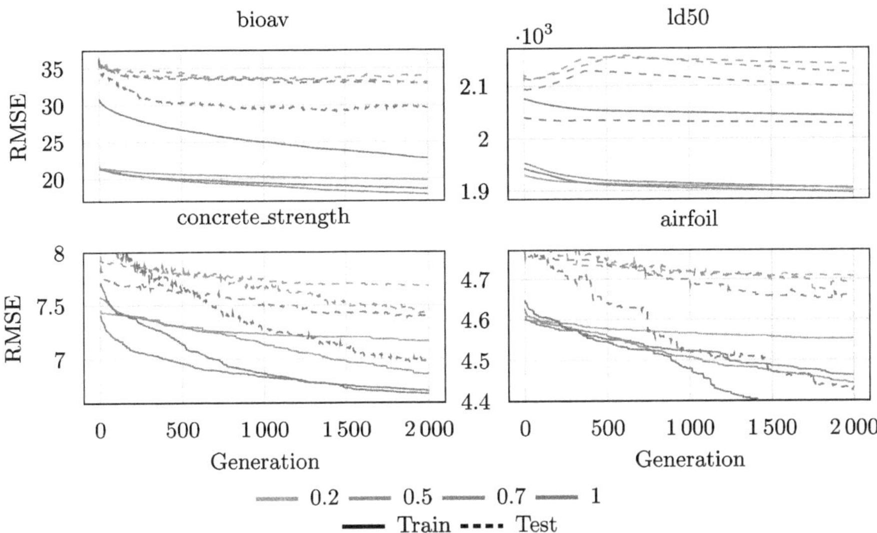

Fig. 7. Comparing the RMSE performance of sizes of the added layers, expressed as percentage of the original depth of the network.

an evolved NEVO-GSPT model. TensorNEAT is run for 1000 generations with a population size of 100, using a maximum of 627 nodes and 626 connections, with 10 individuals per species and a survival threshold of 0.2. For the initial population of TensorNEAT, there must be at least as many connections and nodes as input features, hence the higher values when compared to NEVO-GSPT. SLM is executed for 100 generations using networks with a maximum of 3 layers, starting with 32 initial neurons and adding 4 neurons per layer per GSM mutation. Note the shift in increasing the number of neurons per mutations and decreasing the number of generations is necessary due to limitations in SLM's implementation (PyTorch Genesis [26]) creating significant computational overhead when adding only neuron per layer per generation. We therefore adjust the parameters to ensure a fair comparison. Figure 8 presents the test RMSE of the different algorithms across all datasets. The p-values of the Wilcoxon signed-rank test are provided in the Supplementary Material.

As observed in Fig. 8, NEVO-GSPT is the only algorithm that consistently ranks among the top performers across all tasks. This highlights the effectiveness of population-based training combined with geometric mutation in achieving both strong convergence and robust generalization.

Comparison of Program Size. The novelty of this work also lies in the introduction of a DGSM operator for NE, designed to generate more compact NNs. Figure 9 compares the number of neurons produced by NEVO-GSPT with those obtained by two other NE techniques, namely TensorNEAT and SLM, across all datasets.

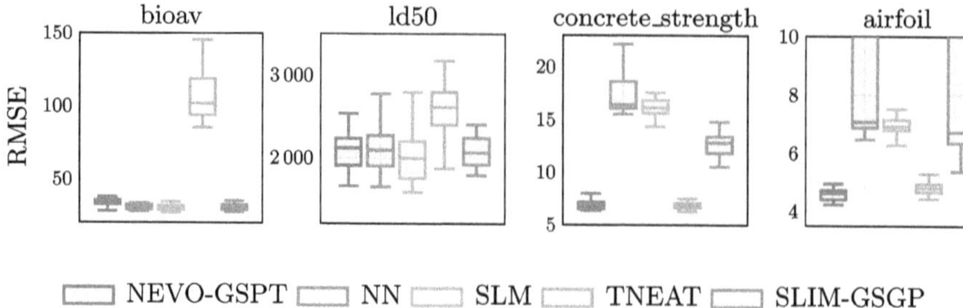

Fig. 8. Comparison of the test performance of NEVO-GSPT against a set of baslines, presented as RMSE.

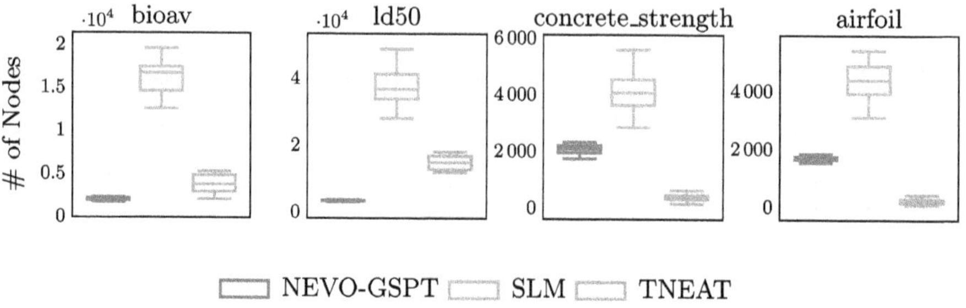

Fig. 9. Comparison of the number of nodes of the individuals evolved by NEVO-GSPT against a set of baselines.

As shown in all test cases, NEVO-GSPT consistently produces smaller NNs than SLM, demonstrating the effectiveness of the proposed DGSM operator. Moreover, in two out of four datasets, NEVO-GSPT also generates smaller networks than TensorNEAT, a state-of-the-art algorithm in NE.

Empirical Analysis of the Running Time. To assess the computational efficiency of NEVO-GSPT, we conducted 30 independent runs comparing runtime performance against standard backpropagation on an Intel Core i5-1235U (1.30 GHz, 16GB RAM). Over 100 generations/epochs, NEVO-GSPT averaged 14.58 s $\pm$ 0.60 s while backpropagation required 5.12 s $\pm$ 0.42 s. However, this comparison requires important contextualization: each NEVO-GSPT generation evaluates 100 individuals in the population, whereas backpropagation represents a single model training run. When examining individual evaluation time, NEVO-GSPT's inflate/deflate operations complete in 0.0006 s $\pm$ 0.00008 s compared to 0.0031 s $\pm$ 0.0006 s for backpropagation, demonstrating approximately 5× faster individual evaluations. It is worth noting that direct runtime comparisons with GPU-accelerated implementations such as PyTorch Genesis (used by SLM) and TensorNEAT may not be entirely equitable, as our CPU-based implementation operates under different computational constraints. Nevertheless, NEVO-GSPT's ability to explore 100 architectural solutions simultaneously within approximately 15 s on CPU hardware represents an advancement for NE, particularly in environments where GPU access may be limited.

6 Conclusions

This paper introduced NEVO-GSPT, a novel NE algorithm that addresses two fundamental challenges in neural architecture search: the computational expense of exploring architectural spaces and the lack of clear mappings between structural changes and their effects on network behavior. A key innovation of our work is the introduction of DGSM for NE. This novel operator enables controlled reduction of network complexity by removing previously added components, addressing the bloating problem that affects geometric semantic approaches while maintaining the favorable fitness landscape properties that facilitate efficient search. The computational efficiency of NEVO-GSPT represents another significant contribution. By maintaining networks as linked lists of perturbation components and evaluating only newly added or removed components, our method reduces the cost of fitness evaluation compared to approaches that must train entire networks from scratch. This efficiency makes it possible evolving large populations at a computational cost measured in minutes rather than GPU-days.

Experimental results across four regression benchmarks demonstrate that NEVO-GSPT consistently ranks among the top performers when compared against standard neural networks, SLM, TensorNEAT, and SLIM-GSGP. This competitive performance is achieved while producing significantly more compact networks than SLM and, in several cases, more compact than TensorNEAT.

Future work should focus on extending NEVO-GSPT to other application domains traditionally associated with NN and NE, such as classification, computer vision, and natural language processing.

Acknowledgments. This work was supported by national funds through FCT (Fundação para a Ciência e a Tecnologia), under the project -UID/04152/2025 - Centro de Investigação em Gestão de Informação (MagIC)/NOVA IMS   (https://doi.org/10.54499/UID/04152/2025)   -   (2025-01-01/2028-12-31)   and   UID/PRR/04152/2025 (https://doi.org/10.54499/UID/PRR/04152/2025) (2025-01-01/ 2026-06-30).

This publication is based upon work from COST Action Randomised Optimisation Algorithms Research Network (ROAR-NET), CA22137, supported by COST (European Cooperation in Science and Technology).

References

1. Archetti, F., Lanzeni, S., Messina, E., Vanneschi, L.: Genetic programming for computational pharmacokinetics in drug discovery and development. Genet. Program Evolvable Mach. **8**(4), 413–432 (2007)
2. Assunção, F., Lourenço, N., Machado, P., Ribeiro, B.: Denser: deep evolutionary network structured representation. Genet. Program Evolvable Mach. **20**(1), 5–35 (2019)
3. Bonin, L., Rovito, L., De Lorenzo, A., Manzoni, L.: Cellular geometric semantic genetic programming. Genet. Program Evolvable Mach. **25**(1), 8 (2024)
4. Castelli, M., Manzoni, L., Gonçalves, I., Vanneschi, L., Trujillo, L., Silva, S.: An analysis of geometric semantic crossover: a computational geometry approach. In: 8th International Conference on Evolutionary Computation Theory and Applications, pp. 201–208 (2016)
5. Castelli, M., Trujillo, L., Vanneschi, L., Silva, S., Z-Flores, E., Legrand, P.: Geometric semantic genetic programming with local search. In: Proceedings of the 2015 Annual Conference on Genetic and Evolutionary Computation, pp. 999–1006 (2015)
6. Castelli, M., Vanneschi, L., Silva, S.: Prediction of high performance concrete strength using genetic programming with geometric semantic genetic operators. Expert Syst. Appl. **40**(17), 6856–6862 (2013)
7. Custode, L.L., Iacca, G., De Falco, I., Scafuri, U., Della Cioppa, A.: Model-free communication federated neuroevolution. ACM Trans. Evol. Learn. Optim. (2025). https://doi.org/10.1145/3745032
8. Dong, X., Liu, Y., Dai, J.: Application of fully connected neural network-based pytorch in concrete compressive strength prediction. Adv. Civil Eng. **2024**(1) (2024). https://doi.org/10.1155/2024/8048645
9. Elsken, T., Metzen, J.H., Hutter, F.: Neural architecture search: a survey. J. Mach. Learn. Res. **20**(1), 1997–2017 (2019)
10. Farinati, D., Pietropolli, G., Vanneschi, L.: A study on the dynamics and effectiveness of the deflate geometric semantic mutation. IEEE Trans. Evol. Comput. 1 (2025). https://doi.org/10.1109/TEVC.2025.3611226
11. Ghosh, A., Jana, N.D., Mallik, S., Zhao, Z.: Designing optimal convolutional neural network architecture using differential evolution algorithm. Patterns **3**(9), 100567 (2022). https://doi.org/10.1016/j.patter.2022.100567, https://www.sciencedirect.com/science/article/pii/S2666389922001787
12. Gonçalves, I., Silva, S., Fonseca, C.M.: Semantic learning machine: a feedforward neural network construction algorithm inspired by geometric semantic genetic programming. In: Portuguese Conference on Artificial Intelligence, pp. 280–285. Springer (2015). https://doi.org/10.1007/978-3-319-23485-4_28
13. Gonçalves, I., Silva, S., Fonseca, C.M., Castelli, M.: Unsure when to stop? In: Proceedings of the Genetic and Evolutionary Computation Conference. ACM (2017). https://doi.org/10.1145/3071178.3071328
14. Huang, H., Peng, J., Li, K., Zhou, J., Sun, Z.: Efficient GPU-accelerated training of a neuroevolution potential with analytical gradients. arXiv:2507.00528 (2025)
15. McDermott, J., et al.: Genetic programming needs better benchmarks. In: GECCO 2012, Proceedings of the 14th Annual Conference on Genetic and Evolutionary Computation, pp. 791–798. ACM, New York, NY, USA (2012). https://doi.org/10.1145/2330163.2330273

16. Miikkulainen, R., et al.: Evolving deep neural networks. In: Artificial Intelligence in the Age of Neural Networks and Brain Computing, pp. 293–312 (2018)
17. Moraglio, A., Krawiec, K., Johnson, C.G.: Geometric semantic genetic programming. In: Coello, C.A.C., Cutello, V., Deb, K., Forrest, S., Nicosia, G., Pavone, M. (eds.) Parallel Problem Solving from Nature. LNCS, vol. 7491, pp. 21–31. Springer (2012). https://doi.org/10.1007/978-3-642-32937-1_3
18. Nguyen, Q.U.: Examining semantic diversity and semantic locality of operators in genetic programming. Ph. D. thesis, University College Dublin, Ireland (2011). http://ncra.ucd.ie/papers/Thesis_Uy_Corrected.pdf
19. Pietropolli, G., Camerota Verdù, F.J., Manzoni, L., Castelli, M.: Parametrizing GP trees for better symbolic regression performance through gradient descent. In: Proceedings of the Companion Conference on Genetic and Evolutionary Computation, pp. 619–622 (2023)
20. Pietropolli, G., Farinati, D., Manzoni, L., Castelli, M., Silva, S., Vanneschi, L.: Introducing crossover in SLIM-GSGP. In: European Conference on Genetic Programming (Part of EvoStar), pp. 103–119. Springer (2025). https://doi.org/10.1007/978-3-031-89991-1_7
21. Pietropolli, G., Manzoni, L., Paoletti, A., Castelli, M.: On the hybridization of geometric semantic GP with gradient-based optimizers. Genet. Program Evolvable Mach. **24**(2), 16 (2023)
22. Real, E., Aggarwal, A., Huang, Y., Le, Q.V.: Regularized evolution for image classifier architecture search. In: Proceedings of the AAAI Conference on Artificial Intelligence, vol. 33, pp. 4780–4789 (2019)
23. Risi, S., Stanley, K.O.: A unified approach to evolving plasticity and neural geometry. In: The 2012 International Joint Conference on Neural Networks (IJCNN), pp. 1–8. IEEE (2012)
24. Risi, S., Tang, Y., Ha, D., Miikkulainen, R.: Neuroevolution: Harnessing Creativity in AI Agent Design. MIT Press, Cambridge, MA (2025). http://www.cs.utexas.edu/users/ai-lab?risi:book25
25. Rosenfeld, L., et al.: Slim_gsgp: a python library for non-bloating GSGP. In: Genetic and Evolutionary Computation Conference (GECCO 2025), p. 9. ACM, Malaga, Spain (2025). https://doi.org/10.1145/3712256.3726398
26. Santos, F.J., Gonçalves, I., Castelli, M.: Neuroevolution with box mutation: an adaptive and modular framework for evolving deep neural networks. Appl. Soft Comput. **147**, 110767 (2023)
27. Silva, S., Vanneschi, L.: Operator equalisation, bloat and overfitting: a study on human oral bioavailability prediction. In: Rothlauf, F. (ed.) Genetic and Evolutionary Computation Conference, GECCO 2009, pp. 1115–1122. ACM (2009)
28. Stanley, K.O., D'Ambrosio, D.B., Gauci, J.: A hypercube-based encoding for evolving large-scale neural networks. Artif. Life **15**(2), 185–212 (2009)
29. Stanley, K.O., Miikkulainen, R.: Evolving neural networks through augmenting topologies. Evol. Comput. **10**(2), 99–127 (2002)
30. Suganuma, M., Kobayashi, M., Shirakawa, S., Nagao, T.: Evolution of deep convolutional neural networks using cartesian genetic programming. Evol. Comput. **28**(1), 141–163 (2020)
31. Suganuma, M., Shirakawa, S., Nagao, T.: A genetic programming approach to designing convolutional neural network architectures. In: Proceedings of the Genetic and Evolutionary Computation Conference, pp. 497–504 (2017)
32. Vanneschi, L.: SLIM_GSGP: the non-bloating geometric semantic genetic programming. In: Giacobini, M., Xue, B., Manzoni, L. (eds.) Genetic Programming, pp.

125–141. Springer Nature Switzerland, Cham (2024). https://doi.org/10.1007/978-3-031-56957-9_8

33. Vanneschi, L., Farinati, D., Rasteiro, D., Rosenfeld, L., Pietropolli, G., Silva, S.: Exploring Non-bloating Geometric Semantic Genetic Programming, pp. 237–258. Springer Nature Singapore, Singapore (2025). https://doi.org/10.1007/978-981-96-0077-9_12

34. Wang, L., Zhao, M., Liu, E., Sun, K., Cheng, R.: Tensorized neuroevolution of augmenting topologies for GPU acceleration. In: Proceedings of the Genetic and Evolutionary Computation Conference, pp. 1156–1164 (2024)

35. White, C., et al.: Neural architecture search: insights from 1000 papers. arXiv preprint arXiv:2301.08727 (2023)

Multi-action Tangled Program Graphs for Multi-task Reinforcement Learning with Continuous Control

Quentin Vacher[(✉)] [iD], Nicolas Beuve [iD], Mickaël Dardaillon [iD], and Karol Desnos [iD]

INSA Rennes, CNRS, IETR – UMR 6164, Univ Rennes, 35000 Rennes, France
{Quentin.Vacher,Nicolas.Beuve,Mickael.Dardaillon,
Karol.Desnos}@insa-rennes.fr

Abstract. Over the past few decades, machine learning has been widely used to learn complex tasks. Reinforcement Learning (RL), inspired by human behavior, is a great example, as it involves developing specific behaviours for specific tasks. To further challenge algorithms, Multi-Task RL (MTRL) environments have been introduced, requiring a single model to learn multiple behaviors.

The Tangled Program Graph (TPG) algorithm is a Genetic Programming (GP) algorithm designed for discrete MTRL environments. Recently, the MAPLE algorithm has been proposed, as another GP algorithm that achieves high results in single task continuous RL environments. A variation of the TPG is proposed alongside MAPLE, named Multi-Action TPG (MATPG) that aggregates MAPLE agents, and creates a control flow to activate them. Initially tested on single task RL environments only, MATPG achieved similar results to MAPLE.

In this work, we present a new benchmark based on the MuJoCo *Half Cheetah* from Gymnasium. This benchmark features five distinct obstacles that are randomly positioned in front of the agent, each of which demands a unique behavior. This benchmark serves as a use case for MATPG, to prove its ability as a GP solution for continuous MTRL environments. Our experiments demonstrate its superiority in this multi-task use case when combined with lexicase selection. Furthermore, we examine the interpretability of the evolved graph, revealing that the decision flow of the model is fully interpretable.

Keywords: Multi-task Learning · Reinforcement Learning · Tangled Program Graphs · Continuous Control

1   Introduction

Intelligence can be measured by the ability to solve diverse tasks. Artificial intelligence has explored learning multiple tasks with single models, particularly in RL, where agents reason over long-term consequences, solving the temporal credit assignment. In MTRL, this challenge extends across tasks, requiring deep environmental understanding.

L. Manzoni et al. (Eds.): EuroGP 2026, LNCS 16521, pp. 259–274, 2026.
https://doi.org/10.1007/978-3-032-23005-8_16

Several MTRL benchmarks capture these challenges [11]. In continuous control, many use the Multi-Joint dynamics with Contact (MuJoCo) suite [24] from Gymnasium [25], by modifying body parts or controlling multiple agents [33]. Most solutions employ deep learning, specifically Deep RL (DRL) [8,9,20,23], using transfer learning to share knowledge. While DRL performs well, its architectures are complex and opaque, producing large, uninterpretable models [4,12,18].

Alternative machine learning architectures, such as GP, solve RL tasks while producing compact and interpretable solutions. Algorithms like Linear GP (LGP) [2], Tree-Based GP (TGP) [15], and Cartesian GP (CGP) [19] handle continuous control, while TPG [13] excels in discrete MTRL [14]. In [29], MATPG was proposed as a continuous adaptation of TPG. However, it does not outperform its simplified version, Multi-Action Programs through Linear Evolution (MAPLE), on single continuous control tasks.

We evaluate the effectiveness of MATPG in continuous-control MTRL settings, comparing it against its simplified version, MAPLE, under both tournament and lexicase selection [7,10]. Tournament selection is the default method for MAPLE and MATPG in [29], while lexicase selection, originally designed for multi-task learning in classification and regression, encourages task-specific elites. Our experiments demonstrate that both MATPG and MAPLE with lexicase selection outperform tournament selection in producing agents capable of solving individual tasks. Furthermore, MATPG proves highly effective in continuous-control MTRL environments, achieving significantly better performance than the other three configurations when evaluated on all tasks simultaneously with lexicase selection.

To assess these algorithms, we designed a MTRL benchmark based on the MuJoCo *Half Cheetah* environment, shown in Fig. 1, extended with five independent obstacles that an agent must overcome. This benchmark is modular, meaning that learning one task does not enable the agent to overcome another without explicit learning. It has been verified that the five obstacles can be learned independently by a single agent, making it a robust testbed for evaluating MATPG and MAPLE. All the code and results used in this study are available on [28].

Section 2 reviews prior work on MTRL environments and GP algorithms applied to RL. Section 3 provides a detailed description of the MATPG algorithm. Our proposed benchmark, based on a customized *Half Cheetah*, is presented in Sect. 4, followed by the experimental setup in Sect. 5, including the selection algorithms and parameter settings. Results are reported in Sect. 6, and Sect. 7 explores the policy of the best MATPG agent, highlighting its interpretability. Finally, Sect. 8 concludes the paper and discusses the findings.

2   Related Work

To motivate the need for a new MTRL benchmark, we provide a review of existing benchmarks in RL in Sect. 2.1. In this work, we challenge the proposed benchmark using the MATPG algorithm, which allows explicit reasoning over multiple tasks. A complementary review of GP algorithms applied to RL is presented in Sect. 2.2, leading up to the development of MATPG.

2.1 Multi-task Reinforcement Learning for Continuous Control

Recent advances in Reinforcement Learning (RL) span diverse domains including robotics, games, autonomous systems, and NLP [31]. This has driven development of DRL algorithms for continuous control like Soft Actor-Critic (SAC) [9], Proximal Policy Optimization (PPO) [23], and Twin Delayed Deep Deterministic (TD3) [8]. Continuous control benchmarks, primarily in Gymnasium [25], often use the MuJoCo physics engine [24]. This benchmark

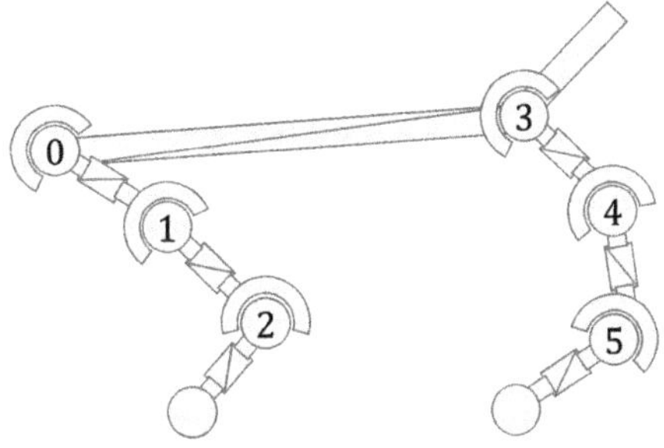

Fig. 1. Scheme of the *Half Cheetah* environments from the MuJoCo suite of Gymnasium.

comprises different robotic tasks that aim to reproduce real-world physics simulations. Thus, the MuJoCo benchmark is highly effective for scalable, realistic, single-task continuous control with RL.

To evaluate multi-task capabilities, benchmarks requiring simultaneous task learning have emerged, leading to specialized MTRL algorithms. IMPALA [6] learns 57 Arcade Learning Environment (ALE) games with discrete actions, while continuous control MTRL uses MuJoCo, modifying *Half Cheetah* morphology [11,33] or learning six tasks simultaneously with single-task performance.

However, DRL suffers from interpretability issues due to its Deep Learning (DL) architecture [18]. For instance, a standard MTRL network with two 400-unit layers exceeds 100k multiplyâĂŞaccumulate (MAC) operations, complicating understanding and increasing complexity [33]. Alternative architectures like GP address this by evolving compact, interpretable policies.

2.2 Genetic Programming for Reinforcement Learning

As in many other areas, GP [15] has been used to learn RL tasks. Genetic Programming (GP) can significantly improve the interpretability [4] and reduce the computational complexity of a solution by orders of magnitude [27] compared to DRL. In [30], the LGP [2] and TGP [15] representations are used to learn different discrete tasks in the Gymansium library [25] such as the *CartPole* or the *Lunar Lander*, as well as some continuous tasks, such as *Hopper* or the *Bidepal Walker*. In [21], LGP ant CGP [19] representations are trained independently on eight MuJoCo environments. Compared to state-of-the-art DRL algorithms, the GP solutions achieve equivalent or better results on the tasks with three or fewer actions. However, when the number of actions increases, DRL solutions become superior.

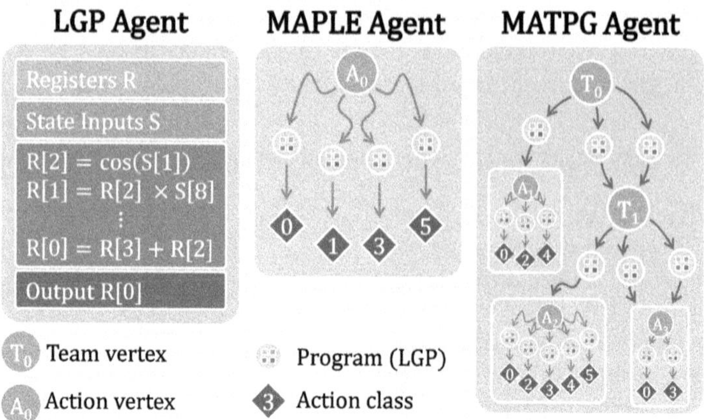

Fig. 2. Examples of LGP, MAPLE and MATPG agents for an environment with six continuous actions. An action vertex uses all the output values of its programs to take the continuous actions, while a team vertex only follow the path of the program outputting the highest value.

For discrete MTRL, TPG evolved a graph of LGP programs, initially achieving Deep Q-Network (DQN)-level performance on ALE with single Atari game, and later learning multiple Atari games with one model [13,14]. It was subsequently extended to continuous RL by replacing discrete actions with an LGP program outputting continuous values [1]. To scale to many actions, two related approaches are proposed in [29]: MAPLE and MATPG employing multiple LGP programs, one per action. MAPLE is a flat multi-program policy, while MATPG integrates this into the hierarchical structure of TPG.

On MuJoCo benchmarks, MAPLE generally outperformed the continuous TPG and matched MATPG, but with a smaller model size. The authors note that standard MuJoCo tasks do not leverage key strength of MATPG: multi-task learning [29]. This paper therefore investigates a dedicated MTRL use case for MATPG. Before doing so, the MATPG algorithm is presented with more details in the next section.

3   Background: Multi-action Tangled Program Graphs

Semantics of MATPG. MATPG [13,29] consists of three main entities forming a direct graph: teams, actions, and programs. Note that MAPLE is a specific case of MATPG without teams. Figure 2 shows an example of LGP, MAPLE and MATPG agent. Each team and action vertex contains multiple edges with programs corresponding to independent LGP. Team programs point to another vertex, while action programs point to an action class. Execution begins at a designated team, called root team. This team activates all its programs with the same state inputs from the observation of the environment. The destination with the highest program output is followed. If it points to another team, the process repeats. If it points to an action vertex, that vertex activates its programs. Each program outputs a continuous value defining the final action for its class. Missing action classes use a default action. The actions taken are then sent to the environments.

Mutation Process. The graph structure evolves dynamically. Each generation creates a fixed number of roots by duplicating and mutating the survivors of natural selection process. Mutation can add or remove edges, or mutate existing ones. When mutating an edge, either its program or destination can be mutated. If the edge led to an action vertex, that vertex can be mutated instead of the program.

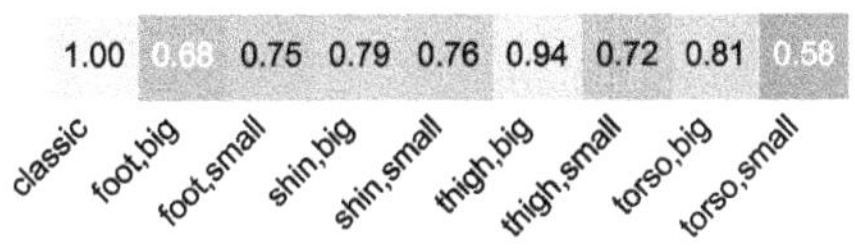

Fig. 3. Scores achieved by maple agents trained on classic *Half Cheetah* but tested with modified body size. Scores are normalized by the unmodified score of 3622.

An action vertex is a MAPLE agent which is mutated in a similar fashion to team vertex, with the additional constraint that than any action class can appear only once in the action class edges.

Sub-population Evolution. MATPG uses two sub-populations: one third MATPG agents and two thirds independent MAPLE agents. During the evolution of an MATPG agent, edge destinations can be changed to MAPLE agents. This approach was used in [29], where authors claimed MATPG alone struggles to learn due to high complexity. Therefore, the learning is divided with MAPLE agents learning behaviors, and MATPG agents learning how to combine them.

As mentioned in Sect. 2.2, MATPG did not surpass MAPLE in [29]. The hypothesis is that the original MuJoCo benchmark lacks multi-task requirements where MATPG should excel. Thus, a MTRL application is needed to measure the efficiency of MATPG.

4   Use Case: Customized Half Cheetah

To measure MATPG efficiency, the benchmark must be an MTRL environment. Tasks must be challenging enough to be independent, where mastering one doesn't imply success in others without further training. They must also be accessible enough for MAPLE agents to succeed individually. In this section, we introduce a new benchmark for MTRL in continuous control, using the *Half Cheetah* from the MuJoCo suite. Inspired by [26], it comprises five distinct obstacle tasks that the *Half Cheetah* must overcome.

The two benchmarks discussed in Sect. 2.1 fail to meet our requirements. The *Half Cheetah* body-size variation benchmark exhibits excessive task correlation, as evidenced by the performance of a MAPLE agent trained without any variations on *Half Cheetah* and tested on this benchmark, see Fig. 3. The benchmark

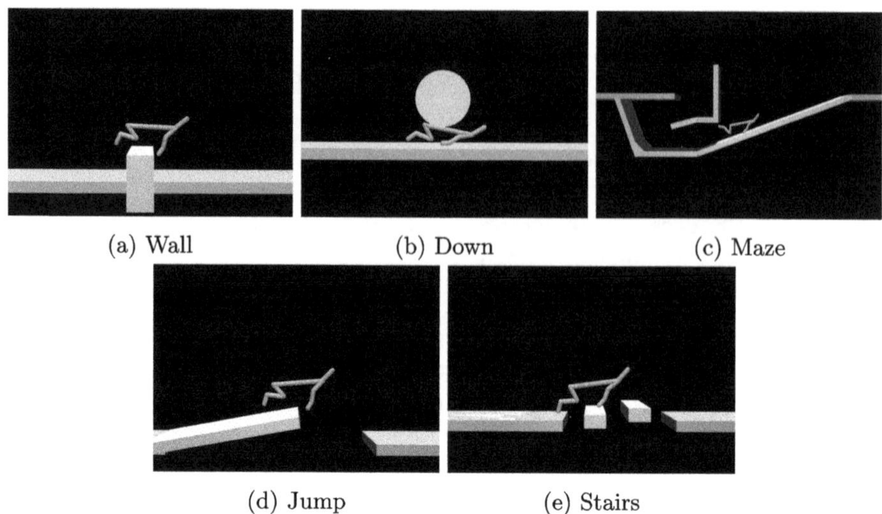

(a) Wall   (b) Down   (c) Maze

(d) Jump   (e) Stairs

Fig. 4. The different obstacles used in the benchmark for the *Half Cheetah*.

requiring multi-task learning across the MuJoCo suite is unsuitable due to its potential excessive complexity and the need for task-specific, independent knowledge. Additionally, unlike real-world multi-task control where tasks involve the same body in different environments, multiple MuJoCo environments introduce distinct input and action spaces, and body morphologies.

Our proposed benchmark strikes a balance: the tasks are independent enough to make achieving a high score challenging for an agent encountering them for the first time, yet they retain similarity through shared *Half Cheetah* behaviors. This setup also enables evaluation in single episodes featuring sequential obstacles.

4.1   The Obstacles

Figure 4 shows the different obstacles used in the benchmark:

- **Wall:** A wall that the *Half Cheetah* needs to climb over.
- **Down:** A cylinder that creates a small tunnel through which the *Half Cheetah* can only pass by getting down.
- **Maze:** A maze where the *Half Cheetah* needs to fall, then go backward, and finally climb a slope. The objective is not the slope itself, so the friction on the slope is maximized to help the *Half Cheetah*.
- **Jump:** As with the previous obstacle, the objective is the jump itself, so the friction on the small slope beforehand is increased to help the *Half Cheetah* gain speed.
- **Stairs:** A set of stairs with holes that the *Half Cheetah* needs to avoid.

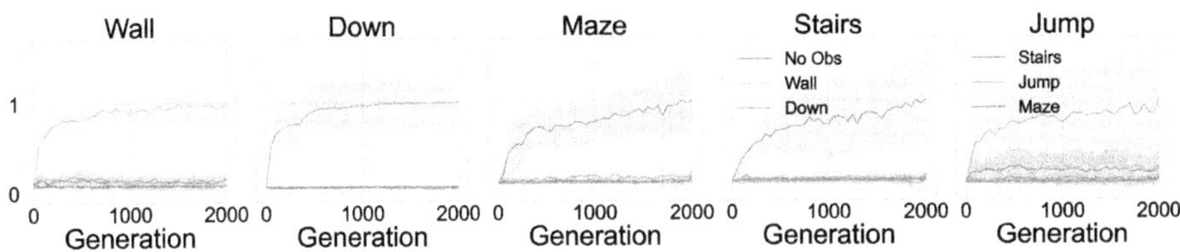

Fig. 5. Validation score achieve on every obstacle, by populations of MAPLE agents trained on each obstacle individually and one on no obstacle. The main curve is the mean score over five seeds, and the shaded area is the standard deviation.

4.2 The Environment

The environment is based on the *Half Cheetah-v5* from Gymnasium [25]. In this environment, the agent performs six actions, applying torques to each of the six joints of the *Half Cheetah*, as shown in Fig. 1. To learn these actions, the sensor inputs are composed of 17 values representing position and velocity information. The default length of an episode is 1000 steps. The objective is to move the *Half Cheetah* as far to the right as possible.

To implement obstacles, the environment is divided into 10-meter sections. When the *Half Cheetah* enters a section, the center of the obstacle is randomly placed between the fourth and fifth meter. This randomness of the position is important to ensure no overfitting. Two observations are added to the standard 17: the identifier of the current obstacle and its center position. While in the obstacle section, the distance to this center is provided, even if negative.

Two additional modifications from the original MuJoCo task were made. First, to prevent the *Half Cheetah* from falling on its back, the episode ends if the z-orientation of the torso leaves $[-2.5, 2.5]$, to avoid undesirable behavior, particularly in the *Maze*. Second change is on the reward, originally divided in two sub rewards. The forward reward, rewarding the agent for going as fast as possible to the right. The control cost, penalizing the agent for applying high torque to the agent motors. While control cost makes sense for pure speed maximization, it becomes less necessary here and can create strong local minima if tasks aren't learned quickly, thus only the forward reward is used.

4.3 Initial Results

Based on the experimental setup detailed later in Sect. 5.2, MAPLE agents are trained on each obstacle individually using five seeds. Figure 5 presents the results for each task. At each validation step, agents are tested on all tasks to assess untrained learning. The results clearly show they do not learn untrained tasks. This observation also holds for populations trained without obstacles.

266   Q. Vacher et al.

To further validate, the best agents from each training run are tested ten times on each obstacle. The correlation matrix in Fig. 6 summarizes these results. Since obstacles like *Maze* require more time steps than *Jump*, raw scores aren't directly comparable. The matrix uses normalized scores, dividing each score by the average achieved by populations trained on that specific obstacle. The correlation matrix reveals complete task independence: agents fail on unseen tasks. This contrasts with Fig. 3, where such generalization does occur.

Thanks to this training, a coefficient can be calculated for each obstacle to indicate its level of difficulty. Since each task is independent, the difficulty cannot be mea-

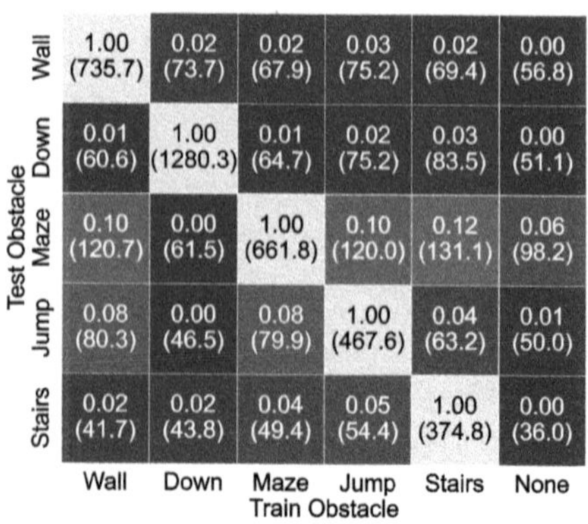

Fig. 6. Correlation matrix of the nomalized score. The x-axis is the obstacle the MAPLE agents have been trained on, while the y-axis is the normalized scores reached on each obstacle. The normalized score presented is obtained by dividing the score of the cell by the score of the trained obstacle, the original score is shown within the brackets.

sure solely by the final score. It is more relevant to base the difficulty on how quickly the task can be learned. Additionally, a higher instability between seeds indicates a higher difficulty. The metric is based on the Area Under the Curve (AUC) of the training [17], corresponding to the following equation:

$$\text{For each task } t, \quad \text{AUC}_t = \frac{\sum_{i=1}^{n}(\mu_{i,t} - \sigma_{i,t})}{\mu_{n,t}} \tag{1}$$

where n is the number of generations, $\mu_{i,t}$ and $\sigma_{i,t}$ represent the average and standard deviation of the score at generation i for task t. The AUC corresponds to the area under the curve below the shaded region in Fig. 5. For obstacles *Wall*, *Down*, *Maze*, *Jump*, and *Stairs*, the AUC values are 29.7, 26.0, 20.6, 16.3, and 10.4, respectively. Normalizing by the highest score yields coefficients of 1.00, 0.88, 0.69, 0.55, and 0.35, where higher values indicate easier tasks. This ranking is supported by one seed failing entirely on the *Stairs* obstacle.

In addition to training with MAPLE, experiments used the state-of-the-art DRL algorithm SAC [9] on the same obstacles. Each obstacle was trained with five independent repetitions, with results shown in Fig. 7, normalized by MAPLE scores. SAC consistently learned the *Down* obstacle quickly. *Maze* and *Stairs* generally required over one million steps, except one *Maze* seed learning faster. For *Jump*, only one seed succeeded, and none on *Wall*.

Although SAC learns efficient behaviors for *Down* and *Maze*, it struggles with new strategies needed for obstacles like *Wall*. While not designed for MTRL, when trained across all obstacles simultaneously, it solved only two versus three when trained separately, supporting the independence of the obstacles.

5   Experimental Setup

This section describes the experimental setup of the study. The parametrization mainly follows [29]. Section 5.1 focuses on selection methods, the main change from the original MATPG paper, presenting both tournament and lexicase selection. The parametrization is then detailed in Sect. 5.2.

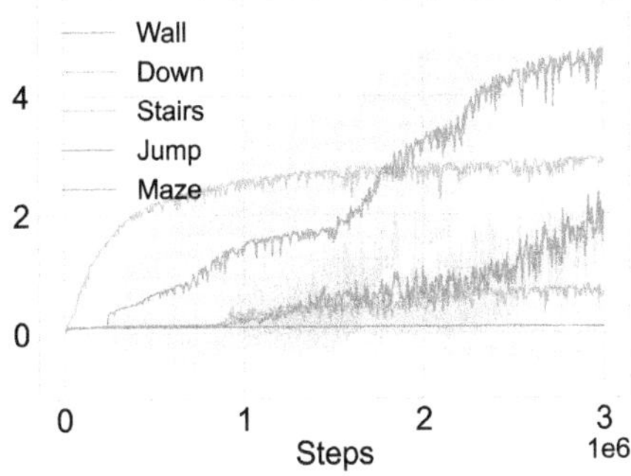

Fig. 7. Validation score achieve on every obstacle, by the SAC. The main curve is the mean score over five seeds, and the shaded area is the standard deviation.

5.1   Selection Methods

Tournament Selection [7] is the algorithm originally used in MATPG. It is widely used and particularly suited for single-task environments. The process for each generation is as follows:

- A small proportion of top-performing agents, called *elites*, are copied directly to the next generation.
- The remaining population is randomly divided into fixed-size subsets. From each subset, the best agent, the winner is selected for reproduction.
- Winners undergo mutation and crossover to generate offspring.
- The next generation consists of elites and the new offspring.

Lexicase Selection [10] was originally designed for classification tasks to promote agents capable of learning multiple classes. The process to select each survivor is:

- A random task order is generated.
- For each task in order, only the top-performing agents are retained.
- After all tasks, if one agent remains, it is selected. If multiple remain, one is chosen randomly.

This repeats until the desired number of survivors is achieved. Survivors generate offspring for the next generation, while others are discarded.

Lexicase selection was proposed for discrete integer scores where top performers often tie. Epsilon-lexicase [16] is a variant for regression problems with continuous scores, where exact equality is rare. In epsilon-lexicase, selection is relaxed: all agents with scores in $[\max(score) \times \epsilon, \max(score)]$ are retained, where ϵ is computed from median absolute deviation. This allows more flexible selection. In our context, epsilon-lexicase is more appropriate due to continuous scores. However, since scores are very high in RL, we multiply the epsilon coefficient by 0.1 to limit its impact.

5.2 Parametrization

To achieve the same solution as in [29], we use the same training framework based on the open source Gegelati library [5]. This C++ library originally implemented TPG and was extended for MAPLE.

We use the same hyperparameters as [29] for MATPG training. The instruction set is $H = \{+, -, \times, \div, \max, \exp, \log, \sin, \cos, \tan, \text{modulo}\}$. Instructions include constants multiplied by their results, with constants mutation based on the mutation in AutoML-Zero [22]. For MAPLE agents on single tasks in Sect. 4.3, identical hyperparameters with tournament selection are used: 1000 agents and MAPLE proportion of 1. For MATPG, values are set to 1500 agents with 0.67 MAPLE proportion, leading to 1000 MAPLE and 500 MATPG agents.

The number of episodes per generation indicates how many times the process needs to be repeated to achieve an accurate score. A value of three per generation is chosen for training, leading to fifteen episodes when five obstacles are used.

6   Experiments

In this Section, the evaluation of MATPG is done based on the experimental setup established in Sect. 5. An ablation study is done to evaluate the performances. In a way, MAPLE is a simplification of MATPG, thus both algorithms are tested with either tournament and lexicase selection. Videos showing the behaviors are available in the supplementary materials.

We show through an ablation study the results of MATPG compared to MAPLE, with either the lexicase or the tournament selection. In addition to evaluating the scalability of the different solutions in overcoming obstacles, the tests are conducted not only on the full set of five obstacles but also on subsets comprising two, three, and four obstacles. The selection of tasks is based on their difficulty, as computed in Sect. 4.3. Specifically, for the two-obstacle scenario, the two easiest obstacles are used for training. This approach is incrementally extended up to the full set of five obstacles.

Figure 8 presents the training performance of the four evaluated approaches across environments with two to five obstacles. In all scenarios, MATPG combined with lexicase selection consistently outperforms both MATPG with tournament selection and MAPLE under either selection method. For the two-obstacle environment, MATPG with tournament selection shows slightly better performance than MAPLE, though the difference is not statistically significant. As the number of obstacles increases, the performance of MATPG with tournament selection aligns closely with that of MAPLE. These results indicate that the original MATPG approach [29] alone struggles to solve MTRL tasks with independent objectives. However, integrating lexicase selection enables MATPG to effectively perform multi-task learning in continuous control environments.

Statistical significance is assessed following prior work [3], using a two-tailed Welch's t-test [32], which is appropriate for small samples with unequal variances. With $N = 10$ independent runs per method, we apply a Bonferroni correction

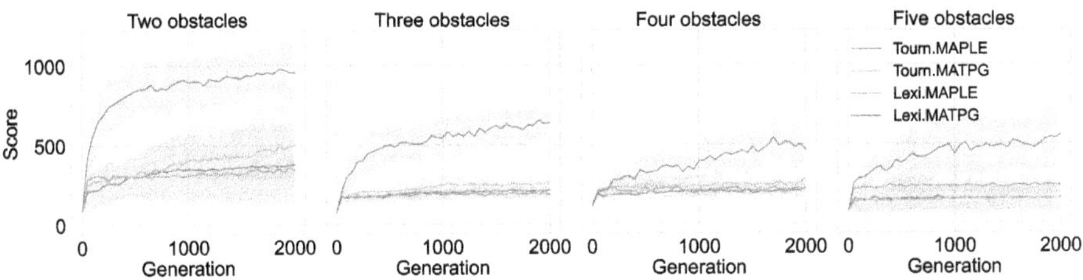

(a) Results over generations. The main curve is the mean score over ten seeds, and the shaded area is the standard deviation.

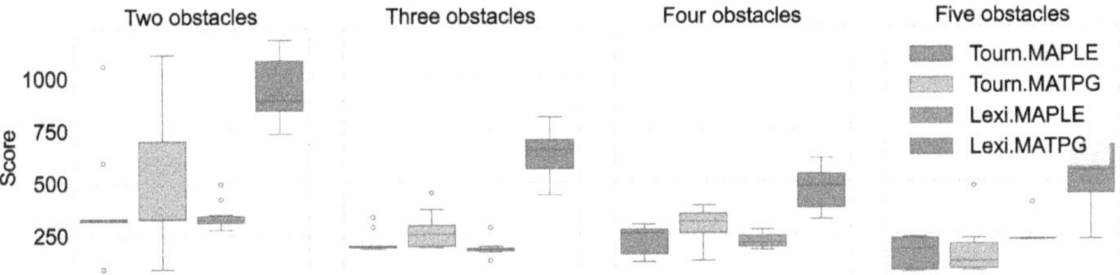

(b) Box plots showing the performances of the best agent of each seed for each solution at the end of the training.

Fig. 8. Validation score achieve on average on five episodes with all trained obstacles showing in random order. Tourn. and Lexi. respectively mean tournament and lexicase selection.

Table 1. Best normalized score on each obstacle, when trained for five obstacles. The normalized score is obtained by dividing the score by the average score obtained by the single obstacle training in Sect. 4.3. The score presented is the average score among ten seeds plus or minus the standard deviation.

Algo.	Selection	Wall	Down	Maze	Jump	Stairs
Best score by any agent within the whole population						
MAPLE	Tourn.	$0.39 \pm (0.3)$	$0.27 \pm (0.2)$	$0.18 \pm (0.2)$	$0.21 \pm (0.1)$	$0.36 \pm (0.3)$
MATPG	Tourn.	$0.60 \pm (0.3)$	$0.17 \pm (0.2)$	$0.22 \pm (0.2)$	$0.23 \pm (0.1)$	$0.63 \pm (0.3)$
MAPLE	Lexi.	$0.73 \pm (0.2)$	$0.72 \pm (0.2)$	$\mathbf{0.70 \pm (0.2)}$	$0.22 \pm (0.0)$	$0.79 \pm (0.2)$
MATPG	Lexi.	$\mathbf{0.81 \pm (0.1)}$	$\mathbf{0.94 \pm (0.2)}$	$0.69 \pm (0.2)$	$\mathbf{0.31 \pm (0.1)}$	$\mathbf{0.99 \pm (0.3)}$
Score by the best agent on episodes with all obstacles						
MAPLE	Tourn.	$0.24 \pm (0.2)$	$0.25 \pm (0.2)$	$0.12 \pm (0.1)$	$0.13 \pm (0.1)$	$0.21 \pm (0.2)$
MATPG	Tourn.	$0.30 \pm (0.2)$	$0.15 \pm (0.2)$	$0.13 \pm (0.2)$	$0.12 \pm (0.1)$	$0.30 \pm (0.3)$
MAPLE	Lexi.	$0.12 \pm (0.2)$	$0.60 \pm (0.2)$	$0.15 \pm (0.2)$	$0.10 \pm (0.0)$	$0.10 \pm (0.1)$
MATPG	Lexi.	$\mathbf{0.46 \pm (0.3)}$	$\mathbf{0.91 \pm (0.2)}$	$\mathbf{0.24 \pm (0.3)}$	$\mathbf{0.14 \pm (0.1)}$	$\mathbf{0.45 \pm (0.4)}$

to control the family-wise error rate, yielding an adjusted significance threshold of $\alpha_{\mathrm{corr}} = 0.005$. Practical significance is measured using Cohen's d, and, we consider $d > 2$ to indicate a sufficiently large effect under limited-run settings. The results in Table 2 report the p-values and corresponding Cohen's d for Lexicase MATPG against all solutions. Across all comparisons, we observe $p < 0.005$ and $d > 2$, demonstrating statistically significant and practically substantial improvements. Therefore, the performance gains are unlikely to be due to random variation.

Table 1 presents the results obtained by each solution for individual obstacles. The first part of the table reports, for each training, the best score achieved by any agent within its entire population, that is, each score can be given by a distinct agent within the population that specialized in the task. Here, both MATPG and MAPLE perform significantly better when using lexicase selection compared to tournament selection, highlighting the ability of lexicase selection to naturally produce elite agents for each task. Notably, MATPG consistently achieves equal or higher scores per obstacle than MAPLE when lexicase selection is employed.

Table 2. Welch's t-test results. $\alpha_{\mathrm{corr}} = 0.005$ with Bonferroni correction ($N = 10$). Cohen's $d > 2$ indicates a very large effect.

	p-value	Cohen's d
Tourn. MAPLE	0.000	2.68
Tourn. MATPG	0.000	2.40
Lexi. MAPLE	0.001	2.20

The second part of the table evaluates the best agent from each population on the individual obstacles. The agent is selected based on performance across episodes containing all obstacles. In an ideal scenario, MATPG with lexicase selection would achieve the same results in both parts of the table. The observed discrepancy indicates that the method does not fully exploit the subpopulation generated by MAPLE. Nevertheless, even under this non-optimal condition, MATPG with lexicase selection outperforms all other approaches, including MAPLE with lexicase selection, across every obstacle. These findings suggest that lexicase selection, regardless of the underlying algorithm, can produce elite agents capable of mastering individual tasks, whereas MATPG is necessary to consolidate this ability into a single agent capable of solving multiple obstacles simultaneously.

7 Interpretability of the Best Agent

In this section, the interpretability of the best MATPG agent is analyzed. This agent achieves high accuracy on the *Wall*, *Down*, and *Stairs* obstacles, with normalized scores of 0.74, 0.95, and 0.77, respectively. However, its performance decreases on the *Maze* and *Jump* obstacles, where it reaches scores of 0.32 and 0.16.

To better understand its behavior, the topology of the agent is illustrated in Fig. 9. The agent consists of two teams vertices, each containing three programs, and five action vertices, where each action vertex includes six programs. Five of the programs connect to the different action vertices, while the remaining connect the two teams. The team vertices, action vertices, and programs are

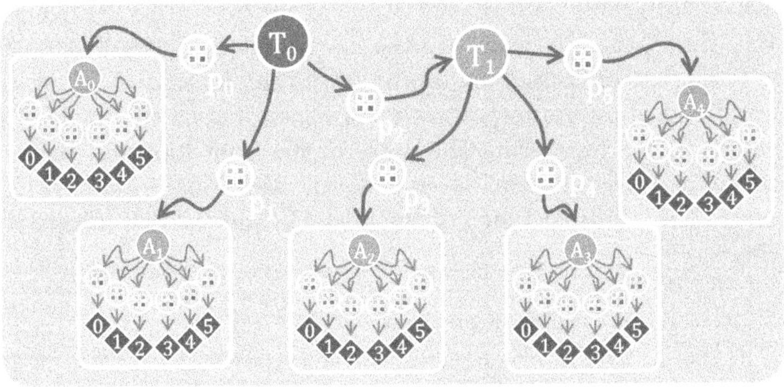

Fig. 9. Topology of the best MATPG agent on the five obstacles.

labeled in Fig. 9, and the detailed behavior of the context programs is described in Algorithm 1. The activation behavior of the agent in response to specific obstacle types can be summarized as follows:

- **Jump** and **Stairs**: Under these conditions, program p_0 consistently produces the highest output value, resulting in the activation of action vertex A_0.
- **Wall**: When encountering a wall obstacle, program p_1 dominates, thereby activating action vertex A_1.
- **Down** and **Maze**: For these obstacle types, program p_2 yields the highest output, leading to the activation of team vertex T_1.
- **Maze (within T_1)**: Once the team vertex T_1 is activated, encountering a maze obstacle causes program p_4 to produce the maximum output, which in turn activates action vertex A_3.
- **Down (within T_1)**: This obstacle exhibits a more complex dynamic. When the agent is positioned far from the obstacle center, programs p_3 and p_4 alternately dominate, producing a cyclic activation pattern lasting between two and four steps per program. As the agent approaches the center of the obstacle, program p_5 becomes predominant, activating action vertex A_4.

In summary, the *Jump* and *Stairs* obstacles are handled by A_0, the *Wall* by A_1, and the *Maze* by A_3. The *Down* obstacle is addressed collaboratively by A_2, A_3, and A_4. The instructions used by each program fully explain this. In short, the traversal paths are mostly due to the state variables s_{17} and s_{18}, which correspond respectively to the obstacle identifier index and the distance to the obstacle, making the following analysis deterministic.

Algorithm 1. Programs used by the teams of the best MATPG agent, see Fig. 9. Each line shows the Program p_i that is computed, depending on the state inputs s_j. The agent only used six inputs from the nineteen of the state inputs. s_3, s_6, s_7 and s_8 are inputs respectively corresponding to the angle of the back shin, the angle of the front shin, the angle of the front foot and the velocity of the x-coordinate of the front tip. s_{17} and s_{18} are additional inputs corresponding respectively to the identifier index of the obstacle and the distance to the center of the obstacle.

$$p_0 \leftarrow \tan(-0.751904 \times (s_{17} + s_6))$$
$$p_1 \leftarrow -7.19471 \times \cos(-0.114644 \times s_8) - 6.002294 \times \frac{s_7}{s_{17}}$$
$$p_2 \leftarrow -0.064495 \times \log(-0.416072 \times \sin(-0.855615 \times s_6))$$
$$p_3 \leftarrow \tan((s_3 - \tan(s_{17})) \times 0.727467)$$
$$p_4 \leftarrow \tan(0.727467 \times (\tan(0.583892 \times \cos(-0.719604 \times s_{18})) + \tan(s_{17})))$$
$$p_5 \leftarrow \sin(0.064883 \times \cos(s_{18}))$$

8 Discussion and Conclusion

In this work, we designed a new MTRL benchmark to assess MATPG efficiency compared to MAPLE. Built on the *Half Cheetah* environment from MuJoCo, it features five distinct obstacles that the agent must overcome. These tasks are designed to satisfy two key properties. First, each task is mutually independent, ensuring that solving one or the original locomotion task does not provide sufficient information to solve the others. Second, all tasks remain learnable by a single MAPLE agent, establishing a baseline that verifies feasibility without requiring specialized methods. Finally, the benchmark is readily extensible, allowing future work to incorporate additional obstacles depending on experimental needs. A current limitation is that agents lack explicit obstacle structure information. Future work could investigate richer task representations by incorporating structural features to enable knowledge sharing.

We evaluate both MAPLE and MATPG under tournament selection, their default configuration, and lexicase selection, which promotes task specialization. Lexicase selection produces populations with improved coverage of task-specific behaviors in both algorithms. However, while MAPLE with lexicase evolves agents solving tasks independently, only MATPG with lexicase achieves significantly higher performance when a single agent faces all obstacles simultaneously. Analysis shows MATPG preserves the inherent interpretability of GP with clear and human-readable decision paths and obstacle-specific behaviors. Although lexicase selection performs better, it requires prior task knowledge. Developing evolutionary techniques without such requirements represents a promising future direction.

Acknowledgements. This research was funded, in whole or in part, by the Agence Nationale de la Recherche (ANR), grant ANR-22-CE25-0005-01. A CC BY license is applied to the AAM resulting from this submission, in accordance with the open access conditions of the grant.

References

1. Amaral, R., Ianta, A., Bayer, C., Smith, R.J., Heywood, M.I.: Benchmarking genetic programming in a multi-action reinforcement learning locomotion task. In: Proceedings of the Genetic and Evolutionary Computation Conference Companion, pp. 522–525 (2022)
2. Brameier, M.F., Banzhaf, W.: Basic concepts of linear genetic programming. Springer (2007)
3. Colas, C., Sigaud, O., Oudeyer, P.Y.: A hitchhiker's guide to statistical comparisons of reinforcement learning algorithms. arXiv preprint arXiv:1904.06979 (2019)
4. De La Torre, C., et al.: Evolution of inherently interpretable visual control policies. In: Proceedings of the Genetic and Evolutionary Computation Conference, pp. 358–367 (2025)
5. Desnos, K., Sourbier, N., Raumer, P.Y., Gesny, O., Pelcat, M.: Gegelati: lightweight artificial intelligence through generic and evolvable tangled program graphs. In: Workshop on Design and Architectures for Signal and Image Processing (14th edn), pp. 35–43 (2021)
6. Espeholt, L., et al.: Impala: Scalable distributed Deep-RL with importance weighted actor-learner architectures (2018). https://arxiv.org/abs/1802.01561
7. Fang, Y., Li, J.: A review of tournament selection in genetic programming. In: International Symposium on Intelligence Computation and Applications, pp. 181–192. Springer (2010)
8. Fujimoto, S., van Hoof, H., Meger, D.: Addressing function approximation error in actor-critic methods (2018). https://arxiv.org/abs/1802.09477
9. Haarnoja, T., Zhou, A., Abbeel, P., Levine, S.: Soft actor-critic: off-policy maximum entropy deep reinforcement learning with a stochastic actor (2018). https://arxiv.org/abs/1801.01290
10. Helmuth, T., Spector, L., Matheson, J.: Solving uncompromising problems with lexicase selection. IEEE Trans. Evol. Comput. **19**(5), 630–643 (2014)
11. Henderson, P., et al.: Benchmark environments for multitask learning in continuous domains. CoRR **abs/1708.04352** (2017). http://arxiv.org/abs/1708.04352
12. Jorgensen, S., et al.: Policy search through genetic programming and LLM-assisted curriculum learning. ACM Trans. Evol. Learn. (2025)
13. Kelly, S., Heywood, M.I.: Emergent tangled graph representations for Atari game playing agents. In: Genetic Programming: 20th European Conference, EuroGP 2017, Amsterdam, The Netherlands, 19–21 April 2017 Proceedings 20, pp. 64–79. Springer (2017)
14. Kelly, S., Heywood, M.I.: Multi-task learning in Atari video games with emergent tangled program graphs. In: Proceedings of the Genetic and Evolutionary Computation Conference, pp. 195–202 (2017)
15. Koza, J.R., Rice, J.P.: Automatic programming of robots using genetic programming. In: AAAI. vol. 92, pp. 194–207 (1992)
16. La Cava, W., Spector, L., Danai, K.: Epsilon-lexicase selection for regression. In: Proceedings of the Genetic and Evolutionary Computation Conference 2016, pp. 741–748 (2016)
17. Mesbahi, G., Panahi, P.M., Mastikhina, O., White, M., White, A.: K-percent evaluation for lifelong RL. arXiv preprint arXiv:2404.02113 (2024)
18. Milani, S., Topin, N., Veloso, M., Fang, F.: Explainable reinforcement learning: a survey and comparative review. ACM Comput. Surv. **56**(7), 1–36 (2024)

19. Miller, J., Turner, A.: Cartesian genetic programming. In: Proceedings of the Companion Publication of the 2015 Annual Conference on Genetic and Evolutionary Computation, pp. 179–198 (2015)
20. Mnih, V., et al.: Human-level control through deep reinforcement learning. Nature **518**(7540), 529–533 (2015)
21. Nadizar, G., Medvet, E., Wilson, D.G.: Naturally interpretable control policies via graph-based genetic programming. In: European Conference on Genetic Programming (Part of EvoStar), pp. 73–89. Springer (2024)
22. Real, E., Liang, C., So, D.R., Le, Q.V.: AutoML-Zero: evolving machine learning algorithms from scratch (2020). https://arxiv.org/abs/2003.03384
23. Schulman, J., Wolski, F., Dhariwal, P., Radford, A., Klimov, O.: Proximal policy optimization algorithms (2017). https://arxiv.org/abs/1707.06347
24. Todorov, E., Erez, T., Tassa, Y.: Mujoco: A physics engine for model-based control. In: 2012 IEEE/RSJ International Conference on Intelligent Robots and Systems, pp. 5026–5033 (2012). https://doi.org/10.1109/IROS.2012.6386109
25. Towers, M., et al.: Gymnasium: a standard interface for reinforcement learning environments. arXiv preprint arXiv:2407.17032 (2024)
26. Tunyasuvunakool, S.: dm_control: Software and tasks for continuous control. Softw. Impacts **6**, 100022 (2020)
27. Vacher, Q., et al.: Hybrid genetic programming and deep reinforcement learning for low-complexity robot arm trajectory planning. In: 16th International Conference on Evolutionary Computation Theory and Applications, pp. 139–150 (2024)
28. Vacher, Q., Beuve, N., Dardaillon, M., Desnos, K.: Eurogp 2026 artifacts (2026). https://github.com/gegelati/EUROGP-2026
29. Vacher, Q., et al.: Maple: Multi-action programs through linear evolution for continuous multi-action reinforcement learning. In: Proceedings of the Genetic and Evolutionary Computation Conference, pp. 1062–1071 (2025)
30. Videau, M., Leite, F.A., Teytaud, O., Schoenauer, M.: Multi-Objective genetic programming for explainable reinforcement learning. In: Lecture Notes in Computer Science. Lecture Notes in Computer Science, vol. 13223, pp. 278–293. Springer International Publishing, Madrid, Spain (2022). https://doi.org/10.1007/978-3-031-02056-8_18, https://inria.hal.science/hal-03886307
31. Wang, X., et al.: Deep reinforcement learning: a survey. IEEE Trans. Neural Netw. Learn. Syst. **35**(4), 5064–5078 (2022)
32. Welch, B.L.: The generalization of 'student's'problem when several different population varlances are involved. Biometrika **34**(1–2), 28–35 (1947)
33. Xu, Z., Wu, K., Che, Z., Tang, J., Ye, J.: Knowledge transfer in multi-task deep reinforcement learning for continuous control (2020). https://arxiv.org/abs/2010.07494

Reducing Computational Overhead
in Biomedical Image Segmentation
via Active Learning and PCA-Based
Diversity Filtering in CGP

Yuri Lavinas[1,2]([✉])[iD], Nathaniel Haut[4][iD], Sylvain Cussat-Blanc[1,2,3][iD],
and Wolfgang Banzhaf[4][iD]

[1] University Toulouse Capitole, Toulouse, France
`yuri.lavinas@ut-capitole.fr`
[2] IRIT - CNRS UMR5505, Toulouse, France
[3] Institut Universitaire de France, Paris, France
[4] Michigan State University, East Lansing, France

Abstract. In this work, we propose an Active Learning (AL) framework for Cartesian Genetic Programming (CGP) to evolve data-efficient programs for biomedical image segmentation. Active Learning enables the dynamic selection of training data by identifying the most informative images to include during evolution. To further reduce computational cost, we introduce a filtering stage that selects a diverse subset of candidate images using Principal Component Analysis (PCA) before uncertainty evaluation. We also investigate several strategies for selecting the initial training data: the most typical image, cluster-based, and random, and analyse their influence on convergence and model diversity. Our results show that filtering substantially reduces computational overhead while preserving data diversity, and that the choice of both initialization and uncertainty metrics significantly impacts convergence speed and overall performance. The proposed approach demonstrates that integrating Active Learning and filtering into CGP leads to faster convergence, improved performance, and more efficient utilization of training data.

Keywords: cartesian genetic programming · biomedical data · data sampling · active learning · data filtering

1 Introduction

The field of biomedical image analysis has experienced significant progress with the usage of Neural Networks, particularly Deep Learning (DL) techniques, displaying a similar trend that happened in the broader domain of Computer Vision (CV). Deep Neural Networks exhibit remarkable performance across various image classification and segmentation tasks within medical disciplines, such as dermatology, radiology, and pathology [16,17], occasionally even surpassing human experts. Nonetheless, DL methods encounter two primary limitations. Firstly, they are regarded as black-box systems, as the decision-making processes of deep Neural Networks often lack interpretability, a difficult challenge in

L. Manzoni et al. (Eds.): EuroGP 2026, LNCS 16521, pp. 275–291, 2026.
https://doi.org/10.1007/978-3-032-23005-8_17

critical domains like medicine. Secondly, DL approaches necessitate substantial amounts of annotated data, a resource-intensive and costly task. Annotating data is typically carried out by busy experts, which reduces the chances of acquisition of larger datasets for training purposes.

Recent work has demonstrated that Cartesian Genetic Programming (CGP) is an effective approach to address the above-mentioned limitations inherent in DL [13,29], with results competitive to DL. CGP evolves solutions based on a set of mathematical functions that process given inputs to produce an expected output. The phenotype of a CGP program can be represented as a graph which often is evolved using a $(1+\lambda)$ Evolutionary Algorithm [34]. One of the main benefits of CGP is the use of a fixed-length integer-based genome to encode the functional graphs, which mitigates the bloat effect encountered in many tree GP approaches, allowing evolution to design small and interpretable solutions [44]. Particularly in image processing tasks, the CGP function library can contain CV functions that are combined and optimized in order to construct tailored image processing pipelines suitable for specific objectives.

Minimizing the amount of data necessary to evolve effective programs for CV tasks remains a key challenge. Our goal in this work is to improve the data efficiency of CGP in biomedical image segmentation. It is now well established that Active Learning (AL) effectively samples the most informative data points across different domains to improve training efficiency in machine learning tasks [18,19,23,37]. In image processing [11,28,31], AL methods can enhance performance by using information gathered during evolution (or training) to iteratively refine the dataset. The main idea is to identify useful images from a larger pool, leading to programs of comparable performance with less training data [12], while also helping evolution avoid overfitting [6].

Our recent works [31,32] showed that building a small set of data samples in a training set increases the convergence speed of CGP and leads to potentially more diverse solutions. In another contribution [30], we performed an in-depth analysis of the impacts of Active Learning in CGP, including a discussion of the frequency of images sampled and the potential impacts of the highly frequent images in the search dynamics of CGP. We investigated the experimental results in terms of performance; convergence speed; program size; and on how to use information about the frequency of sampled images to improve CGP performance.

To further improve efficiency of the algorithm, here we introduce a filtering stage that precedes uncertainty evaluation. Because computing uncertainty is computationally expensive–requiring each model to be evaluated on all available data–we propose filtering the dataset to retain only diverse and informative samples. This is achieved using Principal Component Analysis [20] (PCA) which projects our dataset into a lower-dimensional space to select a predefined number of biomedical images that maximize diversity, that is images that are located in a sparse region of the projected space. This step reduces computational overhead while maintaining representative coverage of the data distribution.

We also address the selection of initial data points for training. The initial subset of images strongly influences early-stage convergence and model diversity. To determine how the initial data distribution affects CGP performance, we evaluated three initialization strategies: the most typical image, a cluster-based

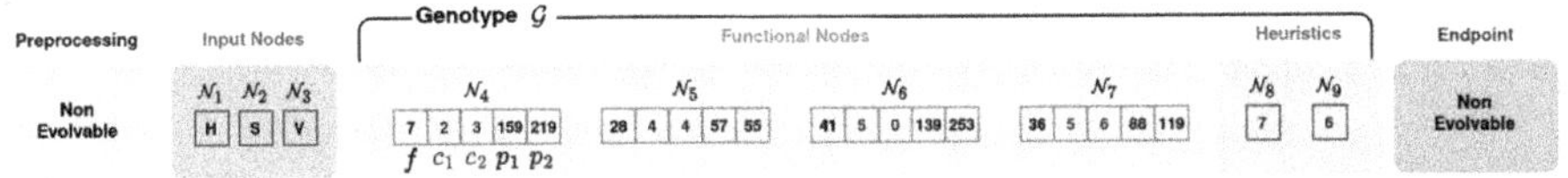

Fig. 1. The genotype of CGP.

strategy, and a random approach. PCA-based initialization ensures diversity across main data modes, cluster-based initialization ensures balanced coverage of the dataset's structure, and random selection serves as a baseline for comparison. Our contributions can be summarized as follows:

(1) We introduce a filtering mechanism based on PCA to reduce computational cost and retain diverse, informative images before uncertainty estimation.
(2) We evaluate three initialization strategies for selecting initial training data (PCA-based, cluster-based, and random) and analyse their influence on convergence and performance.
(3) We show that the number of images required for uncertainty estimation can be greatly reduced without loss of generalizability.

The paper is organized as follows: Sect. 2 introduces the necessary background. Section 3 explains relevant concepts. Section 4 gives the experimental setup. Section 5 presents the experimental results of our analysis. Finally, Sect. 6 concludes the paper and discusses further research.

2 Preliminaries

Most work in the area of Computer Vision tasks uses blackbox approaches, such as artificial Deep Neural Networks. However, these models have proven to be difficult for humans to analyse and interpret [13,36]. One way to complement these blackbox methods is to use methods that are inherently explainable. The class of Genetic Programming (GP) approaches is among such methods. Here we highlight CGP, an evolutionary computation algorithm that evolves easier-to-interpret programs (as with most GP variants), in comparison to DL models [1, 9,13,21,39,44].

2.1 Cartesian Genetic Programming

CGP is a GP variant [34] specialized in evolving graph genotypes. Such graphs are often direct and acyclic and are indexed by Cartesian coordinates. Evolution defines how to connect the nodes of the graphs and the function of each node. CGP has been successfully applied in multiple domains [1,9,39]. Specifically, CGP has been applied in Computer Vision tasks such as the controlling of agents to play ATARI games [44], and in image processing tasks, such as biomedical image segmentation and object detection in robotics [13,21].

Table 1. Description of the function library used in CGP.

Function	Arity	Function	Arity
Max	2	Min	2
Mean	2	Add	2
Subract	2	Bitwise_not	1
Bitwise_or	2	Bitwise_and	2
Bitwise_and_mask	2	Bitwise_xor	2
sqrt	1	pow2	1
exp	1	log	1
median_blur	1	gaussian_blur	1
laplacian	1	sobel	1
robert_cross	1	canny	1
sharpen	1	gabor	1
abs_diff	1	abs_diff2	2
fluo_tophat	1	ref_diff	1
erode	1	dilate	1
open	1	close	1
morph_gradient	1	morph_tophat	1
morph_blackhat	1	fill_holes	1
remove_small_objects	1	remove_small_holes	1
threshold	1	threshold_at_1	1
distance_transform	1	distance_transform_and_thresh	1
inrange_bin	1	inrange	1

CGP generally employs the $(1+\lambda)$ Evolutionary Algorithm (EA), although any other evolutionary algorithm could be used. Initially, a population of λ individuals is randomly generated and evaluated on the problem in question. Then, evaluation is conducted by first generating the programs from the graphs and measuring their performance on the task considered. The solution with the highest performance is maintained to the following generation step, influencing the next λ individuals created via mutation. This process is repeated until a termination criterion is reached. For more information on this algorithm, see [9, 34, 44].

GP has been widely used in the biomedical domain, and as Khan et al. stated, GP is often used in classification of cancerous cells, see this survey for more information [27]. In particular, one can find GP contributions that involve feature extraction [3, 4, 8, 26, 41] or image classification [2, 5, 15, 45].

2.2 Dynamic Data Sampling

Dynamic data sampling is one of the most important components of AL methods and its effects have been studied for different Machine Learning algorithms [14, 23, 40]. AL is frequently used in DL in the context of processing biomedical data [18, 35] with most of the work on AL focusing on finding the metric that leads models to the highest performance [18]. Interestingly, some papers suggest that random sampling is a strong baseline [11, 28].

In the domain of GP, the efficacy of Active Learning for symbolic regression tasks has been widely demonstrated in diverse research contributions, from works that focus on reducing the number of evaluations to create a GP program [19] while creating smaller, balanced datasets by recursively keeping the most 'meaningful' exemplars [42] to studies on improving the rate and consistency at which well-performing solutions are found while reducing the required number of training samples [22,23]. For classification problems, Hamida et al. [6,24] showed how different sampling methods studied over the years affect the performance of GP. Yet, the combination of AL and GP methods in the biomedical domain remain lacking. But recently, we were able to show that AL [31,32] methods can provide performance improvements in biomedical image processing.

Here, we use uncertainty as the base for dynamic data sampling, in a similar manner as discussed by Nguyen et al. [25]. That is, given models trained in a dataset d, where d is a subset of the whole dataset D available, each image in D but not in d is assigned an uncertainty value based on the model's behaviour. The image with the highest uncertainty is added to d and removed from D. The new dataset D is then given to the model for training.

2.3 CGP Implementation

The CGP implementation we use is based on [13], a modular Cartesian Genetic Programming system to generate programs for computer vision tasks. This system introduced the notion of non-evolvable nodes which are functions not subjected to optimization in the syntactic graph. These nodes are positioned at the end of the program generated, that then are directly connected to the output.

Figure 1 shows how our CGP implementation works. This genotype is a sequence of integers known as genes (each one represented as a box containing a single integer) that are organized into nodes. A node is composed of a single function drawn from the function library, at least one connection, and optional parameters. In this implementation, the end nodes of CGP are connected to a fixed endpoint. This endpoint is useful since it allows CGP to obtain insight given by a CV expert. We use the Watershed Transform [7] as the endpoint. Thus, our CGP has 2 outputs, corresponding to the mask and markers necessary for the Watershed Transform endpoint. Finally, we use image processing functions mostly from OpenCV [10] and Scikit-image [43], which apply programs directly to images, see Table 1. Only a few nodes, the "active" nodes, are used, as they are actually connected to the output of the program graph. Other nodes with no connections to the output are called "inactive" nodes. The outputs of a program can come from any node and are determined by the evolutionary process. Table 1 lists the basic functions used in our function library and their related arities. These operators are functions from the OpenCV Python package and are fixed for all CGP variants compared.

3 Uncertainty-Aware Sampling in Cartesian GP

The **Uncertainty-Aware Sampling** CGP (UAS-CGP) template we propose for instantiating and designing sampling method variants is shown in Algorithm 1, **red** indicate the new introduced steps in relation to CGP. The main difference to standard CGP is that instead of using a fixed training dataset during evolution, our template uses a smaller dataset that is selected during evolution, given the different metrics (explained below). We sample the subset of images to be part of the training dataset used in evolution via sampling mechanisms.

Algorithm 1. Uncertainty-Aware Sampling in CGP (UAS-CGP)

1: Initialize $(1 + \lambda)$ CGP.
2: **Select** initial data point.
3: **while** *Stopping criterion is not met* **do**
4:     **Evolve** CGP with the training dataset.
5:     **Update** elite solution.
6:     **Generate** γ solutions by mutating the elite solution δ times.
       Filter η data points.
       Calculate uncertainty *metric* on the images not in the training set.
       Add the image with the highest uncertainty *metric* to the training dataset.
7:     **Calculate** fitness of the elite and the γ solutions.
8:     **Update** elite solution.
9: **end while**

3.1   Initial Image Selection

The reason for selecting initial points is that choosing these initial data is important because it shapes the early stages of model evolution: the size and diversity of this subset influences both convergence speed and generalization.

Typical (PCA-based): The dataset is projected using PCA and the image selected is the one that minimizes the pairwise distances between all images points given the orthogonal distance to the principal components. This idea is to have diversity in the selected subset and reduce over-representation of any particular data region.

Cluster (PCA-based): The dataset is projected using PCA and then we use Kmeans [33] to cluster images. A Knee Plot based on the silhouette Score is conducted to verify into how many clusters the images group together. Then, one image per cluster is picked randomly. This idea is to have diversity that represents all possible image types thus representing all regions of the data space.

Random selection (baseline): For comparison, we also include a random selection strategy, where samples are chosen uniformly at random from the dataset.

3.2   Filtering Images

Now, the rationale for filtering the dataset before computing the uncertainty metric is that this is a computationally expensive task, as each model must be evaluated on all data points. The intuition behind a filtering stage before calculating uncertainty is to first reduce the dataset size to a diverse subset of η candidate images. This subset contains samples that differ from those already used for training, reducing redundancy and focusing uncertainty evaluation on informative regions of the data space.

We employ PCA-based filtering to measure diversity. The dataset is projected into a lower-dimensional space using PCA, and η images are selected that minimize pairwise euclidean distances among themselves given all images points in this reduced space in relation to the images already in use. This ensures that the retained images are well distributed along the major axes of variance while keeping computational cost manageable. This procedure selects images that are more unique (they can be outliers) in relation to the images already in use, providing a diverse set of images to be sampled from.

3.3 Sampling Mechanisms

Uncertainty-based AL utilizes a group of diverse set of γ models to search different areas of the search space. These diverse models allow us to estimate uncertainty via their disagreement to select new training data. The idea is that selecting data where disagreement is high will lead to the selection of data that will be most informative to the current models in training. [23].

To calculate uncertainty in CGP, our $(1 + \lambda)$ Evolutionary Algorithm (EA) forms a group of γ CGP models by applying mutations to the current elite solution. The intuition behind using this group is that slightly different models can help AL identify informative regions of the data space, leading to the selection of more training samples that have features these reference models process differently. By maintaining multiple variants of the elite, we also introduce additional diversity of candidate solutions, a property not typically present in the standard $(1 + \lambda)$ EA commonly used in CGP. Moreover, if any of the variant models outperform the current elite, the highest-performing variant replaces the current elite as the new elite model. The updated elite is then used to generate segmentation masks and to compute all metrics for UAS-CGP.

For uncertainty estimation, we employ the same Weighted Uncertainty (WU), that doubles the uncertainty measurement if one pixel is labeled as 0 and the other pixel is labeled with a non-zero value. This assumes higher uncertainty should be assigned if models disagree on whether a pixel is foreground (non-zero) or background [30].

$$\text{Weighted} = \sum_{i,j}^{\text{Pixels}} k = \begin{cases} 2, & \text{if } m_{1(i,j)} \neq m_{2(i,j)} \text{ and } m_{1(i,j)} = 0 \text{ or } m_{2(i,j)} = 0, \\ 1, & \text{if } m_{1(i,j)} \neq m_{2(i,j)} \text{ and } m_{1(i,j)} \neq 0 \text{ and } m_{2(i,j)} \neq 0, \\ 0, & \text{otherwise.} \end{cases}$$

Random Uniform sampling is used as baseline. No uncertainty is calculated and one image is selected using uniformly distributed values at each step.

3.4 Evaluation Metric and Termination Criterion

We follow the work in [38] that defines average precision $AP = TP/(TP + FP + FN)$, where TP means true positives, FP means false positives and FN means false negatives. We use AP as our fitness function with a threshold of 0.5 to determine the true positives of the predicted mask.

Our goal is to study the effect of sampling of images from the dataset during evolution. Thus, we have a different number of images processed by UAS-CGP and standard CGP, at each generation. Given the different sizes of the datasets used during evolution, we use the number of images processed during evolution as the termination criterion. This criterion does not discriminate if there is a repetition of data points in the training set.

4   Experimental Setup

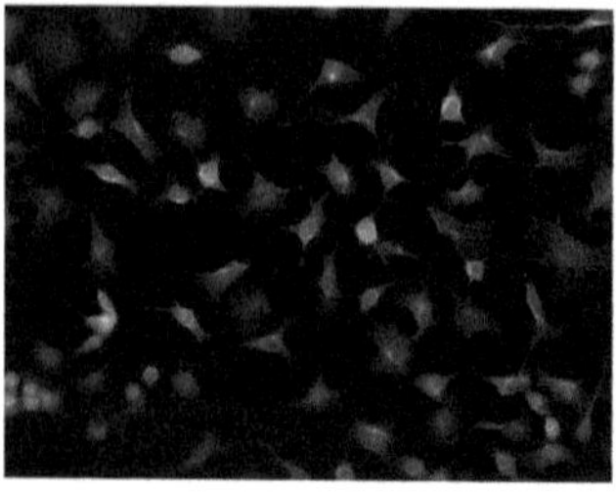

Fig. 2. A target image from the CELL-POSE dataset.

Table 2. Parameters used.

Parameter	Value
Number of nodes	30 nodes
Offspring size, λ	4
Inputs	2, α-tubulin and DAPI channels
Outputs	2, mask and markers
Mutation of function nodes	0.15
Mutation of outputs nodes	0.20
Images processed	1,000,000
Training generations	100
independent executions	30

We compare UAS-CGP with tradicional CGP. UAS-CGP builds the dataset for training using the dynamic data sampling methods (Sect. 3) and standard CGP uses all 89 images available in the training set. We run all CGP variants with the parameters shown in Table 2, the same as in [13], but more work is needed if we want to tune the parameters for the CGP variants. UAS-CGP adds new data to the training dataset every 100 generations. The evolutionary budget is set to 1,000,000 images processed for all CGP variants. For statistical purposes, 30 independent runs are done for each variant.

4.1   Performance Comparison

For prototyping, we use the CELLPOSE dataset [38] which consists of 100 images of fluorescent-labeled protein of cultured neuroblastoma cells with phalloidin FITC and DAPI nuclear stain. For a fair comparison, we follow the work in [38], where the data is split into 89 images for training and 11 for testing. Figure 2 shows a target image from this dataset used for testing. The state-of-the-art performance in this dataset is an AP of 0.93 by DL and 0.89 by CGP [13].

4.2   Reproducibility

For reproducibility purposes, relevant data and code are available at: https://shorturl.at/qQL2u.

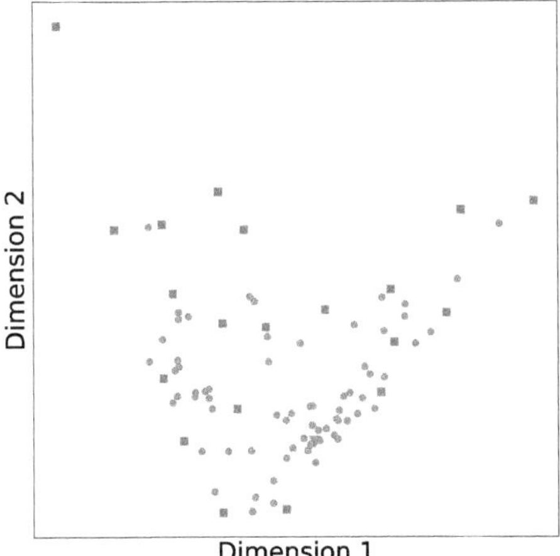

Fig. 3. Example of diverse filtering. The red triangle denotes the most typical point; green squares indicate the $\eta = 20$ selected samples maximizing diversity. (Color figure online)

5 Experimental Comparison

Here, we show a comparative analysis based on dimensionality reduction and initialization strategies. The goal is to better understand the structure of the feature space and how data diversity influences subsequent selection and evaluation processes. We use PCA to reduce the dimensionality of the pixels representations while keeping most of the variance. Retaining 50 components preserves approximately 80% of the total variance, suggesting that the relevant information is concentrated in a relatively low-dimensional subspace. This finding supports earlier observations that compact latent embeddings are often sufficient to describe perceptually meaningful structure.

The choice of data points plays a crucial role in ensuring representative coverage of the underlying feature space. Figure 3 depicts an example of a diverse initialization, where the red triangle denotes the most typical point, and green markers indicate the $\eta = 20$[1] samples selected by the diversity-oriented criterion. The result demonstrates that the selection method effectively spans distinct regions of the space, mitigating redundancy among chosen samples.

Figure 4a presents the results of clustering analysis following dimensionality reduction with PCA. The number of clusters was found to be approximately six, based on the silhouette analysis. However, the post-PCA clustering visualization (for visualization reasons showing the two main components, see Fig. 4a) reveals that the data primarily form two dominant groups, while the remaining four clusters appear as smaller, less cohesive subsets. Therefore, we decided to start evolution with six images, one randomly picked from each of the clusters when

---

[1] Values associated with the parameter analysis in Sect. 5.1.

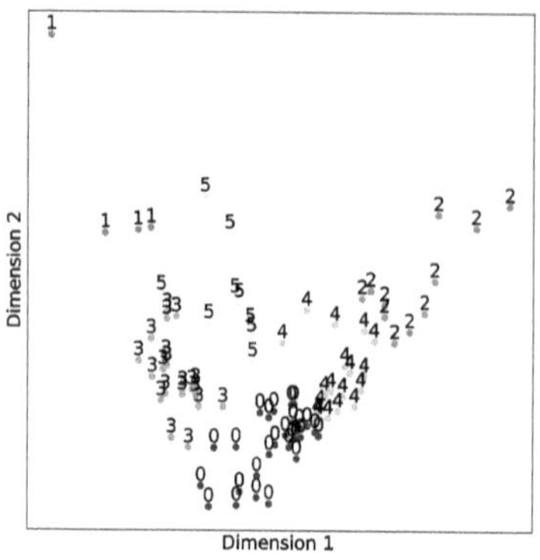

(a) Clustering after PCA with 50 components (only the two most important are shown) showing the six clusters.

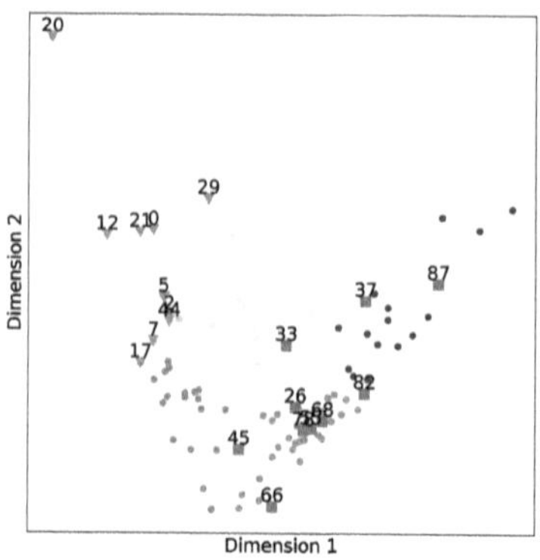

(b) Most (red triangles) and least (blue squares) frequently selected samples, as identified in Lavinas et al. [30].

Fig. 4. The two-dimensional PCA projection reveals two main groups and four minor, outlier-like clusters. To our surprise, there is some connection between sampling images during evolution and the PCA results.

using the cluster initialization method. The other methods start evolution with only one image instead.

A re-examination of the observations reported in Lavinas et al. [30], that found out that more images are selected as the number of cells increases. The most frequently selected images (0, 20, and 21) have 175, 175, and 174 cells, respectively. In contrast, the least frequently selected images (78, 45, and 63) have only 20, 52, and 21 cells, which is much fewer. Looking at Fig. 4b we can see that a distinctive structure exposed by PCA: there are two axes that align with the samples most and least frequently selected images. This suggests that projecting the images into a lower dimensional space can highlight implicit data structures that influence the evolutionary process.

5.1   Impact of the Hyperparameters

In this section, we perform a simple analysis of the impact of the key hyperparameters introduced in this work: (1) ensemble size, (2) number of mutations, (3) number of diverse samples, and (4) initialization strategy. When analysing each hyperparameter, all others are fixed at appropriate values based on preliminary experiments. The default configuration is: ensemble size: $\gamma = 100$, 1 elite+ 99 new models, number of mutations: $\delta = 1$, diverse sampling: $\eta = 20$, and initialization strategy: typical data point. Accordingly, we do not consider possible interactions between different hyperparameter settings in this analysis. Future work should include a more systematic hyperparameter tuning process to better understand the joint effects and potential interactions among these parameters.

Table 3. Higher fitness with ensemble size of 50 or 100.

# models	Mean (Std)
5	0.816 (0.052)
50	0.827 (0.032)
100	**0.827 (0.036)**

Table 4. No clear impact of the mutation on fitness.

Mutations	Mean (Std)
1	0.832 (0.027)
2	0.823 (0.036)
5	0.834 (0.027)
10	0.833 (0.031)

Table 5. Better performance with diversity of $\eta = 20$.

Diverse	Mean (Std)
10	0.82 (0.024)
20	**0.832 (0.027)**
30	0.821 (0.036)
40	0.823 (0.029)

The impact between parameters is not considered here; we will address it in future work.

First, we examine how varying the number of models affects the overall search progress. Our goal is to assess whether performance improvements come from the complementary behaviour of models and to clarify the extent to which model diversity contributes to performance gains. We evaluate three ensemble configurations: (1) four new individuals mutated from the current elite, resulting in an ensemble of $\gamma = 5$ models; (2) forty-nine new individuals plus the elite, forming an ensemble of $\gamma = 50$ models; and (3) ninety-nine new individuals plus the elite, yielding an ensemble of one $\gamma = 100$ models. The comparison across ensemble sizes indicates only a minor difference in overall performance, see Table 3, with the best results obtained using ensembles of size 50 or 100. Although a size of 50 lead to smaller standard deviation values, we adopt an ensemble size of $\gamma = 100$ models, as this setting can lead to more reliability by estimating uncertainty with a higher number of models.

Second, we investigate how the number of mutations influences the balance between diversity (or distance to the elite model) and consistent characteristics across models. We studied four values of successive mutations: (1) $\delta = 1$; (2) $\delta = 2$; (3) $\delta = 5$; (1) $\delta = 10$. Table 4 reveals no substantial performance differences across mutation settings. Thus, we selected the smallest mutation values to favour faster evolutionary iterations.

Lastly, the objective of this analysis is to assess the impact of sampling diversity on the search process. By controlling the number of diverse images introduced at each iteration, we aim to quantify how diversity influences the landscapes of the search space while reducing the number of samples required for a reliable uncertainty estimation. This analysis shows that a reduced diverse subset of points can still preserve the main characteristics of the dataset. We evaluate four configurations of sampling diversity: (1) $\eta = 10$, (2) $\eta = 20$, (3) $\eta = 30$ and (4) $\eta = 40$ diverse images. As shown in Table 5, performance improves slightly when $\eta = 20$ diverse images are sampled (about 20% of the dataset), the setting we adopt in the following experiments.

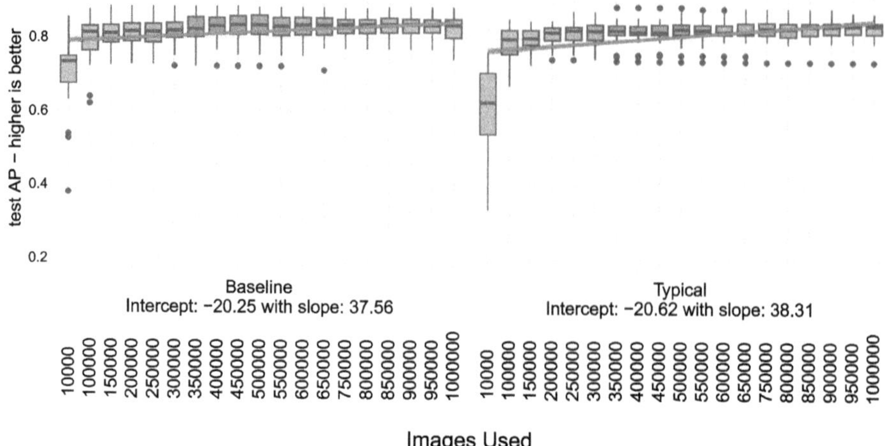

Fig. 5. The baseline, on the left, seems to converge half-way through the search while the Typical strategy, on the right, continues to marginally improve.

5.2   Performance Impact of Initialization Strategies

We investigate how variations in sampling methods influence this process, since the initialization phase can significantly impact search performance. Accordingly, we seek to determine whether certain initialization strategies can facilitate more efficient and consistent search progress. We analyse the impact that different initial sampling strategies have on search performance. From our hyperparameter search experiments, we chose the following setup: $\gamma = 100$, 1 elite+99 new models, number of mutations: $\delta = 1$, diverse sampling: $\eta = 20$, together with the Weighted Uncertainty estimation (with the exception of Random, that uses uniformly sampled values), thus, having the initialization strategy as our only variable. Table 6 shows that the comparison of initialization strategies indicates that the typical strategy achieves the highest overall performance. Therefore, we select the Typical strategy as the initial image selection mechanism. We compare such configuration of UAS-CGP with our baseline.

5.3   Performance Comparison Against the Baseline

The baseline method is a simplified version of the approach presented in [30]. In this version, we remove the islands concept to focus on a single evolutionary setup. As in the original work, the initial image is selected randomly, the training set is build given uniformly sampled values on all available images (no filtering is employed). Here, we use an ensemble of 100 models, the same configuration shown to perform the best in Sect. 2, with the performance of 0.827 (0.035).

As we can see in Table 6, the Typical strategy achieved higher performance of 0.832 (0.027), a higher AP than our baseline, but the student t-test showed no statistically significant difference between the performances with a confidence

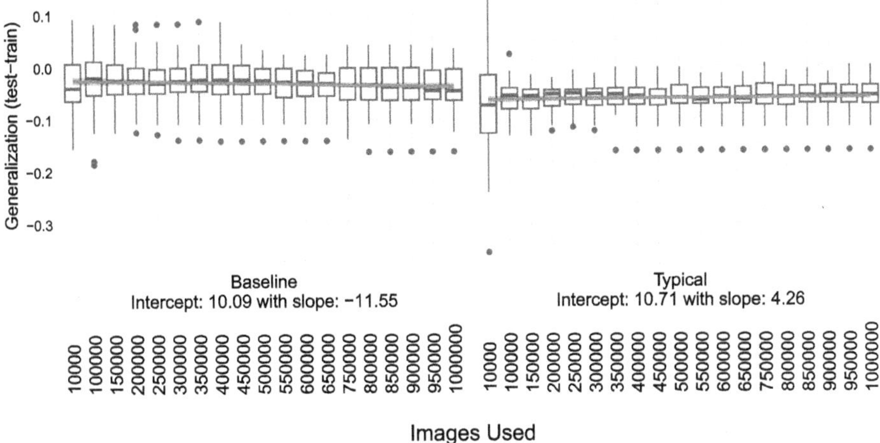

Fig. 6. We plot regression lines in red, with the corresponding confidence intervals shown in green. The baseline, on the left, exhibits a decline in generalization over the course of evolution, indicating potential overfitting. Typical, on the right, gradually improving their generalization as evolution progresses. (Color figure online)

Table 6. Typical strategy leads to higher increments in performance. However, without statistical difference from the baseline.

Strategy	Mean (Std)
Cluster+WU	0.828 (0.031)
Typical+WU	**0.832 (0.027)**
Random+Uniform	0.829 (0.024)
Baseline+Uniform	0.827 (0.036)

level of $\alpha = 0.05$ and p-value $= 0.4986$. The regression analysis, visualized with red lines and green confidence intervals in Fig. 5, shows that the performance of the Typical strategy ($intercept = -20.25$ and $\beta = 37.56$) increases faster than the convergence of the baseline ($intercept = -20.62$ and $\beta = 38.31$).

Now, to examine the relationship between the number of images used during evolution and the generalization ability of the models across initialization strategies, we look at the regression analysis shown by red lines and green confidence intervals in Fig. 6. For the baseline method, the regression coefficients ($intercept = 10.09$ and $\beta = -11.55$) indicates that generalization decreases, suggesting that the baseline may be overfitting. In contrast, the Typical strategy ($intercept = 10.71$ and $\beta = 4.25$) displays the opposite relationship between generalization and the number of images used. This implies that, for these methods, higher generalization is associated with exposure to a broader and more diverse subset of images. The stronger effect observed in the Random strategy suggests that stochastic sampling fosters more varied training dynamics, potentially improving robustness at the cost of slower convergence. Overall, these results highlight a clear distinction between strategies that concentrate on fewer, more specific images (baseline) and those that maintain or expand data diversity during training (Typical). The positive trends in the latter group align with improved generalization performance observed elsewhere in the analysis, reinforcing the importance of maintaining sampling diversity throughout evolution.

Finally, we compared the number of active nodes across different initialization strategies. The results indicate that the baseline, mean (std): 10.133 (2.507), and the Typical strategy, 11.033 (2.058), generate programs of similar size, suggesting that initialization choice does not substantially affect program complexity in this context. Now looking at the execution time, the baseline line, 0.047 (0.021), is clearly feast in comparison to the Typical strategy, 0.077 and (0.022).

## 6	Discussion

We investigated evolving biomedical image segmentation programs using Cartesian Genetic Programming (CGP) enhanced with Active Learning (AL). By dynamically selecting the most informative images, growing training datasets improved both performance and convergence speed of the algorithm.

A key contribution is the filtering stage prior to uncertainty evaluation. Using PCA to select a diverse subset of candidate images reduced computational cost while preserving data coverage, enabling more efficient CGP evolution. Similarly, initial training data selection significantly influenced early convergence. PCA-based and cluster-based strategies provided balanced coverage and diversity, improving early learning dynamics compared to random initialization.

Building on our previous observation that training datasets can be reduced without significant performance loss, we further show that the number of points used for uncertainty estimation can also be significantly reduced. By precomputing a PCA projection before evolution, we are able to reduce the number of points from 89 to 20 (about 20% of the dataset) while maintaining comparable performance. This reduction in data points substantially lowers computational overhead, enabling more efficient evaluation of uncertainty without sacrificing the diversity or informativeness of the sampled data.

Finally, this work advances interactive learning in biomedical image analysis by combining CGP with Active Learning. The system can guide human experts to annotate the most informative images or regions based on uncertainty, focusing on rare cell types, shapes, or colors. Further work should include uncertainty-guided augmentation, and co-evolution of datasets and models promise to further improve efficiency, interpretability, and adaptability in real-world biomedical imaging tasks.

Acknowledgments. This project used compute resources from the CALMIP project P21049.

References

1. Ahmad, A.M., Khan, G.M., Mahmud, S.A., Miller, J.F.: Breast cancer detection using cartesian genetic programming evolved artificial neural networks. In: Proceedings of the 14th Annual Conference on Genetic and Evolutionary Computation, GECCO '12, pp. 1031–1038. Association for Computing Machinery, New York (2012). https://doi.org/10.1145/2330163.2330307
2. Ain, Q.U., Al-Sahaf, H., Xue, B., Zhang, M.: Genetic programming for automatic skin cancer image classification. Expert Syst. Appl. **197**, 116680 (2022)
3. Ain, Q.U., Al-Sahaf, H., Xue, B., Zhang, M.: Automatically diagnosing skin cancers from multimodality images using two-stage genetic programming. IEEE Trans. Cybern. **53**(5), 2727–2740 (2023). https://doi.org/10.1109/TCYB.2022.3182474
4. Ain, Q.U., Xue, B., Al-Sahaf, H., Zhang, M.: Genetic programming for multiple feature construction in skin cancer image classification. In: 2019 International Conference on Image and Vision Computing New Zealand (IVCNZ), pp. 1–6. Institute of Electrical and Electronics Engineers (IEEE), Piscataway (2019). https://doi.org/10.1109/IVCNZ48456.2019.8961001
5. Alkhaldi, E., Salari, E.: Ensemble optimization for invasive ductal carcinoma (IDC) classification using differential cartesian genetic programming. IEEE Access **10**, 128790–128799 (2022). https://doi.org/10.1109/ACCESS.2022.3228176
6. Ben Hamida, S., Hmida, H., Borgi, A., Rukoz, M.: Adaptive sampling for active learning with genetic programming. Cogn. Syst. Res. **65**, 23–39 (2021)
7. Beucher, S., Meyer, F.: The morphological approach to segmentation: the watershed transformation. In: Mathematical Morphology in Image Processing, pp. 433–481. CRC Press (2018)
8. Bi, Y., Xue, B., Zhang, M.: Genetic programming-based discriminative feature learning for low-quality image classification. IEEE Trans. Cybern. **52**(8), 8272–8285 (2022). https://doi.org/10.1109/TCYB.2021.3049778
9. Biau, J., Wilson, D., Cussat-Blanc, S., Luga, H.: Improving image filters with cartesian genetic programming. In: IJCCI, pp. 17–27 (2021)
10. Bradski, G.: The OpenCV library. Dr. Dobb's J. Softw. Tools (2000)
11. Casanova, A., Pinheiro, P.O., Rostamzadeh, N., Pal, C.J.: Reinforced active learning for image segmentation. In: International Conference on Learning Representations (2020). https://openreview.net/forum?id=SkgC6TNFvr
12. Cohn, D.A., Ghahramani, Z., Jordan, M.I.: Active learning with statistical models. J. Artif. Intell. Res. **4**, 129–145 (1996). https://doi.org/10.1613/jair.295
13. Cortacero, K., et al.: Evolutionary design of explainable algorithms for biomedical image segmentation. Nat. Commun. **14**(1), 7112 (2023)
14. Csiba, D., Richtárik, P.: Importance sampling for minibatches. ArXiv arxiv:1602.02283 (2016). https://api.semanticscholar.org/CorpusID:15471954
15. De La Torre, C., et al.: Evolved and transparent pipelines for biomedical image classification. In: Xue, B., Manzoni, L., Bakurov, I. (eds.) Genetic Programming, pp. 173–189. Springer, Cham (2025)
16. Deng, S., Zhang, X., Yan, W., Chang, E.I.C., Fan, Y., Lai, M., Xu, Y.: Deep learning in digital pathology image analysis: a survey. Front. Med. **14**, 470–487 (2020)
17. Esteva, A., et al.: Deep learning-enabled medical computer vision. NPJ Dig. Med. **4**(1), 5 (2021)
18. Gaillochet, M., Desrosiers, C., Lombaert, H.: Active learning for medical image segmentation with stochastic batches. Med. Image Anal. **90**, 102958 (2023)

19. Gathercole, C., Ross, P.: Dynamic training subset selection for supervised learning in Genetic Programming. In: Davidor, Y., Schwefel, H.P., Männer, R. (eds.) Parallel Problem Solving from Nature – PPSN III, pp. 312–321. Springer, Heidelberg (1994)
20. Greenacre, M., Groenen, P.J., Hastie, T., d'Enza, A.I., Markos, A., Tuzhilina, E.: Principal component analysis. Nat. Rev. Methods Primers **2**(1), 100 (2022)
21. Harding, S., Leitner, J., Schmidhuber, J.: Cartesian Genetic Programming for Image Processing, pp. 31–44. Springer, New York (2013). https://doi.org/10.1007/978-1-4614-6846-2_3
22. Haut, N., Banzhaf, W., Punch, B.: Active learning improves performance on symbolic regression tasks in StackGP. In: Proceedings of the Genetic and Evolutionary Computation Conference Companion, pp. 550–553. GECCO '22, Association for Computing Machinery, New York (2022). https://doi.org/10.1145/3520304.3528941
23. Haut, N., Punch, B., Banzhaf, W.: Active learning informs symbolic regression model development in genetic programming. In: Proceedings of the Companion Conference on Genetic and Evolutionary Computation, GECCO '23 Companion, pp. 587–590. Association for Computing Machinery, New York (2023). https://doi.org/10.1145/3583133.3590577
24. Hmida, H., Hamida, S.B., Borgi, A., Rukoz, M.: Sampling methods in genetic programming learners from large datasets: a comparative study. In: Angelov, P., Manolopoulos, Y., Iliadis, L., Roy, A., Vellasco, M. (eds.) Advances in Big Data, pp. 50–60. Springer, Cham (2017)
25. Hüllermeier, E.: How to measure uncertainty in uncertainty sampling for active learning. Mach. Learn. **111**, 89–122 (2021). https://api.semanticscholar.org/CorpusID:221726816
26. Huml, M., Silye, R., Zauner, G., Hutterer, S., Schilcher, K., et al.: Brain tumor classification using AFM in combination with data mining techniques. BioMed Res. Int. **2013** (2013)
27. Khan, A., Qureshi, A.S., Wahab, N., Hussain, M., Hamza, M.Y.: A recent survey on the applications of genetic programming in image processing. Comput. Intell. **37**(4), 1745–1778 (2021)
28. Kirsch, A., van Amersfoort, J., Gal, Y.: BatchBALD: efficient and diverse batch acquisition for deep bayesian active learning. In: Wallach, H., Larochelle, H., Beygelzimer, A., d' Alché-Buc, F., Fox, E., Garnett, R. (eds.) Advances in Neural Information Processing Systems, vol. 32. Curran Associates, Inc. (2019). https://proceedings.neurips.cc/paper_files/paper/2019/file/95323660ed2124450caaac2c46b5ed90-Paper.pdf
29. Lavinas, Y., Cortacero, K., Cussat-Blanc, S.: Evolving graphs with cartesian genetic programming with lexicase selection. In: Proceedings of the Companion Conference on Genetic and Evolutionary Computation, GECCO '23 Companion, pp. 1920–1924. Association for Computing Machinery, New York (2023). https://doi.org/10.1145/3583133.3596402
30. Lavinas, Y., Haut, N., Punch, W., Banzhaf, W., Cussat-Blanc, S.: Adaptive sampling of biomedical images with cartesian genetic programming. In: Affenzeller, M., et al. (eds.) Parallel Problem Solving from Nature - PPSN XVIII, pp. 256–272. Springer, Cham (2024)
31. Lavinas, Y., Haut, N., Punch, W., Banzhaf, W., Cussat-Blanc, S.: Data sampling via active learning in cartesian genetic programming for biomedical data. In: 2024 IEEE Congress on Evolutionary Computation (CEC) 2024

32. Lavinas, Y., Haut, N., Punch, W., Banzhaf, W., Cussat-Blanc, S.: Dynamically sampling biomedical images for genetic programming. In: Proceedings of the Genetic and Evolutionary Computation Conference Companion. GECCO '24. Association for Computing Machinery, New York (2024). https://doi.org/10.1145/3638530.3654202
33. Lloyd, S.: Least squares quantization in pcm. IEEE Trans. Inf. Theory **28**(2), 129–137 (1982). https://doi.org/10.1109/TIT.1982.1056489
34. Miller, J.F.: An empirical study of the efficiency of learning boolean functions using a cartesian genetic programming approach. In: Proceedings of the 1st Annual Conference on Genetic and Evolutionary Computation. GECCO'99, , vol. 2, pp. 1135–1142. Morgan Kaufmann Publishers Inc., San Francisco (1999)
35. Nath, V., Yang, D., Landman, B.A., Xu, D., Roth, H.R.: Diminishing uncertainty within the training pool: active learning for medical image segmentation. IEEE Trans. Med. Imaging **40**(10), 2534–2547 (2021). https://doi.org/10.1109/TMI.2020.3048055
36. Rudin, C.: Stop explaining black box machine learning models for high stakes decisions and use interpretable models instead. Nat. Mach. Intell. **1**(5), 206–215 (2019)
37. Settles, B.: Active learning literature survey. Computer Sciences Technical Report 1648, University of Wisconsin–Madison (2009)
38. Stringer, C., Wang, T., Michaelos, M., Pachitariu, M.: Cellpose: a generalist algorithm for cellular segmentation. Nat. Methods **18**(1), 100–106 (2021)
39. Suganuma, M., Shirakawa, S., Nagao, T.: Designing convolutional neural network architectures using cartesian genetic programming, pp. 185–208. Springer, Singapore (2020). https://doi.org/10.1007/978-981-15-3685-4_7
40. Tokdar, S.T., Kass, R.E.: Importance sampling: a review. Wiley Interdisc. Rev. Comput. Stat. **2** (2010). https://api.semanticscholar.org/CorpusID:14552197
41. Vanneschi, L., Farinaccio, A., Giacobini, M., Mauri, G., Antoniotti, M., Provero, P.: Identification of individualized feature combinations for survival prediction in breast cancer: a comparison of machine learning techniques. In: Pizzuti, C., Ritchie, M.D., Giacobini, M. (eds.) Evolutionary Computation, Machine Learning and Data Mining in Bioinformatics, pp. 110–121. Springer, Heidelberg (2010)
42. Vladislavleva, E., Smits, G., den Hertog, D.: On the importance of data balancing for symbolic regression. IEEE Trans. Evol. Comput. **14**(2), 252–277 (2010). https://doi.org/10.1109/TEVC.2009.2029697
43. van der Walt, S., et al.: The scikit-image contributors: scikit-image: image processing in Python. PeerJ **2**, e453 (2014). https://doi.org/10.7717/peerj.453
44. Wilson, D.G., Cussat-Blanc, S., Luga, H., Miller, J.F.: Evolving simple programs for playing atari games. In: Proceedings of the Genetic and Evolutionary Computation Conference, GECCO '18, pp. 229–236. Association for Computing Machinery, New York (2018). https://doi.org/10.1145/3205455.3205578
45. Xia, T., Cosgrove, J., Alty, J., Jamieson, S., Smith, S.: Application of classification for figure copying test in Parkinson's disease diagnosis by using cartesian genetic programming. In: Proceedings of the Genetic and Evolutionary Computation Conference Companion, GECCO '19, pp. 1855–1863. Association for Computing Machinery, New York (2019). https://doi.org/10.1145/3319619.3326822

Using Monte Carlo Tree Search to Enhance Search Space Exploration in Cartesian Genetic Programming

Christina Berghegger[1,2]([✉]) [iD], Camilo De La Torre[1,2] [iD],
Sylvain Cussat-Blanc[1,2,3] [iD], Yuri Lavinas[1,2] [iD], and David Simoncini[1,2] [iD]

[1] University Toulouse Capitole, Toulouse, France
[2] IRIT - CNRS UMR5505, Toulouse, France
christina.berghegger@ut-capitole.fr
[3] Institut Universitaire de France, Paris, France

Abstract. With the need for enhanced interpretability, Cartesian Genetic Programming has proven to be a promising alternative to widely used black-box approaches in fields such as image processing or program synthesis. Yet, the approach often fails to reach state of the art performance due to fitness landscape characteristics which make it difficult to efficiently explore the search space. We extend Multimodal Adaptive Graph Evolution, a multi-type variant of Cartesian Genetic Programming, with Monte Carlo Tree Search, a search heuristic known for its capability to overcome these difficulties. This approach outperforms the classic $1+\lambda$ Evolutionary Algorithm, and shows a significant increase in performance, particularly for problems where $1+\lambda$ guided search tends to get trapped in local optima or neutral fitness plateaus. We show how search can be guided to explore deceptive search paths and reach optimal solutions by traversing low-fitness search regions. Our contribution illustrates the need for problem- and fitness landscape-specific search strategies in Cartesian Genetic Programming with the aim to efficiently explore the search space.

Keywords: Genetic Programming · Graph-based Genetic Programming · Cartesian Genetic Programming · Monte Carlo Tree Search · Search Space Exploration · Program Synthesis

1 Introduction

The inherent interpretability of Genetic Programming (GP) makes it a particularly appealing approach for solving tasks in domains where understanding the decision making process is crucial both for advancing fundamental knowledge, and for ensuring trust and fairness. Cartesian Genetic Programming (CGP) [11] is a graph-based variant of GP where algorithmic building blocks, such as functions or mathematical operators, are assembled into a directed acyclic graph. The graph is evolved through genetic operators, typically mutation, with the goal of

L. Manzoni et al. (Eds.): EuroGP 2026, LNCS 16521, pp. 292–304, 2026.
https://doi.org/10.1007/978-3-032-23005-8_18

improving the fitness of candidate solutions to a given problem. CGP has many attributes that enhance interpretability, such as fixed-length representation of genomes and the notion of active graph: only the subset of the nodes connected to the output contribute to the final solution. While these characteristics make CGP an interesting approach in various domains, such as biomedical imaging [2], game playing [15] or the evolution of control policies for visual tasks [6], the difficulty to evolve solutions often results in inferior performance compared to the state of the art. A variety of work on how to improve evolution has been proposed, ranging from extensive studies on the crossover operator in CGP [9] to modified mutation operators aiming at a more efficient exploration of the search space by influencing the mutation probability and favoring inactive nodes [3]. We reason that the problem of evolvability is likely dependent on certain characteristics of the fitness landscape, which represents how solution quality varies across the search space. Ruggedness, neutrality and deceptiveness are such emergent characteristics determined by the topology of the fitness landscape. Ruggedness is influenced by the number and distribution of local optima. Neutrality refers to the size of networks of connected solutions with equal fitness. Deceptiveness occurs when paths that appear to improve fitness misguide the search, leading it toward regions without the global optimum.

By the example of Multimodal Adaptive Graph Evolution (MAGE), we want to explore how to increase evolvability under specific fitness landscape characteristics. MAGE was proposed as a variant of CGP that can handle multiple data types [5], making it more suitable for real-world applications than the classic implementation of CGP, able to handle a single data type only. In a comprehensive comparison on the program synthesis benchmark *PSB2* [8], MAGE shows higher performance compared to other multi-type approaches in terms of the number of problems that were solved, and the consistency with which solutions were found. Still, the majority of the problems remains unsolved. MAGE, in its current implementation, employs a $1+\lambda$ Evolutionary Algorithm (EA), which is widely adopted as the standard evolutionary process for CGP. At each iteration, the current elite parent generates λ offspring, and the best-performing individual, including the parent, is selected as the new elite. This strategy, however, is not designed to efficiently explore the search space or escape local minima and may perform poorly on neutral, rugged and deceptive fitness landscapes.

We hypothesized that replacing the $1+\lambda$ EA with a search strategy that can better explore the search space and escape local minima would improve the performance of MAGE on *PSB2* instances where it currently struggles to find good solutions. To test our hypothesis, we implemented a Monte Carlo Tree Search (MCTS) into MAGE as an alternative search strategy to the $1+\lambda$ EA. Our results confirm that MAGE's poor performance on some instances is indeed related to characteristics of the fitness landscape, and that performance improves when search space exploration is enhanced with MCTS. In previous work, MCTS-inspired mutation operators to expand programs have been proposed with the goal of overcoming deceptiveness, tested on a set of benchmark symbolic regression problems [10]. To the best of our knowledge, MCTS has not yet been applied for exploring search spaces moving between complete solutions, corresponding to functioning graphs in the field of Graph-based Genetic Programming. In this contribution, we propose an MCTS-inspired Evolutionary Algorithm as an alternative to the classic *1+λ* EA that is able to move in and explore the search space more efficiently when characteristics like ruggedness, neutrality, and deceptiveness, exist.

2   Monte Carlo Tree Search

Monte Carlo Tree Search (MCTS) [14] is a probabilistic search heuristic that guides the exploration of the search space by measuring the expected performance from traversing a node (a location in the search space), which is achieved by simulating future states that can be reached from it. MCTS is composed of four main stages: selection, expansion, simulation and backpropagation. Each search iteration begins at the root and traverses the tree using the *Upper Confidence Bounds Applied To Trees* (UCT) criterion. At each node, the child that maximizes the UCT score is selected as the next node, calculated by its visit count since the start of the search, and the mean reward over all iterations that traversed this node. The selection is defined as:

$$Q_i = \frac{V_i}{n_i} \tag{1}$$

$$U_i = Q_i + c \cdot \sqrt{\frac{\ln(N)}{n_i}} \tag{2}$$

where Q_i is the average reward of child node i, V_i is the total accumulated reward (value) of node i, n_i is the number of visits to node i, N is the number of visits to the parent node, and c is the exploration constant. The exploration constant balances nodes that yield good average fitness and nodes that were visited less often. Once a terminal node is reached during the UCT selection process, this node is expanded by creating child nodes. A child node is then simulated by sampling subsequent moves until a final state is reached given a stop criterion. The final backpropagation process updates statistics of all child nodes that have been traversed.

Compared to *1+λ* EA, MCTS offers two main advantages to address the challenge of efficiently traversing the search space in GP. First, it can escape local optima and leave plateaus due to UCT controlling the exploration of the search space. Second, MCTS can exploit deceptiveness by accepting mutations with inferior fitness that can lead to high performing solutions in the long term.

3   MAGE-MCTS Integration

In MAGE-MCTS, the root of the search tree corresponds to the initial genome of a functional graph. Each subsequent node is a mutated genome of its parent node. At each iteration, the tree is traversed calculating the UCT score from the nodes' visit count and average fitness. The simulation of leaf nodes is done by progressively executing a defined number of random mutations to determine which is the highest performing solution reachable in the near future. Finally, the backtracking phase updates statistics of nodes in terms of their visit count and their fitness. The integration is visualized in Fig. 1 and the code is available online[1]

In this work, we craft a binary search tree. A leaf node is expanded by creating one new solution through mutation, and by passing over the genome of the leaf node itself. With this setup we ensure the best current solution is maintained, while forcing search to explore mutations independently of their instant fitness.

3.1   Comparison to the State of the Art

MAGE was evaluated and compared against PushGP [12,13] and Mixed-Type Cartesian Genetic Programming (MT-CGP) [7] on the Second Program Synthesis Benchmark (*PSB2*) in previous work [5]. The benchmark is composed of 25 program synthesis problems from various sources, such as coding katas or college courses. 25 independent runs were performed for each problem both with MAGE and MT-CGP. MAGE outperforms MT-CGP both in terms of the number of problems with at least one complete solution (zero error on train and test data), as well as the consistency in finding complete solutions (percentage of complete solutions out of all runs). MAGE and MT-CGP solved 9 and 5 out of 25 problems respectively, with MAGE having a higher consistency in all problems that were solved with both approaches. Compared to the results obtained with PushGP, published together with the *PSB2* benchmark suite by the authors, MAGE solves less problems but obtains higher consistency in five out of the seven problems where both methods found solutions.

We applied the MAGE-MCTS implementation on all *PSB2* problems to compare the results with the classic MAGE implementation adopting a *1+λ* EA. With this comparison, we determine if MAGE can benefit from extended exploration capabilities resulting in a higher number of complete solutions. Furthermore, we examine the search behavior of independent runs to check the ability of MAGE-MCTS to navigate deceptive landscapes.

3.2   Hyperparameter Tuning

For the experiment, we performed hyperparameter tuning for MCTS-specific hyperparameters: the exploration constant for UCT-selection and the rollout depth during the simulation phase. All other parameters not specific to MCTS, such as the length of the genome and mutation rate, remain the same as for the previous evaluation of MAGE with a *1+λ* EA. Parameters were tuned with the hyperparameter optimization framework Optuna [1] for a total of 1500 trials (combination of hyperparameters). Trials were evaluated on three problems of the *PSB2* suite: *Indices of Substring*, *Fuel Cost*, and *FizzBuzz*.

---

[1] https://github.com/chrissi-b/mage-mcts-binary.

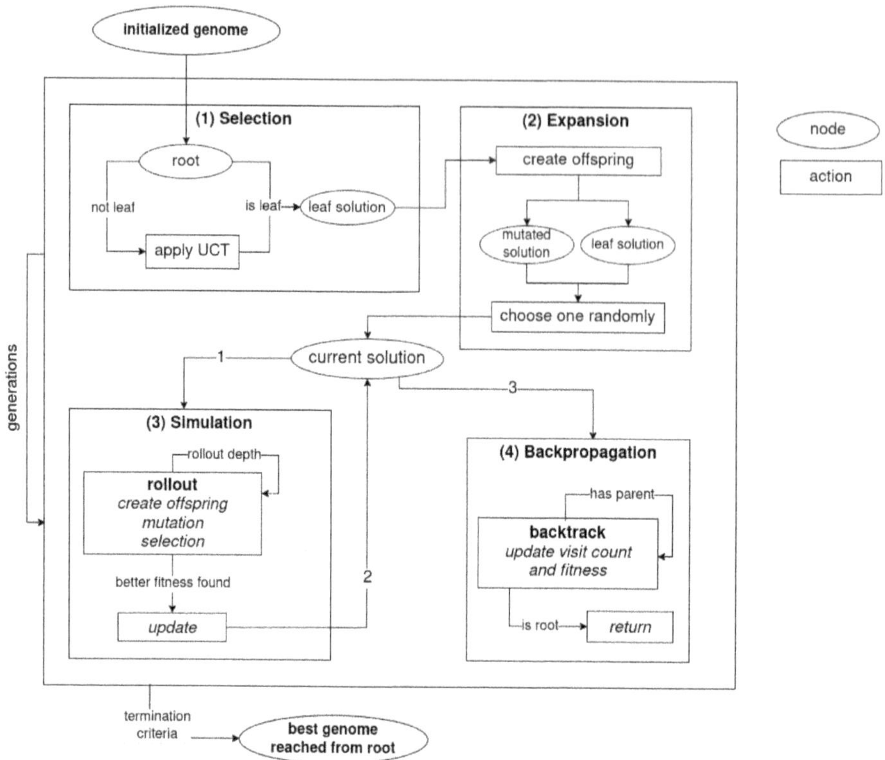

Fig. 1. Visualization of the integration of MCTS and MAGE. An initialized genome, corresponding to a functioning graph, serves as root for MCTS, and is optimized over a defined number of generations undergoing the 4 stages of MCTS: After selecting a leaf node with UCT, it is expanded with a mutated genome and its own state. One of the two is picked for simulation (arrow 1). The node is updated if a better genomes if found during rollout (arrow 2). After the rollout is finished (arrow 3), backpropagation starts at the selected leaf node and recursively updates statistics of the parent node until reaching the root. Ellipses correspond to nodes, rectangles correspond to actions.

These problems were selected because they show big differences in how difficult and long it was for MAGE to find complete solutions. *Indices of Substring* can be classified as a rather easy problem, with 80% of all runs resulting in a complete solution after on average 9,666 evaluations. Complete solutions for *FizzBuzz* were found in 64% of all runs at a relatively late stage, after on average 110,570 evaluations. Finally, *Fuel Cost* was completely solved rather early, after on average 32,160 evaluations, but in only 20% of all runs. To account for the stochastic nature of MAGE, each problem was evaluated over three independent runs with different starting genomes. The mean over each problem was calculated and returned as trial result, and the Optuna study was configured as a multi-objective optimization minimizing each

Table 1. Hyperparameter tuning for MCTS with a random offspring and rollout strategy. For each trial, each hyperparameter is suggested by Optuna from a specified range. Ranges include start and end values.

Hyperparameter	Type	Range	Steps
Exploration constant	Int	(2, 10)	1
Rollout depth	Int	(5, 250)	5

of the three problems. The possible value ranges and corresponding steps for both hyperparameters are visualized in Table 1. Each problem run had a budget of 100,000 evaluations, corresponding to half of the evaluations of the final comparison, and a maximum wall time of 2 h, whichever was reached first.

After tuning, the Pareto front was composed of three trials obtaining the best results over all problems (Table 2), with one hyperparameter configuration being able to solve two out of the three problems with zero error. It reached error values (the lower the better) of 0.0, 0.0 and 0.60 for *Indices of Substring*, *Fuel Cost* and *FizzBuzz*, respectively. Given this higher consistency in solving two out of three problems, we use this configuration in our following experiments and the values used are: 2 for the exploration constant and 100 for the rollout depth.

Table 2. Pareto front after hyperparameter tuning. Parameters are provided as a tuple (*exploration constant, rollout depth*). The third trial was used for the final experiment as it was able to solve two out of three problems with 0 error.

Parameters	Indices of Substring	Fuel Cost	FizzBuzz
(2, 125)	0.0	5.42	0.36
(2, 150)	0.0	0.26	0.52
(2, 100)	0.0	0.0	0.60

4   Results

With the tuned hyperparameters, we completed 25 independent runs for each of the 25 problems contained in the *PSB2* benchmark suite. The number of evaluations was increased to 200,000 as this limit was also set for the previous evaluation of MAGE [5]. Table 3 shows the mean (std) values of the final error on the test data, the percentage of complete solutions and the mean number of evaluations required to find such complete solution for MAGE with a $1+\lambda$ EA and with a MCTS integration. In this context, a complete solution has zero error on the train and test test.

Table 3. Final results for MAGE with a *1+λ* EA and with a MCTS integration. The final fitness is computed on the test set and averaged over 25 runs, standard deviation is provided in parentheses. Bold results indicate superior fitness, the asterisk indicates statistical difference ($p < 0.05$). *Complete* denotes the percentage of runs with zero error on the train and test set. *Evals* indicates the mean number of evaluations to find the complete solution.

Problem	$1+\lambda$ EA			MAGE-MCTS		
	final fitness	complete	evals	final fitness	complete	evals
Basement	0.16 (0.63)	80 %	21650	**0.01 (0.04)**	80 %	35245
Bouncing Balls	57.02 (16.27)	0 %	-	**50.85 (14.88)**	0 %	-
Bowling	14.6 (7.57)	0 %	-	**11.57 (1.29)**	0 %	-
Camel Case	**0.0 (0.0)**	100 %	5300	**0.0 (0.0)**	100 %	2065
Coin Sums	4.64 (11.55)	44 %	122550	**1.08 (0.32)***	0 %	-
Cut Vector	3764.33 (324.01)	0 %	-	**3719.59 (149.47)**	0 %	-
Dice Game	0.09 (0.06)	0 %	-	**0.04 (0.01)***	0 %	-
Find Pair	**3337.89 (155.78)**	0 %	-	3348.26 (169.77)	0 %	-
FizzBuzz	**0.1 (0.13)***	64 %	110570	0.39 (0.32)	12 %	124331
Fuel Cost	2639.57 (2680.36)	20 %	32160	**1.20 (3.66)***	76 %	42139
GCD	**16.04 (43.09)***	16 %	79720	23.94 (106.79)	44 %	79304
Indices of Sub.	0.28 (0.79)	88 %	9660	**0.0 (0.0)**	100 %	4110
Leaders	507.57 (549.89)	40 %	18890	**<0.001 (0.0)***	96 %	33496
Luhn	5.58 (0.32)	0 %	-	**5.47 (0.20)***	0 %	-
Mastermind	1.21 (0.1)	0 %	-	**1.19 (0.03)**	0 %	-
Middle Char.	1.31 (0.31)	0 %	-	**1.14 (0.39)***	0 %	-
Paired Digits	15.68 (1.05)	0 %	-	**14.92 (0.54)***	0 %	-
Shopping List	26.24 (5.4)	0 %	-	**20.21 (1.31)***	0 %	-
Snow Day	6.19 (3.04)	0 %	-	**4.56 (1.39)***	0 %	-
Solve Boolean	**0.1 (0.05)***	4 %	150860	0.15 (0.04)	0 %	-
Spin Words	3.82 (1.36)	0 %	-	**2.62 (0.79)***	0 %	-
Square Digits	**5.69 (0.7)***	0 %	-	6.30 (0.18)	0 %	-
Subst. Cipher	**10.59 (0.3)**	0 %	-	10.70 (0.27)	0 %	-
Twitter	**3.23 (1.95)**	0 %	-	4.09 (1.09)	0 %	-
Vector Dist.	47.53 (14.95)	0 %	-	**28.49 (5.60)***	0 %	-

It is clear that using MAGE-MCTS leads to better performance (the lower the better) in many of the problems. We can observe that the average final test fitness is superior in 17 out of 25 problems with MAGE-MCTS. In 11 out of the 17 problems, the more exploratory approach MCTS leads MAGE to achieve performance values that are statistically different, determined with a Mann-Whitney U test and a p-value threshold of 0.05. On the other hand, for 4 out of the 7 problems where the *1+λ* EA, focusing on exploitation, performs better, values are statistically different. The *Camel Case* problem was equally and completely solved with both approaches.

The two easiest problems for *1+λ* EA, *Camel Case* and *Indices of Substring*, were solved with less than half of the evaluations with MAGE-MCTS compared to *1+λ* EA. Solutions for problems where the number of complete solutions significantly increased with MAGE-MCTS (*GCD*, *Fuel Cost* and *Leaders*) were generally found later, which can be explained by the exploratory nature of MCTS and the additional budget used to explore different paths of the search space.

With MCTS-guided search, the consistency in moving towards an optimal solution increases, as Table 3 shows when we look at the number of complete solutions. The problems *GCD*, *Fuel Cost*, *Leaders* and *Indices of Substring* highly benefit from exploration, with percentages of complete runs increasing from 16% to 44%, from 20% to 76%, from 40% to 96%, and from 88% to 100%, respectively. Moreover, the standard deviation of the final test fitness in smaller in a majority of cases. For other problems, such as *FizzBuzz*, *Coin Sums*, and *Solve Boolean*, a less exploratory approach seems to be more beneficial to move towards an optimal solution.

For two problems, *Coin Sums* and *GCD*, the mean fitness and the number of complete solutions are in disagreement. The final fitness with MAGE-MCTS is significantly lower for the *Coin Sums* problem, and yet MAGE-MCTS could not solve this problem in any of the runs, while MAGE with a *1+λ* EA found solutions in 44% of all runs. The same holds true for *GCD*, where MAGE-MCTS found more complete solutions, but achieved a worse final fitness and a higher standard deviation. After analyzing these results, we verify that one of the runs of MAGE-MCTS is performing much differently than the others and could benefit from more running time. Overall, we understand that this indicates that MAGE-MCTS would benefit from a larger evaluation budget.

4.1   Search Tree Example

The MCTS search tree created during a run appearing in the experiments and solving the *Fuel Cost* problem is visualized in Fig. 2. Starting from the root (upper purple node), two paths are presented: the yellow nodes depict the path always accepting the child node with better fitness (the lower the better), imitating a classic *1+λ* EA. The green and red nodes represent the path leading to the optimal solution (lower purple node), with green nodes depicting child nodes with higher fitness, and red nodes depicting child nodes with lower fitness compared to their neighboring offspring.

MAGE-MCTS had to accept and explore two mutations with inferior fitness to reach the global optimum. This is an example of deceptiveness that MAGE-MCTS was able to overcome with its exploration capabilities. Exploring search regions that, at first, seemed unattractive, allowed this MAGE variant to find the optimum faster. That said, it cannot be ruled out that following the yellow path would have led to an optimal solution at some point.

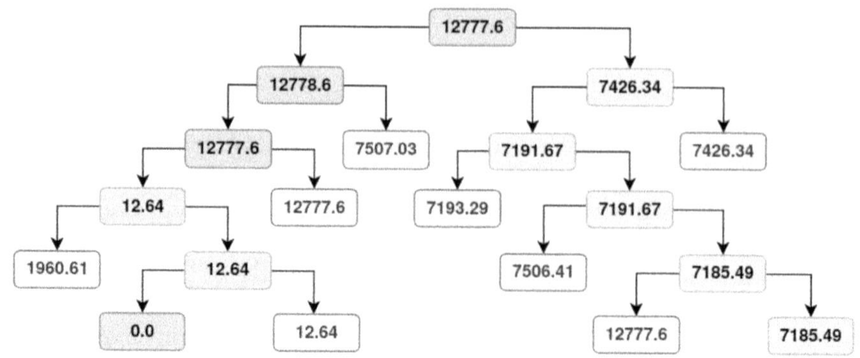

Fig. 2. Example of a search tree generated through MCTS-guided search. Node values correspond the the fitness of the node's genome. The yellow path is built by always choosing the subsequent child node with better fitness. The green and red nodes together build the path to the optimal solution, with green nodes showing child nodes with superior fitness, and red nodes showing child nodes with inferior fitness. Red nodes represent deceptive mutations, where following an initially inferior fitness eventually led to an optimal solution.

4.2   Impact of the Fitness Landscape

Understanding which search heuristics are best capable of traversing certain fitness landscapes is crucial to increase performance. In Fig. 3, we illustrate repeated visits of genomes during the evolutionary process from a $1+\lambda$ EA search instance both on the *Fuel Cost* and *FizzBuzz* problems. Figure 4 shows the corresponding diversity ratios expressed as number of individual genomes out of all genomes visited, with a higher ratio indicating a stronger exploration throughout evolution. Data was collected until visiting 10,000 distinct genomes, reaching a complete solution or having performed 10,000 duplicate visits, whichever was reached first. The best fitness obtained during evolution is visualized in red, the number of duplicate visits (Fig. 3) and diversity ratio (Fig. 4) at the moment of visiting a new genome is visualized in blue. The two problems were picked as examples of two highly different search behaviors with a $1+\lambda$ EA, that can help understand how different problems can differently benefit from an alternative search strategy. Two single instances of each problem are presented below to showcase how analyzing $1+\lambda$ search can provide information on fitness landscape characteristics.

Looking at Fig. 3athat depicts a *Fuel Cost* instance, we can observe that the fitness, shown in red, was stable in the beginning of the evolutionary process. The fact that the number of duplicate genomes remains almost flat and close to zero, as shown by the blue line, indicates a neutral fitness landscape. After exiting the first long plateau and having visited around 250 unique genomes, the number of duplicate genomes increases rapidly to 10,000, while only 200 new unique genomes are explored. The exploration capacity decreased drastically

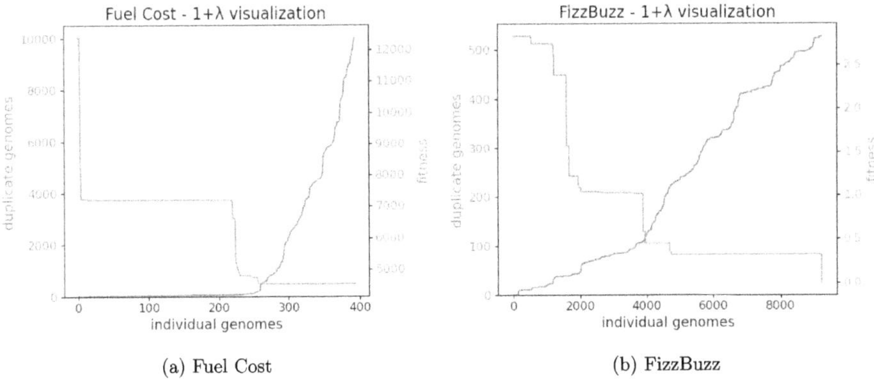

(a) Fuel Cost

(b) FizzBuzz

Fig. 3. Search behavior of a *1+λ* EA search instance for the *Fuel Cost* and *FizzBuzz* problem. The horizontal axis counts the number of unique genomes visited during evolution. The blue line shows how many duplicate visits of any genome it took to reach a number of unique genomes. The red line represents the best fitness of all visited genomes.

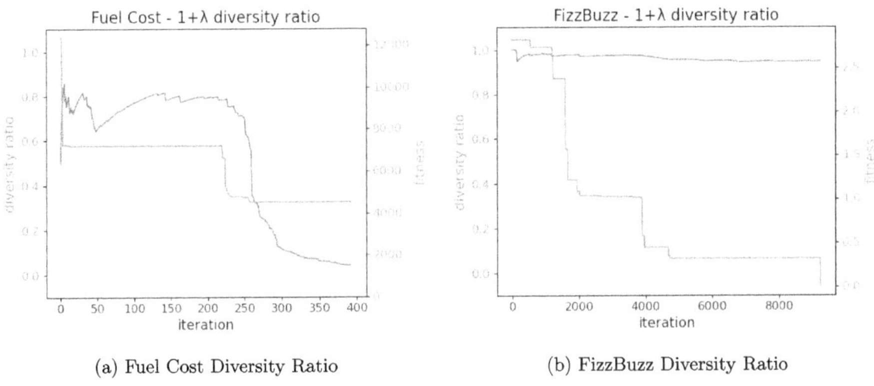

(a) Fuel Cost Diversity Ratio

(b) FizzBuzz Diversity Ratio

Fig. 4. Diversity ratios, shown by the blue line, expressed as number of individual genomes out of all genomes visited during evolution. The horizontal axis shows the number of iterations. The red line represents the best fitness of all visited genomes. (Color figure online)

and the search was redirected to already known neighborhoods, as visualized by the corresponding diversity ratio in Fig. 4a.

For the instance of the *FizzBuzz* problem, visualized in Figs. 3b , we can see that the fitness, shown in red, decreases gradually while the search efficiently moves through unexplored neighborhoods, until finding a complete solution after visiting 9260 distinct genomes. The fitness values are stable during some parts of the search, but we can see that MAGE-MCTS keeps exploring new genomes, as shown by the flat diversity ratio remaining at almost 1 during the entire search, represented by the blue line in Fig. 4b.

In the case of *Fuel Cost*, MAGE-MCTS can guide the search to traverse neutral plateaus faster and escape local optima, while the $1+\lambda$ EA failed to do so. In the case of *FizzBuzz*, the $1+\lambda$ EA allowed for an already efficient search with a constant exploration of new solutions. In this case, MCTS might stop the exploitation of the current search path too early by giving preference to the exploration of unknown neighborhoods, leading to inferior performance.

The numbers shown in Table 4 corroborate the findings presented above: We can see the mean ratio of deceptive mutations to find the best genome averaged over 25 independent runs, which for *FizzBuzz* is much less than for *Fuel Cost*. For *Fuel Cost*, accepting and exploring mutations with inferior fitness was thus more necessary to reach a better final performance. Combined with the wide min – max range, this suggests that *Fuel Cost* exhibits a more rugged and variable landscape, where progress can depend heavily on the starting genome. In contrast, the low deceptiveness ratio for *FizzBuzz* suggests a smoother landscape. Overall, this suggests that we need to design or select problem-specific search heuristics. In the work of De La Torre et al. [5], the authors indicated with the help of Search Trajectory Networks [4], that MAGE might benefit from accepting and exploring variations of solutions with inferior fitness, as they can open up pathways towards optimal solutions. Our analysis strengthens this hypothesis and puts some light into the reasons why.

Table 4. The ratio of deceptive mutations out of all mutations to find the best solution after all generations. The mean, min and max are computed over the same 25 independent runs that are presented in Table 3, column *MAGE-MCTS*.

Problem	*mean* deceptive	*min* deceptive	*max* deceptive
Fuel Cost	21.63%	6.67%	40.00%
FizzBuzz	7.99%	0.0%	19.23%

5 Conclusion

Certain fitness landscape characteristics make it difficult for CGP, a Graph-based GP approach, to evolve. Local optima, neutral fitness plateaus, and deceptive fitness paths misleading search, can result in an inefficient exploration of the search space and poor performance compared to other state of the art approaches.

In this contribution, we extended Multimodal Adaptive Graph Evolution, a multi-type variant of CGP, with MCTS as an alternative to the $1+\lambda$ EA strategy usually employed. The classic $1+\lambda$ EA highly focuses on exploitation and, by design, finds it hard to escape local optima or to explore deceptive search paths. On the contrary, MCTS is able to balance exploitation and exploration, thus guiding the search into unknown regions of the search space to overcome difficult fitness landscape characteristics.

Comparing both search methods on the *PSB2* program synthesis benchmark, MAGE-MCTS outperformed MAGE with a *1+λ* EA on the majority of the problems. By the example of two problems, we illustrate that the benefit of using MCTS is highly dependent on the fitness landscape. We analyzed the search behavior of both MAGE variants and we argue that problems with difficult fitness landscape characteristics, such as long neutral plateaus and low diversity of visited genomes, highly benefit from MCTS. Instead, when a high diversity of genomes can be maintained during the search with a *1+λ* EA, and the neutral degree is low, MCTS hinders search.

With the aim of maintaining high performance of CGP on a variety of applications, this stresses the need to dispose of search strategies adapted to different fitness landscape characteristics. In future work, we want to test different implementations of MAGE-MCTS, and study alternative EAs to analyze their behavior and performance when exploring the search space. We also aim to determine in which landscape MCTS deteriorates performance and the reasons why. With a knowledge base of how to rapidly determine characteristics of a fitness landscape, and then adopt a search strategy best adapted to these specific characteristics, we believe performance of MAGE can be increased and sped up.

Acknowledgments. This project used compute resources from the CALMIP projects P21049 and P24042.

Disclosure of Interest. The authors have no competing interests to declare that are relevant to the content of this article.

References

1. Akiba, T., Sano, S., Yanase, T., Ohta, T., Koyama, M.: Optuna: a next-generation hyperparameter optimization framework. CoRR **abs/1907.10902** (2019). http://arxiv.org/abs/1907.10902
2. Cortacero, K., et al.: Evolutionary design of explainable algorithms for biomedical image segmentation. Nature commun. **14**(1), 7112 (2023). https://doi.org/10.1038/s41467-023-42664-x
3. Cui, H., Pätzel, D., Margraf, A., Hähner, J.: Weighted mutation of connections to mitigate search space limitations in cartesian genetic programming. In: Proceedings of the 17th ACM/SIGEVO Conference on Foundations of Genetic Algorithms, pp. 50–60. FOGA '23, Association for Computing Machinery, New York, NY, USA (2023). https://doi.org/10.1145/3594805.3607130
4. De La Torre, C., Cussat-Blanc, S., Wilson, D., Lavinas, Y.: On search trajectory networks for graph genetic programming. In: Proceedings of the Genetic and Evolutionary Computation Conference Companion, pp. 1681–1685. GECCO '24 Companion, Association for Computing Machinery, New York, NY, USA (2024). https://doi.org/10.1145/3638530.3664169, https://doi.org/10.1145/3638530.3664169
5. De La Torre, C., Lavinas, Y., Cortacero, K., Luga, H., Wilson, D.G., Cussat-Blanc, S.: Multimodal adaptive graph evolution for program synthesis. In: Affenzeller, M., et al. (eds.) Parallel Problem Solving From Nature - PPSN XVIII, pp. 306–321. Springer Nature Switzerland, Cham (2024)

6. De La Torre, C., Nadizar, G., Lavinas, Y., Luga, H., Wilson, D., Cussat-Blanc, S.: Evolution of inherently interpretable visual control policies. In: Proceedings of the Genetic and Evolutionary Computation Conference, pp. 358–367. GECCO '25, Association for Computing Machinery, New York, NY, USA (2025). https://doi.org/10.1145/3712256.3726332, https://doi.org/10.1145/3712256.3726332
7. Harding, S., Graziano, V., Leitner, J., Schmidhuber, J.: MT-CGP: mixed type cartesian genetic programming. In: Proceedings of the 14th Annual Conference on Genetic and Evolutionary Computation, pp. 751–758. GECCO '12, Association for Computing Machinery, New York, NY, USA (2012).https://doi.org/10.1145/2330163.2330268, https://doi.org/10.1145/2330163.2330268
8. Helmuth, T., Kelly, P.: PSB2: the second program synthesis benchmark suite. In: Proceedings of the Genetic and Evolutionary Computation Conference, pp. 785–794. GECCO '21, Association for Computing Machinery, New York, NY, USA (2021). https://doi.org/10.1145/3449639.3459285
9. Husa, J., Kalkreuth, R.: A comparative study on crossover in cartesian genetic programming. In: Castelli, M., Sekanina, L., Zhang, M., Cagnoni, S., García-Sánchez, P. (eds.) Genetic Programming, pp. 203–219. Springer International Publishing, Cham (2018)
10. Islam, M., Kharma, N., Grogono, P.: Mutation operators for genetic programming using monte carlo tree search. Appl. Soft Comput. **97**, 106717 (2020). https://doi.org/10.1016/j.asoc.2020.106717, https://www.sciencedirect.com/science/article/pii/S1568494620306554
11. Miller, J.F., Harding, S.L.: Cartesian genetic programming. In: Proceedings of the 10th Annual Conference Companion on Genetic and Evolutionary Computation, pp. 2701–2726. GECCO '08, Association for Computing Machinery, New York, NY, USA (2008). https://doi.org/10.1145/1388969.1389075
12. Spector, L., Klein, J., Keijzer, M.: The push3 execution stack and the evolution of control. In: Proceedings of the 7th Annual Conference on Genetic and Evolutionary Computation, pp. 1689–1696. GECCO '05, Association for Computing Machinery, New York, NY, USA (2005). https://doi.org/10.1145/1068009.1068292
13. Spector, L., Robinson, A.: Genetic programming and autoconstructive evolution with the push programming language. Genet. Program. Evolvable Mach. **3**, 7–40 (2002). https://doi.org/10.1023/A:1014538503543
14. Swiechowski, M., Godlewski, K., Sawicki, B., Mandziuk, J.: Monte carlo tree search: a review of recent modifications and applications. CoRR **abs/2103.04931** (2021). https://arxiv.org/abs/2103.04931
15. Wilson, D.G., Cussat-Blanc, S., Luga, H., Miller, J.F.: Evolving simple programs for playing atari games. CoRR **abs/1806.05695** (2018). http://arxiv.org/abs/1806.05695

Extended Semantics Operator for Genetic Programming: A Semantic-Density Approach to Improve Model Robustness

Sofia Pereira[(✉)] and Leonardo Vanneschi

NOVA Information Management School (NOVA IMS), Universidade Nova de Lisboa,
Campus de Campolide, 1070-312 Libson, Portugal
{spereira,lvanneschi}@novaims.unl.pt

Abstract. The use of semantic awareness in Genetic Programming (GP) has attracted growing attention in recent years. In this framework, the semantics of a program are defined as the vector of outputs it produces on the training cases and loss, commonly used as a fitness measure in GP for supervised learning, corresponds to a distance between the program's semantic vector and the target vector from the dataset. This formulation has an implicit consequence: programs performing well in densely populated output regions tend to receive higher fitness scores and are thus favored during evolution. Conversely, poor performance in sparse regions may go unnoticed, as these areas are underrepresented and contribute little to the overall fitness evaluation. This work aims to mitigate this imbalance by promoting the survival of programs whose performance is more uniform across the entire space of output values. To this end, we introduce the Extended Semantics Operator (ExtenSO), a mechanism that probabilistically expands an individual's semantic vector by generating artificial output values through oversampling. These synthetic values are added in underrepresented areas of the output space, producing a more homogeneous semantic distribution, spanning more diverse output values. Experimental results show that GP enhanced with ExtenSO achieves performance comparable to, or better than, standard GP, while generating smaller individuals. ExtenSO acts as a semantic regularization operator, reducing overfitting and encouraging smoother, more balanced behavior across the output domain. These findings confirm the relevance of semantic awareness in GP and introduce a simple, effective way to exploit it for greater accuracy and robustness.

Keywords: Genetic Programming · Semantic Awareness · Density Estimation · Overfitting

1 Introduction

Genetic Programming (GP) [9] is a flexible evolutionary algorithm that has been successfully applied to a wide range of real-world problems, often producing

L. Manzoni et al. (Eds.): EuroGP 2026, LNCS 16521, pp. 305–320, 2026.
https://doi.org/10.1007/978-3-032-23005-8_19

human-competitive results [12]. Despite these achievements, GP still faces several well-documented limitations, most notably bloat and overfitting [22]. Overfitting, a particularly critical issue in predictive modeling, occurs when evolved individuals perform well on training data but fail to generalize to unseen cases [11]. This behavior indicates that the underlying functional relationships were not fully captured, and instead, the models exploited patterns specific to the training set that do not generalize across the broader data distribution. Bloat, on the other hand, refers to the uncontrolled growth of program size without corresponding fitness improvements, which often leads to computational inefficiency and reduced interpretability [17]. Importantly, previous studies have shown that bloat and overfitting are not necessarily correlated: eliminating one does not necessarily mitigate the other [22,23]. While bloat and overfitting focus primarily on performance and structural aspects, the semantics of GP programs (often used to assess fitness and functional complexity) plays a crucial role in determining generalization performance.

Over the years, several strategies have been proposed to address these issues. One research direction studies this problem from the data-space perspective. In imbalanced or heterogeneous problems, oversampling techniques such as SMOTE and ADASYN [4,7,20] have been used to enrich underrepresented regions of the target space, promoting balanced learning across the target distribution. However, such methods typically operate at an external level, without considering how the evolutionary dynamics or the semantics of GP individuals evolve during the search. More recently, attention has turned toward the semantics of GP individuals, the vector of outputs a program produces over all training instances. Semantic-based methods exploit those outputs to design operators that guide the search more meaningfully. The work on Geometric Semantic GP (GSGP) [10] demonstrated that using operators with known and measurable effects on the semantic space can produce unimodal error surfaces and significantly enhance generalization. Subsequent developments, such as SLIM_GSGP [21], have shown that semantic operators can be implemented efficiently, while maintaining strong generalization properties and producing interpretable models.

However, despite these advances, the internal distribution of semantics, that is, the density and balance of outputs within the space of possible output values, remains largely unexplored. Understanding and leveraging this density could help GP systems avoid over-specialization in "dense" regions of the space of output points and encourage more balanced, generalizable solutions. In this vein, this study introduces the novel *Extended Semantics Operator* (ExtenSO). ExtenSO combines principles from regularization and oversampling, but applies them directly within the semantics of the evolving individuals (internal level) rather than the data (external level). The name ExtenSO (from the Latin *in extenso*, *"in its entirety"*) reflects its goal of encouraging GP to consider the space of the output values as a whole rather than focusing solely on its densest regions. The operator estimates the density of an individual's semantic outputs using Kernel Density Estimation (KDE) [18], effectively quantifying the representational balance of the individual's outputs. It then identifies underrepresented (sparse) regions of the space of output values and temporarily extends the semantics by interpolating new synthetic points in those areas. Fitness is recalculated over this extended semantic vector, penalizing individuals that perform poorly in sparse regions and favoring those with more balanced semantic coverage. By doing so, ExtenSO functions as a stochastic, semantic-aware regularizer. We hypothesize that this mechanism reduces overfitting and improves generalization by encouraging robust, uniformly performing individuals across the complete semantic space.

The remainder of this paper is organized as follows. Section 2 reviews the relevant theoretical background. Section 3 introduces the proposed Extended Semantics Operator and its implementation. Section 4 describes the experimental setup and datasets. Section 5 presents and discusses the obtained experimental results; finally, Sect. 6 concludes the work, also discussing directions for future research.

2   Background

The issues of overfitting and bloat have long been central topics in GP research. Over the years, many researchers have proposed techniques to mitigate these problems, and while several approaches have shown promise, they remain persistent challenges. Overfitting occurs when evolved models perform well on training data but fail to generalize to unseen cases, while bloat refers to the uncontrolled growth of program size without corresponding fitness improvement. Both issues hinder generalization, efficiency, and interpretability, motivating the continued exploration of new strategies to improve GP's robustness and learning behavior.

As for any other machine learning system, a particularly relevant scenario for understanding generalization issues in GP involves imbalanced datasets. The core motivation behind studying such cases lies in the observation that an imbalanced distribution of training samples can bias learning algorithms toward the most frequent, or dense, regions of the input or target space, leading to poor performance in rare but important regions. This imbalance often translates into limited generalization capability, as models fail to adequately learn information from the underrepresented cases. A common solution to this problem is the use of oversampling techniques, such as SMOTE and ADASYN, which aim to rebalance the training data distribution by generating synthetic samples in sparse regions of the data space. By creating new data points in less populated areas (corresponding, for instance, to minority classes in classification problems), these algorithms help models learn from a more balanced representation of the target distribution, mitigating bias toward overrepresented regions in the data space. SMOTE (Synthetic Minority Oversampling Technique) [4] is one of the earliest and simplest methods to address class imbalance. It combines under-sampling of frequent classes with over-sampling of minority classes by generating synthetic instances through interpolation. For each rare instance, SMOTE randomly selects one of its k-nearest neighbors from the minority class and creates a new example whose features are interpolated between the two. This process densifies the sparse regions of the feature space and encourages the model to pay more attention to less frequent cases. When dealing with regression problems, a variant of SMOTE, called SMOTER [20], adapts this strategy to continuous target values. SMOTER identifies rare instances based on a relevance function and then generates synthetic samples by interpolating between a rare observation and one of its k-nearest rare neighbors. The new target value is obtained as a weighted average of the two seed targets, where weights are inversely proportional to their distance from the new point. This results in synthetic examples that preserve local continuity and balance the target space without distorting the underlying relationships. ADASYN (Adaptive Synthetic Sampling Approach) [7] extends these ideas by dynamically adjusting the number of synthetic samples generated according to the local data density. Minority observations located in sparser or more complex regions of the data space generate more synthetic samples, while those in well-represented areas generate fewer. In this way, ADASYN focuses the learning process on the most difficult and underrepresented regions, thereby reducing bias and improving generalization. Over the years, numerous other resampling approaches have been proposed and extensively studied [3]. Most of these methods focus on addressing class imbalance in classification tasks, with considerably less attention given to imbalance in regression problems. Within the GP framework, the exploration of such methods remains scarce, although recent works on GP-based oversampling frameworks for classification [5,6] have demonstrated the significant impact that GP-oriented approaches can have in imbalanced learning. A deeper investigation into imbalance regression and resampling strategies in GP thus remains an open research avenue. Nevertheless, these oversampling methods reveal a powerful principle: by synthetically enriching sparse regions, one can improve balance and generalization. Although originally designed as external preprocessing methods that modify the data distribution, this concept of density-aware learning provides valuable inspiration for internal evolutionary mechanisms such as those used in GP.

Another well-established approach to shaping GP evolution is the use of penalty-based methods. These techniques act internally within the evolutionary process by penalizing the fitness of programs that exhibit undesirable characteristics. One of the most studied examples is parsimony pressure, which introduces a penalty proportional to an individual's size, discouraging excessive growth and indirectly controlling bloat [13]. Such penalization helps guide evolution toward simpler, more interpretable models that better reflect the desired behavior. Unlike oversampling, which modifies the data to rebalance learning, penalty-based approaches modify the fitness function to steer the search toward preferred solution characteristics. These internal mechanisms can have a substantial influence on the GP dynamics and have been shown to impact both model complexity and generalization.

An interesting approach to address imbalanced datasets involves using Kernel Density Estimation (KDE) to estimate target distributions. Recognizing that continuous targets inherently possess meaningful relationships, [25] proposed a distribution smoothing strategy for both labels and features. Their method explicitly accounts for the influence of nearby target values by employing a symmetric kernel function (such as a Gaussian kernel) to smooth label distributions, thereby leveraging the similarity among neighboring targets. Building on this principle, [19] introduced Density-Based Weighting for Imbalanced Regression, which assigns weights to data points according to the rarity of their target values, estimated via KDE. This approach adjusts each sample's contribution to the loss function, granting greater importance to rare targets while reducing the influence of common ones. Although originally developed for Deep Neural Networks, the idea of modeling continuous target densities through KDE is broadly applicable across learning paradigms, including GP. An example of such an application is presented in [15], where synthetic data generation is explored as a strategy to improve extrapolation performance through knowledge distillation (KD). Kernel density estimation (KDE) identifies sparsely sampled regions of the input space, where a pre-trained teacher model (e.g., NN, RF, or GP) generates synthetic samples that are subsequently used to train a student model (e.g., GP). This opens promising research directions for integrating density-aware mechanisms within GP to enhance learning under imbalanced conditions.

The Extended Semantics Operator (ExtenSO) proposed in this study, and discussed in the next section, draws inspiration from these influential ideas, synthetic data generation, fitness penalization and density estimation, and combines them within the context of GP's semantic aware systems. By combining the data-level insight of oversampling with the fitness-level control of regularization and kernel density estimation, ExtenSO aims to produce smoother, more balanced, and ultimately more generalizable evolutionary models.

3 Extended Semantics Operator

As stated beforehand, the Extended Semantics Operator (ExtenSO) is a semantic regularization and oversampling-based operator that penalizes individuals that behave poorly on underrepresented regions of the space of output values. Standard GP operators act exclusively on the syntax of individuals, with no regard for their semantics. Acting on the vector of output values (the semantics of an individual), ExtenSO introduces semantic awareness into the evolution through guided penalization, encouraging individuals that perform consistently across both dense and sparse regions of the output space. This operator originates from the observation that, in regression problems with imbalanced output distributions, algorithms often focus on patterns from denser areas of the space of outputs. By designing an operator that explicitly penalizes individuals performing poorly in sparse semantic regions, we guide evolution toward solutions that generalize better across the entire semantic space, even in the presence of uneven data distributions.

ExtenSO works stochastically, with a predefined probability p_{ext}, acting on the offspring produced by mutation and crossover at each generation.When applied, ExtenSO receives the individual's semantic vector $\mathbf{s}$ and target values $\mathbf{t}$ and estimates the density of the semantic space using *Kernel Density Estimation (KDE)*. KDE is a nonparametric statistical method used to estimate

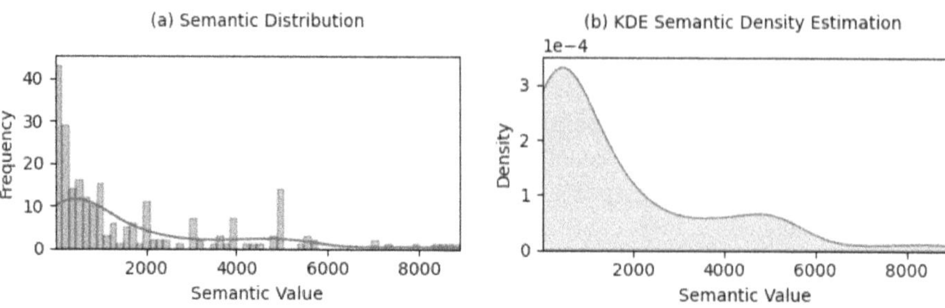

Fig. 1. Graphical representation of the application of Gaussian KDE to a semantic vector. (a) presents the original semantic distribution; (b) presents the respective density estimation through KDE.

a probability density function from a finite set of observations. It operates by placing a smooth kernel (Gaussian, for this study) around each observed point, effectively producing a continuous approximation of how likely it is to encounter values in different regions of a space. Figure 1 presents a visual representation of this process over an arbitrary semantic vector. Formally, the *KDE* for a semantic point s is expressed as [18]:

$$\hat{f}(s) = \frac{1}{nh} \sum_{i=1}^{n} K\left(\frac{s - s_i}{h}\right),$$ (1)

where $K(\cdot)$ is the kernel function, h is the bandwidth parameter controlling smoothness, s_i are the individual's semantic outputs and n the number of training instances, which corresponds to the cardinality of the semantic vector. The selection of the bandwidth parameter h in KDE is non-trivial, as it directly affects the smoothness of the estimated density: a smaller h can lead to an overfitted, noisy estimate, while a larger h may oversmooth important features of the distribution. For the purpose of this study, Scott's *rule of thumb* [16] was adopted for bandwidth selection:

$$h = \sigma \, n^{-1/5},$$ (2)

where σ is the standard deviation of the coordinates of the semantic vector, and n, as in Equation (1) is the number of training instances. Scott's rule provides a data-driven balance between bias and variance in the density estimate, assuming that the underlying distribution is approximately normal. This heuristic ensures that the estimated semantic density is neither too coarse, missing local irregularities, nor too fine, overemphasizing noise in the semantics.

Example. Consider an individual evaluated on a dataset consisting of three training instances. Let the target values corresponding to these instances and the semantics of the individual be as reported in Table 1 (the target is reported in the second column and the semantics in the third one).

Table 1. Target outputs (second column) and semantics of an individual (third column) for three training instances used in the example discussed in the text.

Instance	Target (t_i)	Output (s_i)
1	3.0	2.0
2	6.0	5.0
3	9.0	6.0

The semantic vector (i.e., the vector of the outputs calculated by the individual on the training instances) is $\mathbf{s} = [2.0, 5.0, 6.0]$. Using KDE, we now estimate the density of these outputs. Intuitively, since the last two values (5.0 and 6.0) are close together, they form a dense region, while 2.0 is isolated, forming a sparse region, as illustrated in Fig. 2. The value 2.0 will thus have a smaller estimated density $\hat{f}(s)$, identifying it as a candidate for extension.

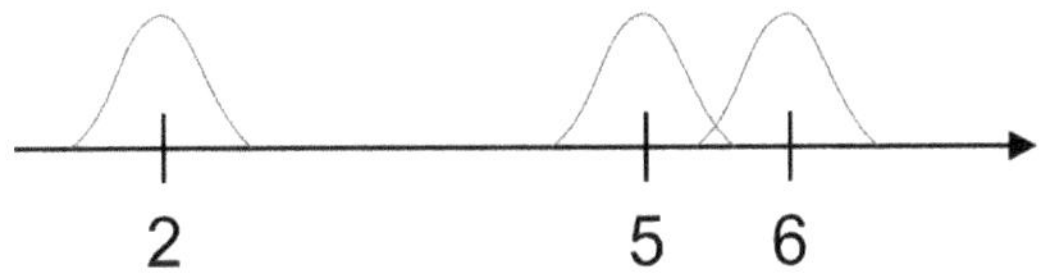

Fig. 2. Graphical representation of the distribution of semantic vector $\mathbf{s} = [2.0, 5.0, 6.0]$ in a 1-dimentional space of output values.

Once the density values are estimated, ExtenSO selects a predefined proportion (ρ_m) of output points for extension, corresponding to the m points with the lowest estimated density. The parameter ρ_m controls the influence of the operator on the extended fitness: larger values increase the contribution of synthetic points, but may also bias the model towards regions dominated by artificially generated output points.

For each of the selected points s_i, ExtenSO finds the closest semantic neighbor s_j and generates a synthetic semantic output by linear interpolation:

$$s_{\text{new}} = s_i + \alpha(s_j - s_i), \quad \alpha \sim U(0, 1). \tag{3}$$

The same interpolation process is applied to the corresponding target values:

$$t_{\text{new}} = t_i + \alpha(t_j - t_i). \tag{4}$$

Continuing from the previous example, the sparse semantic value $s_i = 2.0$ (with corresponding target $t_i = 3.0$) has its nearest neighbor $s_j = 5.0$ (with corresponding target $t_j = 6.0$). Assuming, for instance, $\alpha = 0.5$, the new synthetic pair is:

$$s_{\text{new}} = 2 + 0.5(5 - 2) = 3.5, \qquad t_{\text{new}} = 3 + 0.5(6 - 3) = 4.5.$$

This synthetic pair $(s_{\text{new}}, t_{\text{new}}) = (3.5, 4.5)$ represents a new coordinate of an extended semantic vector that fills the sparse region between output values 2.0 and 5.0, densifying the individual's semantics, along with the corresponding target value. The new synthetic output and target are added to the original vectors, obtaining the following extended vectors:

$$\mathbf{s'} = [2, 5, 6, 3.5], \quad \mathbf{t'} = [3, 6, 9, 4.5].$$

A new fitness value is then computed using these extended vectors. In the case of RMSE, the new fitness is now given by:

$$RMSE = \sqrt{\frac{1}{n+m} \sum_{k=1}^{n+m} (t_k - s_k)^2}, \tag{5}$$

where n is the original number of training instances and m is the number of synthetic points added. Because RMSE penalizes large deviations, individuals that produce high errors in sparse regions will now receive a larger penalty than they did before the extension, while individuals that are already accurate in those regions remain unaffected. In this sense, we say that ExtenSO acts as a semantic-level regularizer, penalizing models that perform inconsistently across their semantic values and favoring those that maintain uniform accuracy. From an evolutionary standpoint, ExtenSO alters the selection pressure of GP in a subtle but meaningful way. Individuals that underperform in underrepresented regions of the space of output values become less competitive, while those that maintain consistent accuracy across both dense and sparse regions are rewarded. This introduces a form of semantic bias that prioritizes robust, well-distributed behavior over localized specialization.

Analogously to regularization in statistical learning, ExtenSO discourages extreme, uneven semantic behavior by penalizing large localized errors. Because the operator is applied probabilistically (it is applied with a probability equal to p_{ext} to each newly generated individual, as mentioned above), it tends to preserve diversity across the population and prevents over-regularization. This ensures that exploration through standard mutation and crossover remains active, while ExtenSO selectively reinforces individuals with more stable and generalizable semantics.In this way, ExtenSO introduces semantic awareness directly into the evolutionary process without altering traditional GP operators. It complements existing mechanisms by possibily promoting smoother fitness

Table 2. Experimental setup with general and algorithm-specific parameters.

Component	Parameter	Value
General Experimental Setup		
Initialization	Method	Ramped Half-and-Half (RHH)
	Max Depth	6
Selection	Method	Tournament selection (size: 2)
	Elitism	Best individual preserved
Function Set		$\{+, -, \times, \div\}$ (protected division as in [9])
Terminal Set		Problem variables only (no constants)
Evolutionary Run	Population	100
	Generations	1500
	Fitness Function	Root Mean Squared Error (RMSE)
Operators	Crossover Probability	0.9
	Mutation Probability	0.1
	Max Tree Depth	17
Algorithm-Specific Parameters		
ExtenSO	p_{ext}	0.25
	ρ_n	0.05

Table 3. p-values of the Wilcoxon Signed-Rank Test relative to the results in Figs. 3 and 4 at termination, for each studied dataset. These p-values compare GP using the ExtenSO operator with Standard GP (StdGP). Values in bold indicate cases where ExtenSO is statistically better than StdGP (p-value < 0.05). The final row represents the number of datasets where ExtenSO is statistically comparable or better than StdGP (with better in parentheses).

Problem	Test Fitness	Overfitting	Individual Size
Toxicity	0.15	**1.50E-02**	**2.40E-02**
Istanbul	0.99	0.42	**3.60E-02**
PPB	0.81	0.25	**3.03E-02**
Bio	0.24	**2.20E-02**	**8.00E-03**
RBSP	0.61	**3.60E-02**	0.81
Concrete	**4.97E-02**	**2.80E-03**	0.62
Summary	**6(1)**	**6(4)**	**6(4)**

landscapes, improving generalization, and providing a lightweight yet effective means to handle imbalance and overfitting in GP.

4   Test Problems and Experimental Setup

The presented experiments are conducted on six real-life symbolic regression problems, commonly used for benchmarking in GP research: Toxicity (characterized by 626 features and 274 samples, and presented in [2]); PPB (626 features, 234 samples, [2]); Istanbul (7 features, 536 samples, [1]); Residential Building Sales Price (RBSP) (107 features, 372 samples, [14]); Human Oral Bioavailability (Bio) (242 features, 359 samples, [2]); Concrete Slump (9 features, 102 samples, [26]). Each dataset is evaluated using Monte Carlo crossvalidation [24] with 30 random splits (70% training, 30% testing). In other words, at each run the datasets were randomly split with a uniform distribution into training and test partitions composed of 70% and 30% of the data, respectively. The same partitions were used across all compared methods in each run to ensure fairness of comparison. All the results reported in the next section are medians over the 30 independent runs.

Table 2 presents a summary of the parameter setting used in this study. The general experimental setup followed parameter settings commonly adopted in related studies (see, for example, [21]), while the ExtenSO algorithm-specific parameters were determined through a preliminary calibration phase, where the final configuration corresponded to the values that yielded the best performance.

5   Experimental Results and Discussion

The obtained experimental results are presented in Figs. 3 and 4. Figure 3 illustrates the evolution of training and test fitness after the application of ExtenSO, as well as overfitting dynamics throughout evolution. For simplicity, and for having a quick intuition, here overfitting is computed as the division between training and test fitness. Figure 4 reports the evolution of individual size, measured as the node count of the best individual in the population on the training set at each generation. Table 3 reports the p-values of the Wilcoxon Signed-Rank Test for comparisons between Standard GP (stdGP in the figures and tables) and GP with ExtenSO (simply ExtenSO in the figures and tables) in terms of test fitness, overfitting and individual size. Bold p-values indicate statistically significant improvement at a significance level of $\alpha = 0.05$, considering the best individual on the training set at generation 1500.

A closer inspection of the evolutionary dynamics (Figs. 3 and 4) suggests that ExtenSO tends to reduce overfitting early in the evolution, promoting smoother convergence. Interestingly, this reduction occurs without degrading performance on unseen data, indicating that ExtenSO encourages the development of individuals that generalize more effectively. Moreover, the operator shows potential in controlling model growth: in most datasets, the GP variant with ExtenSO exhibited a tendency to generate smaller offspring compared to Standard GP. While not as impactful, the studied operator also has a generally positive impact when it comes to model fitness on unseen data, showing an overall evolutionary improvement on all but one of the studied datasets.

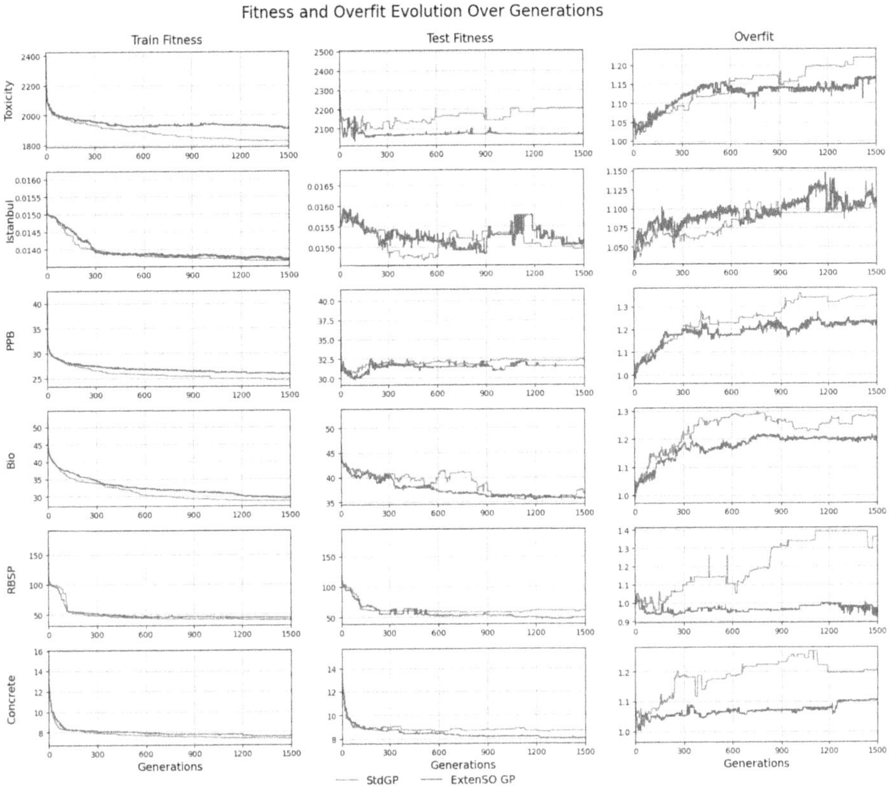

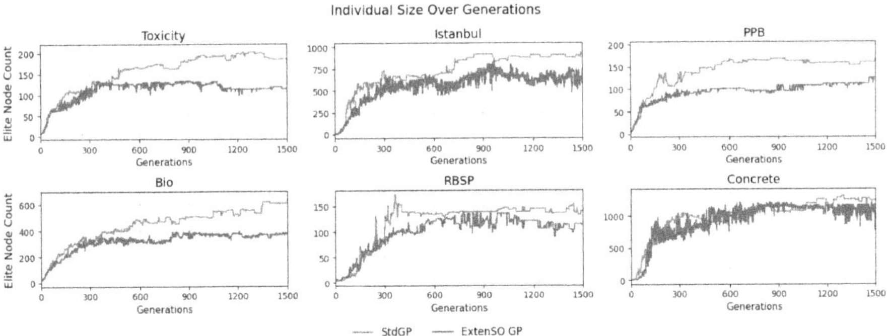

Fig. 3. Comparison between Standard GP (StdGP) and GP with ExtenSO across the studies datasets. Lines show the evolution of the training fitness, test fitness, and overfitting across generations, with median values across 30 runs for each algorithm.

Fig. 4. Comparison between Standard GP (StdGP) and GP with ExtenSO across the studied datasets. Lines show the evolution of the individual size (number of nodes of the best individual on the training set in the population) across generations, with median values across 30 runs for each algorithm.

When analyzing the statistical results in Table 3, it becomes clear that, across all studied datasets, GP with ExtenSO performs on par or better than Standard GP. Specifically, ExtenSO significantly reduced overfitting in four of the six datasets, decreasing the difference between training and test fitness. When it comes to model sizes, it also shows statistically significant improvements in four out of six datasets, maintaining performance in the remaining. Interestingly, this operator also shows the ability to improve model performance on unseen data, performing better than Standard GP in a statistically significant way in terms of test fitness in one dataset and maintaining performance on the other tested cases studied (from which four out of five present an improvement, although not statistically significant, at the end of the median run).

These findings indicate that ExtenSO's semantic dynamics contribute positively to the evolutionary process, improving model generalization, interpretability, and even test fitness, and never compromising or hindering GP performance. This analysis provides a relevant insight: semantic-based mechanisms can meaningfully influence GP evolution, improving both model generalization and complexity. The core principle of densifying the semantic vector, rewarding individuals that perform well across all regions of the space of output values and penalizing those focusing on highly dense areas, leads to smaller models with better overall generalization. This suggests that ExtenSO filters out individuals that would otherwise dominate evolution due to good fitness values in dense output regions, but which in reality lack true generalization capability. By reducing the evolutionary advantage of such individuals, the operator allows more balanced, semantically robust models to propagate, resulting in lower overfitting and better overall performance.

ExtenSO showed the most pronounced improvements in datasets with skewed target distributions, such as PPB (skewness = −0.95), Bio (skewness = −0.63), Toxicity (skewness = 1.52), RBSP (skewness = 1.26), and Concrete (skewness = −1.1, but with a bimodal distribution, with two clearly denser regions). These datasets exhibit strong left or right-tailed distributions with rare, sparse extreme target values. In contrast, datasets with more balanced or near-normal target distributions, such as Istanbul (skewness = −0.08), showed smaller or negligible gains. This pattern suggests that the operator's KDE guided sampling is more effective when the semantic space reflects underlying target sparsity, allowing it to better handle outlier or low-density regions in the data. Figure 5 presents the distribution of the target values for each dataset:

This relationship between dataset skewness and the operator's performance supports the underlying intuition of the ExtenSO mechanism. Datasets with higher target skewness (e.g., PPB, Bio, Toxicity, and RBSP) contain sparse, tail-heavy output regions analogous to "minority areas" in SMOTE terminology. Because the operator estimates semantic densities via KDE and preferentially extends individuals in low-density regions, it enhances the representation of these rare semantic behaviors. This mechanism allows the population to better explore and generalize over underrepresented parts of the output distribution, which explains the stronger improvements observed in these skewed datasets.

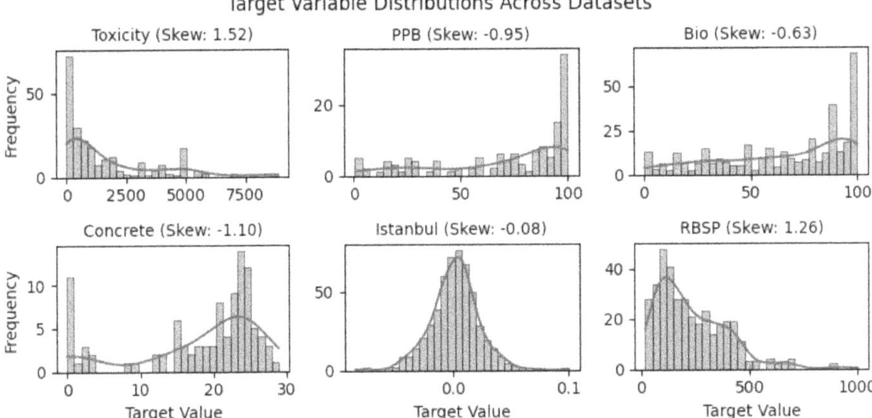

Fig. 5. Distribution of target values for each dataset, with respective skewness presented between parentheses. Skewness is computed as the Fisher-Pearson coefficient of skewness [8].

Conversely, datasets with approximately symmetric target distributions (e.g., Istanbul) offer fewer sparse regions to exploit, leading to results closer to the baseline. These findings suggest that ExtenSO is particularly advantageous for symbolic regression problems characterized by nonuniform or skewed target distributions.

These findings raise a compelling hypothesis regarding the relationship between bloat and semantics. Although previous studies have shown that bloat and overfitting are not directly correlated [17,22], our results suggest that semantic specialization may underlie both phenomena. If a model performs well in dense regions of the output space but poorly in sparse ones, its overall fitness may stagnate due to errors in underrepresented areas, mirroring the early signs of bloat. This leads to an intriguing open question: can bloat sometimes emerge as a symptom of semantic over-specialization, where fitness improvements concentrate on a limited subset of the fitness cases? ExtenSO offers a potential pathway to explore this hypothesis by directly regulating semantic density and exposing the relationship between output distribution, generalization, and structural complexity.

6 Conclusions and Future Work

This paper introduced the Extended Semantics Operator (ExtenSO), designed to enhance the generalization capability of Genetic Programming (GP) by incorporating semantic density awareness into the evolutionary process. We hypothesize that generalization in GP is often hindered by imbalance in the output space, where models tend to overfit dense regions and underperform in sparse ones. ExtenSO addresses this limitation by integrating the principles of oversampling and regularization directly within the semantics of the evolving individuals. By extending the semantics in underrepresented regions and recalculating fitness over the enriched semantic set, the operator effectively densifies sparse areas of the output space. This process encourages GP to evolve individuals that perform consistently across all regions of the target distribution, thereby promoting more robust and generalizable solutions. When compared with Standard GP, GP equipped with ExtenSO achieved comparable or superior results across all studied datasets in terms of test fitness, overfitting, and individual size. Specifically, ExtenSO demonstrated statistically significant improvements in four test cases for overfitting, and model size and produced significantly more general models in one test case, suggesting that semantic regularization can simultaneously enhance generalization and control model complexity.

Despite these promising findings, several directions remain open for future research. A more detailed analysis of ExtenSO's parameterization, particularly the extension probability p_{ext} and the proportion of semantics ρ_n selected for extension, could provide valuable insights into its optimal configuration. Further investigation into additional metrics such as population diversity, functional complexity, and semantic similarity may reveal deeper relationships between semantic regularization and evolutionary dynamics. Moreover, applying ExtenSO to other GP paradigms, particularly Geometric Semantic GP (GSGP), where evolution is inherently guided by semantic transformations, represents an exciting avenue for future exploration. Such extensions could further validate the potential of semantic density regularization as a unifying principle for promoting generalization.

Acknowledgments. This work was supported by national funds through FCT (Fundação para a Ciência e a Tecnologia), under the project - UID/04152/2025 - Centro de Investigação em Gestão de Informação (MagIC)/NOVA IMS - https://doi.org/10.54499/UID/04152/2025 (2025-01-01/2028-12-31) and UID/PRR/04152/2025 https://doi.org/10.54499/UID/PRR/04152/2025 (2025-01-01/ 2026-06-30).

References

1. Akbilgic, O., Bozdogan, H., Balaban, E., Balaban, M.: A novel hybrid RBF neural networks model as a forecaster. Stat. Comput. **24** (2013). https://doi.org/10.1007/s11222-013-9375-7
2. Archetti, F., Lanzeni, S., Messina, E., Vanneschi, L.: Genetic programming for computational pharmacokinetics in drug discovery and development. Genet. Program Evolvable Mach. **8**(4), 413–432 (2007). https://doi.org/10.1007/s10710-007-9040-z
3. Avelino, J.G., Cavalcanti, G.D.C., Cruz, R.M.O.: Resampling strategies for imbalanced regression: a survey and empirical analysis. Artif. Intell. Rev. **57**(4), 82 (2024). https://doi.org/10.1007/s10462-024-10724-3
4. Chawla, N., Bowyer, K., Hall, L., Kegelmeyer, W.: SMOTE: synthetic minority over-sampling technique. J. Artif. Intell. Res. (JAIR) **16**, 321–357 (2002). https://doi.org/10.1613/jair.953

5. Cui, Y., Du, Z., Ge, H., Zou, G., Hou, Y.: Multitree genetic programming with spherical-based operators for synthetic minority over-sampling technique in unbalanced data. Swarm Evol. Comput. **98**, 102126 (2025). https://doi.org/10.1016/j.swevo.2025.102126. https://www.sciencedirect.com/science/article/pii/S2210650225002846
6. Farinati, D., Vanneschi, L.: An empirical study of GM4OS for imbalanced binary classification. SN Comput. Sci. **6**(5), 510 (2025). https://doi.org/10.1007/s42979-025-04048-4
7. He, H., Bai, Y., Garcia, E., Li, S.: ADASYN: adaptive synthetic sampling approach for imbalanced learning. In: Proceedings of the International Joint Conference on Neural Networks. pp. 1322 – 1328 (2008). https://doi.org/10.1109/IJCNN.2008.4633969
8. Kokoska, S., Zwillinger, D.: CRC Standard Probability and Statistics Tables and Formulae: Student Edition. CRC Press, 1 edn. (2000). https://doi.org/10.1201/b16923
9. Koza, J.R.: Genetic programming: on the programming of computers by means of natural selection. MIT Press, Cambridge, MA, USA (1992)
10. Moraglio, A., Krawiec, K., Johnson, C.: Geom. Semant. Genet. Program. **7491**, 21–31 (2012). https://doi.org/10.1007/978-3-642-32937-1_3
11. O'Neill, M., Vanneschi, L., Gustafson, S.: Open issues in genetic programming. Genet. Program. Evolvable Mach. **11**, 339–363 (2010). https://doi.org/10.1007/s10710-010-9113-2
12. Poli, R., Langdon, W., Mcphee, N.: A Field Guide To Genetic Programming (2008)
13. Poli, R., McPhee, N.F.: Parsimony pressure made easy. In: Proceedings of the 10th Annual Conference on Genetic and Evolutionary Computation. pp. 1267–1274. GECCO '08, Association for Computing Machinery, New York, NY, USA (2008). https://doi.org/10.1145/1389095.1389340, event-place: Atlanta, GA, USA
14. Rafiei, M.: Residential building (2015). https://doi.org/10.24432/C5S896, [Dataset]
15. Ramlan, F.W., O'Riordan, C., Kronberger, G., McDermott, J.: Can synthetic data improve symbolic regression extrapolation performance? In: Proceedings of the Genetic and Evolutionary Computation Conference Companion. p. 2548–2555. GECCO '25 Companion, Association for Computing Machinery, New York, NY, USA (2025). https://doi.org/10.1145/3712255.3734356
16. Scott, D.W.: Multivariate Density Estimation: Theory, Practice, and Visualization (1992)
17. Silva, S., Costa, E.: Dynamic limits for bloat control in genetic programming and a review of past and current bloat theories. Genet. Program Evolvable Mach. **10**(2), 141–179 (2009). https://doi.org/10.1007/s10710-008-9075-9
18. Silverman, B.: Density Estimation for Statistics and Data Analysis (02 2018).https://doi.org/10.1201/9781315140919
19. Steininger, M., Kobs, K., Davidson, P., Krause, A., Hotho, A.: Density-based weighting for imbalanced regression. Mach. Learn. **110**(8), 2187–2211 (2021). https://doi.org/10.1007/s10994-021-06023-5
20. Torgo, L., Ribeiro, R., Pfahringer, B., Branco, P.: SMOTE for Regression. **8154**, 378–389 (2013). https://doi.org/10.1007/978-3-642-40669-0_33
21. Vanneschi, L.: SLIM_GSGP: the non-bloating geometric semantic genetic programming. In: Genetic Programming: 27th European Conference, EuroGP 2024, Held as Part of EvoStar 2024, Aberystwyth, UK, April 3–5, 2024, Proceedings. pp. 125–141. Springer-Verlag, Berlin, Heidelberg (2024). https://doi.org/10.1007/978-3-031-56957-9_8

22. Vanneschi, L., Castelli, M., Silva, S.: Measuring bloat, overfitting and functional complexity in genetic programming. In: Proceedings of the 12th Annual Genetic and Evolutionary Computation Conference, GECCO '10. pp. 877–884 (2010). https://doi.org/10.1145/1830483.1830643
23. Vanneschi, L., Silva, S.: Using Operator Equalisation for Prediction of Drug Toxicity with Genetic Programming. In: Lopes, L.S., Lau, N., Mariano, P., Rocha, L.M. (eds.) Progress in Artificial Intelligence, pp. 65–76. Springer, Berlin Heidelberg, Berlin, Heidelberg (2009)
24. Vanneschi, L., Silva, S.: Lectures on Intelligent Systems. Natural Computing Series, Springer (2023). https://doi.org/10.1007/978-3-031-17922-8
25. Yang, Y., Zha, K., Chen, Y.C., Wang, H., Katabi, D.: Delving into deep imbalanced regression (2021), https://arxiv.org/abs/2102.09554
26. Yeh, I.C.: Simulation of concrete slump using neural networks. Constr. Mater. **162**, 11–18 (2009)

Author Index

GPSR Compliance
The European Union's (EU) General Product Safety Regulation (GPSR) is a set
of rules that requires consumer products to be safe and our obligations to
ensure this.

If you have any concerns about our products, you can contact us on

ProductSafety@springernature.com

In case Publisher is established outside the EU, the EU authorized
representative is:

Springer Nature Customer Service Center GmbH
Europaplatz 3
69115 Heidelberg, Germany